ALSO BY PETER WATSON

NONFICTION: IDEAS

*Ideas: A History of
Thought and Invention from
Fire to Freud*

*The Modern Mind:
An Intellectual History of
the Twentieth Century*

*The Great Divide:
Nature and Human
Nature in the Old World
and the New*

*The German Genius:
Europe's Third Renaissance,
the Second Scientific
Revolution, and the Twentieth
Century*

*The Age of Atheists:
How We Have Sought
to Live Since the Death
of God*

*War on the Mind:
The Military Uses and
Abuses of Psychology*

NONFICTION: ART

The Medici Conspiracy

Inside Sotheby's

*From Manet to Manhattan:
The Rise of the Modern
Art Market*

*Wisdom & Strength:
The Biography of a
Renaissance Masterpiece*

The Caravaggio Conspiracy

FICTION

The Nazi's Wife

Crusade

Landscape of Lies

Stones of Treason

Capo

Gifts of War

The Clouds Beneath the Sun

Madeleine's War

CONVERGENCE

The Idea at the Heart of Science

---

*How the Different Disciplines Are
Coming Together to Tell One Coherent,
Interlocking Story, and Making Science the
Basis for Other Forms of Knowledge*

PETER WATSON

Simon & Schuster
New York   London   Toronto   Sydney   New Delhi

For David Henn and David Wilkinson

It is a wonderful feeling to recognize the unity of a complex of phenomena that to direct observation appear to be quite separate things.

—ALBERT EINSTEIN

The history of science teaches us again and again how the extension of our knowledge may lead to the recognition of relations between formerly unconnected groups of phenomena.

—NIELS BOHR

By tracing the arrows of explanation back toward their source, we have discovered a striking convergent pattern—perhaps the deepest thing we have yet learned about the universe.

—STEVEN WEINBERG

We are at a moment of great convergence, when data, science, and technology are all coming together to unravel the biggest mystery yet—our future, as individuals and as a society.

—ALBERT-LÁSZLÓ BARABÁSI

We shall not rest satisfied until we are able to represent all physical phenomena as an interplay of a vast number of structural units intrinsically alike.

—ARTHUR EDDINGTON

Nature is pleased with simplicity.

—ISAAC NEWTON

Everything is made of one hidden stuff.

—RALPH WALDO EMERSON

All of us secretly wish for an ultimate theory, a master set of rules from which all truth would flow.

—ROBERT LAUGHLIN

Reality, in the modern conception, appears as a tremendous hierarchical order of organised entities, leading, in a superposition of many levels, from physical and chemical to biological and sociological systems.
　　　　　　　　　　　　—LUDWIG VON BERTALANFFY

As scientific knowledge advances, previously unrelated phenomena are found to be related.
　　　　　　　　　　　　—AUSTEN CLARK

The universe is orderly. It has certain built-in characteristics that came we know not whence or why but that are determinable and that have not changed during the course of recoverable history.
　　　　　　　　　　　　—GEORGE GAYLORD SIMPSON

Reductionism is the primary cutting tool of science.
　　　　　　　　　　　　—EDWARD O. WILSON

We have inherited from our forefathers the keen longing for unified, all-embracing knowledge.
　　　　　　　　　　　　—ERWIN SCHRÖDINGER

The search for the elementary ingredients making up the universe and the deepest laws governing their interactions may be a search that one day draws to a close. The deeper we look, the simpler and more unified the laws become, and there may well be a limit to this process.
　　　　　　　　　　　　—BRIAN GREENE

Biology presupposes physics but not vice versa.
　　　　　　　　　　　　—RUDOLF CARNAP

Once there was physics and there was chemistry but there was no biology.
　　　　　　　　　　　　—JULIUS REBEK

Mathematics can expose the underlying unity of phenomena that otherwise seem unrelated.
—STEVEN STROGATZ

We live in a world orderly enough that it pays to measure.
—GEORGE JOHNSON

Our everyday activity implies a perfect confidence in the universality of the laws of nature.
—LUCIEN LÉVY-BRUHL

It is now evident that where one discipline ends and the other begins no longer matters.
—PATRICIA CHURCHLAND

In every age there is a turning point, a new way of seeing and asserting the coherence of the world.
—JACOB BRONOWSKI

Science aims both to detect order and to create order.
—JOHN DUPRÉ

There can be no explanation which is not in need of a further explanation.
—KARL POPPER

CONTENTS

Part One

The Most Important Unifying Ideas of All Time

Part Two

The Long Arm of the Laws of Physics

Part Three

"The Friendly Invasion of the Biological Sciences by the Physical Sciences"

Part Four

The Continuum from Minerals to Man

Preface

CONVERGENCE: "THE DEEPEST IDEA IN THE UNIVERSE"

I n early April 1912, the Danish physicist Niels Bohr arrived in the bustling city of Manchester in the north of England. When he had first stepped ashore from Denmark, some months previously, he had never imagined working in the industrial heartland of Britain, where the forest of factory chimneys billowed smoke and soot twenty-four hours a day, and where Market Street was said to be the most crowded in all Europe. Instead, his first destination had been the "mellow and stately" colleges and quadrangles of Cambridge. He had just completed his PhD, in Copenhagen, on the electron theory of metals, and he went to Cambridge to work with J. J. Thomson, the director of the Cavendish Laboratory and the man who, in 1897, had discovered the electron as a fundamental unit of matter, for which he had won the Nobel Prize.

But although Bohr was always very polite about Thomson in his letters home to his fiancée, Margrethe, Niels and "JJ"—as he was invariably known—didn't really hit it off. The Dane, a large-boned, heavyset man, had studied English at school, but his spoken syntax was rather stilted and formal and was hardly helped by the fact that he was trying to polish it by reading *David Copperfield*. Nor did he do himself any favors by attempting to advance his friendship with the director by pointing out several small errors in the other man's work. For his part, the notoriously absentminded JJ took weeks to read Bohr's dissertation, which had been translated from the Danish but by someone who wasn't a physicist. (The phrase "charged particles" had been rendered as "loaded particles.") Thomson, who in fairness was very busy as director of the Cavendish, just didn't seem overly interested in Bohr or his work.

And so when, shortly after Christmas, Ernest Rutherford came to Cambridge to speak at the annual Cavendish dinner—a riotous affair, mixing lectures and sing-alongs—Bohr was entranced. Rutherford was a down-to-earth, broad-shouldered man with a ruddy complexion and a reputation for swearing at experiments when they didn't go according to plan. He was a New Zealander who had done postgraduate work at the Cavendish, and then worked at McGill University in Canada, before returning to Manchester, as professor. Rutherford, who had won the Nobel Prize in 1908, for his investigation of radioactivity, had astonished the world of physics for a second time by discovering the basic structure of the atom in May 1911. He showed that it was a bit like a miniature solar system, with a nucleus of positive charge, surrounded at a great distance by orbiting electrons of equal negative charge. (To put this into context, in the atom the proportions of the nucleus to the electron cloud surrounding it are of the order of a grain of sand in London's Albert Hall. Put another way, if the nucleus were the size of a basketball, the electrons would be about three city blocks away. In real terms, the largest atom is that of caesium, a silvery-gold alkali metal, similar to potassium, discovered in 1860, which is just 0.0000005 millimeters—5×10^{-7}mm—across. It would take 10 million of these atoms laid out side by side to stretch between two points of the serrated edge of a postage stamp.)

After hearing Rutherford, Bohr seems to have decided there and then that he wanted to work with him. He arranged a face-to-face meeting via a friend of his father, who lived in Manchester but had worked in Copenhagen. This was a much more successful relationship than the one with JJ—Rutherford later said that Bohr was the most intelligent man he had ever met.

It was the custom of the Manchester laboratory for all the staff to get together each afternoon, late on, for tea—cakes and bread-and-butter laid out on the lab benches. Rutherford led the discussions, perched on a high wooden stool. The discussions were not confined to physics—everything from theater to politics to the new automobiles were legitimate topics—but it was here that Bohr first tentatively advanced his view that, with the basic structure of the atom now being known, it ought to be possible to further our understanding of the elements. Their different properties, he said, should be related to the way the atom is structured, that structure governing why some elements are metal, say, and others gases, why some are reactive and others inert. He suggested that the radioactive properties of matter stem from the nucleus, while the chemical properties arise from the electrons, on the outside.

It was tidy reasoning but there were problems with it. Matter is both stable and discrete: iron is rigid and hard; other elements are liquids or gases. In chemical reactions, one element interacts with another, to produce a third substance, which is both different from the other two and yet in general stable. However, on Rutherford's model, according to classical physics, no one could understand why the orbiting electrons didn't lose energy and spiral down and collapse into the nucleus. Where did this stability come from?

When Bohr arrived in Manchester, Rutherford had just returned from a conference in Belgium where he had met for the first time both Albert Einstein and Max Planck. Both of them had introduced into physics the concept of the quantum, the idea that energy comes in small discrete packets and is not continuous, as classical physics had it. This was very controversial at the time, but it gave Bohr the idea that would make him famous. As he wrote later, "In the spring of 1912 I became convinced that the electronic constitution of the Rutherford atom was governed throughout by the quantum of action."

After four months in Manchester, Bohr returned to Copenhagen in July to get married. Over the next months he refined his ideas to show that an atom was formed by the successive "binding" of electrons. "One free electron after another would be drawn into the atomic solar system until the number of electrons equalled the charge of the nucleus and the whole system was rendered neutral."[1] But, and this was his real advance, he argued that the binding energy existed as discrete packets—quanta—and so electrons could occupy only *certain stable states* as they orbited the nucleus at different radiuses. Under certain conditions (in chemical reactions, for example), the electron could move between orbits but only by quanta of action, discrete jumps of a minimum size. And so the arrangement of these orbits explained not only the stability of matter but also how the elements differed. The number of electrons in the successive orbits—particularly the outer ones—gave the elements their characteristic properties.

To begin with, Bohr's ideas were, as some historians of science have put it, "intuitive," even "philosophical." Rutherford, a dedicated experimentalist who distrusted theory, was nevertheless supportive of Bohr's efforts and helped him get his ideas into print, in three seminal papers published in 1913. In these papers, known now as "The Trilogy," Bohr explained how the elements fitted into the periodic table, how the electrons were arranged on concentric orbits that related to the element's atomic weight, how one element was related to others, with similar properties, and why some were more reactive than others, depending on the arrangement of electrons in the outermost orbits.

In other words, Bohr had unified physics and chemistry. It was one of the most riveting and important unifications in science and Bohr's "Trilogy" led to the award of the Nobel Prize in Physics in 1922.

Or, it would be truer to say, Bohr had *almost* unified physics and chemistry. At the time of the Nobel ceremony in 1922 there was one uncomfortable, outstanding problem. At that stage a gap in the table of elements occurred at number 72. According to Bohr's theory, the missing element should be similar to zirconium (number 40) and titanium (number 22), the two elements in the same column of the periodic table, rather than resemble the rare earths that occupied the places next to it. But in May 1922 the question of element 72 took a new and dramatic turn. Scientists in France claimed to have discovered a new rare-earth element, which they placed at number 72 in the periodic system.[2] The new element was named celtium, after France. If celtium *was* a rare earth, it would be a major embarrassment for Bohr's theory.

When he had departed his native Copenhagen to travel to Stockholm for the Nobel Prize ceremony he had left two colleagues working on the matter. They were investigating zircon-bearing minerals by X-ray spectrographic analysis. Showing a sense of timing that any theater director would be proud of, the two assistants wired Bohr on the evening immediately before the Nobel ceremony to say that the long-missing element had been found at last and that its chemical properties resembled nothing so much as those of zirconium. The new element was given the name hafnium, for Hafnia, the ancient name of Copenhagen. And so, when Bohr gave his Nobel lecture—as all prize-winners do, on the day after the awards ceremony—he was able to announce this latest result, which did indeed confirm that his theory had successfully unified physics and chemistry.

In the same year that Bohr began his work into the structure of the atom, 1913, Andrew Ellicott Douglass launched his researches, which he wouldn't feel confident enough about publishing until 1928–29. This was the science of dendrochronology, which exploited the links between astronomy, climatology, botany, and archaeology.

In the notebooks of Leonardo da Vinci there is a brief paragraph to the effect that wet years and dry years can be traced in tree rings. The same observation was repeated in 1837 by Charles Babbage—more famous as the man who designed the first mechanical calculators, ancestor of the computer. But Babbage added the notion that tree rings might also be related to other forms of dating. No one took this up for generations, but then Douglass, an American physicist and astronomer, and director of the University of Arizona's Steward Observatory, made a conceptual breakthrough.

His research interest was the effect of sunspots on the climate of the earth, and like other astronomers and climatologists, he knew that, crudely speaking, every eleven years or so, when sunspot activity is at its height, the earth is wracked by storms and rain—one consequence of which is that there is well above average moisture for plants and trees. In order to prove this link, Douglass needed to show that the pattern had been repeated far back into history. For such a project, the incomplete and occasional details about the weather reported in newspapers, say, were woefully inadequate. It was then that Douglass remembered something he had noticed as a boy, an observation familiar to everyone brought up in the countryside. When a tree is sawn through and the top part carried away, leaving just the stump, we see row upon row of concentric rings. All woodsmen, gardeners, and carpenters know, as part of the lore of their trade, that tree rings are annual rings. But what Douglass observed—which no one else had thought through—was that the rings are not of equal thickness. In some years there are narrow rings, in other years the rings are broader. Could it be, Douglass wondered, that broad rings represent what the Bible calls "fat years" (i.e., moist years) and the thin rings represent "lean years"—in other words, dry years?

It was a simple but inspired idea, not least because it could be tested fairly easily. Douglass set about comparing the outer rings of a newly cut tree with official weather records from recent years. To his satisfaction he discovered that his assumption fitted the facts. Next he moved further back. Some trees in Arizona, where he lived, were three hundred years old. If he followed the rings all the way into the pith of the trunk, he should be able to recreate climate fluctuations for his region in past centuries. And that is what he found. Every eleven years, coinciding with sunspot activity, there had been a "fat period," several years of broad rings. Douglass had proved his point: sunspot activity—astronomy—weather and tree growth are related.[3]

But now he saw other uses for his new technique. In Arizona, most of the trees were pine and didn't go back earlier than 1450, just before the European invasion of America. At first Douglass obtained samples of trees cut by Spaniards in the early sixteenth century to construct their missions. Later, he wrote to a number of archaeologists in the American Southwest, asking for core samples of the wood on their sites. Earl Morris, working amid the Aztec ruins fifty miles north of Pueblo Bonito, a prehistoric site in New Mexico, and Neil Judd, excavating at Pueblo Bonito itself, both sent samples. These Aztec "great houses" appeared to have been built at the same time, judging by their style and the objects excavated. But there had been no written calendar in ancient North America, and so no one had been able to place an exact date on the pueblos. Sometime after Douglass received his samples from Morris and Judd, he was able to thank them with a bombshell. "You might be interested to know," he said in a letter, "that the latest beam in the ceiling of the Aztec ruins was cut just nine years before the latest beam from Bonito."[4]

A new science, dendrochronology, had been born, and Pueblo Bonito was the first classical problem it helped to solve. At that point, by overlapping samples from trees of different ages felled at different times, Douglass had an unbroken sequence of rings in southwest America going back first to AD 1300, then to AD 700. Among other things, the sequence revealed that there had been a severe drought, which lasted from 1276 to 1299 and explained why there had been a vast migration at that time by Pueblo Indians, a puzzle that had baffled archaeologists for decades. Botany had resolved one of the prime problems of archaeology.

A third kind of unification took place in the wake of World War II. One of the prime problems in psychology at that time was the number of homeless children in postwar Europe. France,

Holland, Germany, and Russia, in addition to Britain, had all suffered heavy bombing and the disruption of family life that went with it. John Bowlby, a child psychiatrist and psychoanalyst, and head of the Children's Department at the Tavistock Clinic in London, was commissioned in 1949 to write a report for the World Health Organization on the mental health of these homeless children. Preparation of the report gave Bowlby an opportunity to pick the brains of many practitioners, and he visited France, Holland, Sweden, Switzerland, and the United States.

Bowlby's international travels set him on a path that would before long result in the unification of pediatrics, psychoanalysis, ethology—in particular the study of animal behavior seen in an evolutionary context—and the hardening of the idea of the unconscious from a philosophical/psychological concept to a firmly based biological entity. His unification of these disciplines came under ferocious attack at the time from psychoanalysts determined to resist his "biologification" of their discipline. But Bowlby stuck to his guns, and history has vindicated him.

Bowlby's report was written in six months and published in 1951 as *Maternal Care and Mental Health* by the WHO. It was translated into fourteen languages and sold 400,000 copies in its English paperback edition. A second edition, entitled *Child Care and the Growth of Love*, was later published by Penguin.[5]

It was this report that first confirmed for many people the crucial nature of the early months of an infant's life, introducing the key phrase "maternal deprivation" to describe the source of a general pathology of development in children, the effects of which were found to be widespread. The very young infant who went without proper mothering was found to be "listless, quiet, unhappy, and unresponsive to a smile or a coo" and later to be less intelligent, bordering in some cases on the defective. No less important, Bowlby drew attention to a large number of studies which showed that victims of maternal deprivation failed to develop the ability to hold relationships with others, or to feel guilty about their failure. Such children either "craved affection" or were "affectless." Bowlby went on to show that delinquent groups were comprised of individuals who, more than their counterparts, were likely to have come from broken homes, where, by definition, there had been widespread maternal deprivation.

This was quite an achievement on Bowlby's part, but then, in 1951, through Julian Huxley, the eminent biologist, he was introduced to the work of the ethologist Konrad Lorenz, particularly his 1935 paper on imprinting. This is a well-known study now, famously showing that if, at a certain critical stage, young geese are exposed to a stimulus (Lorenz himself in the famous case), they will become "imprinted" on that stimulus. The photographs and film of Lorenz being followed wherever he went by a line of young goslings caught the imagination of everyone who saw it. From then on, Bowlby embraced ethology as a new discipline which could connect with and enrich pediatrics and psychoanalysis and would in time help refine the concept of the unconscious. He was joined at the Tavistock by Mary Ainsworth, a Canadian who moved to London for a time, following her husband's deployment there, and then on to Uganda and finally Baltimore. There she carried out parallel studies, using a variety of observational techniques, and ethological comparisons with other species (such as mother-child interaction in monkeys), to build up their notion of what became famous as "attachment theory."[6]

The significance of this was that it showed how linking one science with another could amplify understanding, different disciplines supporting each other, and lead to new methods of treatment. Bowlby and Ainsworth's alignment of pediatrics and ethology placed the mother-infant bond, and the unconscious motivation that results, on a firm and familiar biological basis and, no less important, situated it in an evolutionary context. According to the Bowlby-Ainsworth theory, attachment was an instinctual response (like imprinting) with the function of binding the infant at a critical period to the mother and vice versa, and in so doing promoting the evolutionary fitness of the offspring.[7]

And, as part of all this, Bowlby said, the child acquires an "internal working model" of itself as either valued and reliable, or as unworthy and incompetent. This was, for Bowlby, the best way to understand the unconscious. "Internal working models" are acquired in the first year of life, well before words, and become less and less accessible to awareness as they become habitual and automatic. This is also because, in mainly dyadic patterns of relating (more or less all that are available at that age), the requirements of reciprocal expectancies are formed exceptionally strongly in such a narrow environment.[8] What had begun life, before Freud, as a purely philosophical/psychological entity now had a firm biological underpinning.

From the Big Bang to Big History

These three examples—involving very different subject matter, spreading across many countries, and extending over decades—jointly introduce the theme of this book.

Convergence is a history of modern science but with a distinctive twist. The twist has been there for all to see, but so far it has not been set out as clearly as it deserves. The argument is that the various disciplines—despite their very different beginnings, and apparent areas of interest—have in fact been gradually coming together over the past 150 years. Converging and coalescing to identify one extraordinary master narrative, one overwhelming interlocking coherent story: the history of the universe. Among its achievements, the intimate connections between physics and chemistry have been discovered. The same goes for the links between quantum chemistry and molecular biology. Particle physics has been aligned with astronomy and the early history of the evolving universe. Pediatrics has been enriched by the insights of ethology; psychology has been aligned with physics, chemistry, and even with economics. Genetics has been harmonized with linguistics, botany with archaeology, climatology with myth—and so on and so on. Big History—the master narrative of the trajectories of the world's great civilizations—has been explained and is being further fleshed out by the interlocking sciences. This is a simple insight but one with profound consequences. Convergence is, as Nobel Prize–winning physicist Steven Weinberg has put it, "the deepest thing about the universe."

This story of the convergence of the sciences—their synthesis, symphysis, and coherence—turns out to offer one timeline of history on which all of the major discoveries that have ever been made can fit. It is not a straight line by any means but a definite line nonetheless, not unlike a very long and complicated backbone, or spine, which curves and is made up of vertebrae of different sizes. I further argue that the order that emerges from this convergence—and the way one science supports another—gives scientific understanding an unrivaled authority as a form of knowledge and that we should therefore expect it to *extend* its reach in the years ahead, into fields not traditionally associated with science. In truth, it is already doing so and we should welcome that fact. The proven interlocking nature of science now helps to guide future research.

Not all the links and overlaps in the story are equally strong. Niels Bohr's amalgamation of physics and chemistry was fundamental, as was the later linking of quantum chemistry to molecular biology, by Linus Pauling and others (chapter 9). In more recent decades, the linking of fundamental particles to the early history of the evolving universe (chapter 11), and the "hardening" of psychology—the links between behavior and brain chemistry, for example—are no less fundamental (chapter 16). The same too goes for the overlaps that have also been revealed between genetics and archaeology, and between genetics and archaeology and language (chapter 12). At other times, the overlaps—while not exactly trivial—are more helpful and intriguing than fundamental. The example of tree-ring chronology is a case in point, as are some of the other scientific dating technologies that have been developed, the potassium/argon method, for example (chapter 12). They show that not just botany but also physics, molecular biology, and genetics can help us reconstruct history. Importantly, the different dating mechanisms are consistent with one another, so that ancient history in particular is now an interdisciplinary branch of science.

But—and this is the underlying point—all the connections and overlaps, all the patterns and hierarchies that have been revealed, whether fundamental or otherwise, dovetail together conceptually. There are no exceptions, no important ones anyway. Scientific discoveries repeatedly come together, in all manner of ways, to support one another, to tell one coherent, interlocking story. In an important sense, and to use another analogy, it is as if this story has its own form of gravity as—like particles in cooling gases—the different chapters come together to form a solid narrative.

That narrative leads from the origins of the universe in a Big Bang 13.8 billion years ago, up through the creation of elementary particles, the formation of the lighter then the heavier chemical elements, the formation of the stars and planets, including our own sun, the evolution of the broad structure of the universe (the way the galaxies are laid out), of the gases that coalesced to comprise the rocks of the earth, how those rocks align in the way that they do, how the earth has aged, how the ice ages have come and gone, why the continents are arranged around the globe as they are, why the oceans circulate the planet in a particular pattern, where and when primitive forms of life developed, how ever more complex molecules and organisms came to be, how sex evolved, why trees and flowers take the form that they do, why leaves are green, why some animals have six limbs and others four, why the plants and animals (including people) are distributed across the earth in the way that they are, how major catastrophes have given rise to widespread myths and shape our beliefs, how accuracy developed and became important, how and why and where science itself emerged, culminating (so far) in humankind and the very different civilizations that populate the globe. Indeed, this one story shows *why* there are different civilizations that populate the globe where they do. The convergence of the sciences helps us explain the greatest single story there could be—Big History.

An Epic Detective Story and a New Dimension

I do not, however, tell the story by beginning at the beginning and ending at the end. It is much more revealing, more convincing, and altogether more thrilling to tell the story as it emerged; as it began to fall into place, piece by piece, chapter by chapter, converging tentatively at first, but then with increasing speed, vigor, and confidence. The overlaps and interdependence of the sciences, the patterns and hierarchies of the discoveries in different fields, the underlying order that they are gradually uncovering, is without question one of the most enthralling aspects—perhaps the most enthralling aspect—of modern science. It is in effect a collective detective story of epic dimensions. The convergence and the emerging order—even a kind of unity—between the sciences is one of the most important and satisfying elements in scientific knowledge, and all the more convincing because nobody went looking for it in the first place.

Nor do I begin, as many science histories do, in ancient Greece, the so-called Ionian Enchantment, or with the discoveries of Copernicus and Galileo, or with the scientific revolution of the seventeenth century. I begin much later, in the 1850s—a crucial decade as I show—because that is when the convergence began, when the interconnections and overlaps between the various disciplines first started to show themselves in two fundamental areas and so added a whole new dimension to science, one that hadn't been fully grasped until then.

It was in the 1850s that the idea of the conservation of energy was first aired, which brought together recent discoveries in the sciences of heat, optics, electricity, magnetism, food, and blood chemistry. Almost simultaneously, Darwin's theory of evolution by natural selection brought together the new sciences of deep-space astronomy, deep-time geology, paleontology, anthropology, geography, and biology. These two theories comprised the first great coming together, meaning that the 1850s was in many ways the most momentous decade in the annals of science, and possibly, as it has turned out, the years which saw the greatest intellectual breakthrough of all time: the realization of the way one science supports another, the beginning of a form of understanding like no other. This was in every way a new era intellectually.

I am not aware that anyone has told the history of science, or the history of the universe, in quite this way before. This is the distinctive twist that, I suggest, sets this science history apart.

I *am* aware that some historians of science, social scientists, and philosophers object to the very idea that there is unity or order in the sciences. But I argue that the story of convergence and the emerging order described in this book speaks for itself, and I address several of their objections in the Conclusion.

The idea that the sciences are linked in some hierarchical way is not new, of course, and is known as reductionism. Although reductionism has been criticized—especially in the last twenty to thirty years, even as the evidence in its favor has grown stronger than ever—for the most part, leading scientists themselves have overridden these objections. Such figures as George Gaylord Simpson, Philip Anderson, Ilya Prigogine, Abdus Salam, Steven Weinberg, and Robert Laughlin (the last five being Nobel Prize winners) have all described themselves as wholehearted reductionists. Edward O. Wilson, the noted sociobiologist, put it this way: "Reductionism is the primary cutting tool of science."

As this book was being finalized, there came the news that researchers had inserted two small silicon chips into the posterior parietal cortex of a tetraplegic individual, ninety-six microscopic electrodes that could record the activity of about a hundred nerve cells at the same time. Based on previous work with monkeys, which guided the researchers to a specific area of the human brain, they found that they could reliably read out where the patient *intended* to move his paralyzed arm by analyzing the differing patterns of these hundred cells. This information was then used—bypassing his damaged spinal cord—to enable him to direct a robot arm either to pick up a beer or move a cursor on a computer screen. The researchers could even predict how fast he wanted to move, and whether he wanted to move his left or right arm. In a related experiment, by showing the activity of a single nerve cell on a screen, the patient was able to modulate the cell activity. The experiment was very specific. One nerve cell, for example, would increase its activity when he imagined rotating his shoulder, and decrease its activity when he imagined touching his nose. The specificity of this experiment, and the fact that it could throw light on the man's *intentions*, not just his actual movements, offers great hope for the future, but from our point of view it takes reductionism to a new level, uniting still further psychology and physics.

The Beauty of Deep Order

That said, there is no final order yet, and there may never be. But the order that has emerged already is impressive enough. Order, in particular spontaneous order, is now a major interest of science (chapter 18).

And of course the story of this book is more than just a narrative—for there are two deeper implications of the order that convergence is producing.

The first is that alluded to earlier. Because the convergence—the emerging order—is so strong, and so coherent, science as a *form* of knowledge is beginning to invade other areas, other systems of knowledge traditionally different from or even opposed to science, and is starting to explain—and advance—*them*. Science is invading—and bringing order to—philosophy, to morality, to history, to culture in general, and even to politics (see chapters 14, 15, and 19). Critics object that this is a form of intellectual imperialism, but our newspapers are peppered every day with reports, for example, of the latest psychological research having a bearing on our honesty, generosity, trustworthiness, proneness to violence, and much else. This genie can't be put back into its bottle.

It is not too much to say that the overall coherence and order revealed by the convergence of the sciences is ushering in a new phase of history. No other form of knowledge has the coherence and order that the converging sciences have brought about. The methods and infrastructure of science are invaluable, are indeed unrivaled aspects of modern democracy, and will shape society in all its manifestations even more in the future than they have in the past, and rightly so. This is a quintessentially contemporary story.

The second aspect of the order that is emerging relates to order itself. Order, the way even inanimate matter spontaneously organizes itself in nature (without, it should be said, any input from a supernatural power), has emerged in recent decades as one of the most important new topics. The very idea that there is a preexisting order *in* nature—a deep order underlying even "chaoplexity" (a mix of chaos and complexity), as appears to be the case—sounds itself very much like a philosophical conundrum as important as any other. Spontaneous order is being explored by physicists, chemists, biologists, and mathematicians and has been found to occur among elementary particles, among molecules, in complex systems, in living things, in the brain, in mathematics, even in traffic. All of which gives an idea of how central the subject now is (chapters 17 and 18). A breakthrough in this area could have breathtaking consequences, not least for our understanding of evolution (chapter 18).

And so there is no other story quite like the one told in this book. Convergence is, as Steven Weinberg says, and without exaggeration, the most fundamental story that could ever be imagined.

Nor, finally, should we overlook the fact that the way the sciences are coming together may offer comfort of a kind. Not quite a religious comfort perhaps, but the converging sciences—the emerging order in nature—certainly appears to offer an intellectual/philosophical *satisfaction*, a form of beauty almost and, yes, for the time being at least, a mystery as to what that underlying order might ultimately mean. In this, the converging sciences sustain their power to *thrill*.

Introduction

"THE UNITY OF THE OBSERVABLE WORLD"

W e begin in the mid-nineteenth century, and in the most unlikely of places. Walking on a beach in Cornwall in 1852, a passerby chanced upon a length of driftwood that had been washed ashore following a recent storm. There was writing on the plank. It read: *"Mary Somerville."* The ship of that name, which had been commissioned in 1834, had plied between Liverpool, India, and China, carrying cotton, tea, and flour. She had foundered on a return journey shortly before.

In that year, 1834, a wealthy Liverpool shipbuilder, William Potter, had asked the real-life Mary Somerville if he could name a merchant ship in her honor and, at the same time, obtain a copy of a bust that had been made of her for use as a figurehead of the ship. The original bust, recently completed, had been carved by Sir Francis Chantrey, the celebrated society portrait sculptor, whose other subjects included such eminences as King George III, King George IV, Prime Minister William Pitt the Younger, President George Washington, and scientists James Watt and John Dalton. The bust had been commissioned by the Fellows of the Royal Society of London and placed in the society's great hall.[1]

There was never any question that, as a woman, Mary Somerville would be *elected* to the Royal Society as a fellow—women were not allowed even to attend lectures there until 1876. But, as the commissioning of the bust and the dedication of the merchantman testify, she had nonetheless made her mark. And though it is unusual, it is by no means unsatisfactory to begin a book about science with an account of a remarkable *woman*, who so admirably introduces our theme.

She was born Mary Fairfax in Jedburgh, on the Scottish borders, in December 1780. Her mother had only just returned from waving off her husband—a naval officer—on a series of voyages from which he would not return until Mary was a girl of eight. During the intervening years, she received no formal education and was allowed "to run wild." When her father eventually returned home, he was alarmed to find that Mary had failed to master the skills of reading, writing, and account-keeping "that would make her a suitable wife" and so sent her away to boarding school, where she was taught dancing, painting, music, cookery, needlework, and "elementary geography." [2]

She had a more serious cast of mind, however, and taught herself algebra, using puzzles set in popular magazines as a way to begin. Mary found that she had a natural aptitude for mathematics. An avid book lover, she had no fortune to speak of but was fortunate in being beautiful, and at twenty-three, she married. She and her husband, Captain Samuel Greig, set up home in London, where he held a commission in the Russian Navy and was Russian consul. They had two sons, but Mary was lonely inside the marriage, and when her husband died suddenly, although she was inconsolable at first, she returned to Scotland. [3] Here, now having a small income deriving from her late husband's position, she was able to cultivate the kind of life she preferred. All the more so after she met her cousin, William Somerville, who soon proposed. This was a much better match. Both held liberal views on politics, religion, and education and both were interested in science. William, a military doctor, had done pioneering work on natural history and ethnological exploration in South Africa (plus some other, more secret, military duties).

"The Most Extraordinary Woman in Europe"

It was now that Mary's intellectual life really began to take off. The couple moved first to Edinburgh. This was the time of the Scottish Enlightenment; many of the men there had liberal views about the role of women, and among the individuals with whom she could share her interest in mathematics were the likes of James Hutton and John Stuart Mill. This was the heyday of the *Edinburgh Review*, one of the best periodicals in Britain, or anywhere, but in the early nineteenth century the reformers of British science had launched a new journal, which focused on mathematical challenges (this was a fashion of the times). The publication was entitled *New Series of the Mathematical Repository*, and in June 1811 Mary was delighted to learn that she had won the Prize Question, for which she received a silver medal with her name engraved on it.[4]

James Secord, the Cambridge-based historian of Victorian science, says that Mary felt "most intensely alive and completely herself" in mathematics. For her, he writes, "the practice of mathematics was a form of theological engagement. . . . For Somerville, the divine transcendence of God's power could most fully be experienced by those who—like herself—understood the language of mathematics." Or, as she herself put it, "These formulae, emblematic of Omniscience, condense into a few symbols the immutable laws of the universe. This mighty instrument of human power itself originates in the primitive constitution of the human mind, and rests upon a few fundamental axioms which have eternally existed in Him who implanted them in the breast of man when He created him after His own image." Mary was from the start interested in how the manifest diversity of the world could be reduced to those few fundamental axioms.

Then she and William moved to London, where they became well known among those with scientific interests: at least twenty-six of their regular friends were Fellows of the Royal Society, "possibly the most distinguished corps that any author ever commanded during a lifetime."[5] Mary Somerville took this in her stride. She was well connected socially but became famous, says Allan Chapman, in his biography, via her letters, by her conversation, and by the fact that everybody in intellectual London knew of this singular woman who had mastered the most abstruse mathematics of the age, and had acquired from her studies a sophisticated grasp of how physical science worked. Sir David Brewster, a physicist and mathematician who was principal of both St. Andrews and Edinburgh universities, described her as "the most extraordinary woman in Europe."[6]

In the long run, two things set her apart, in addition to that grasp of mathematics. Like other Grand Amateurs of the day, she took part in simple experiments, in her case into the connection between magnetism and sunlight.[7] This was in the excited wake of Hans Christian Ørsted's discovery of a connection between magnetism and electricity (see chapter 1), and the results she obtained were interesting enough for William, her husband, himself an FRS, to read her account of them before the Royal Society. The papers were subsequently published in the society's *Philosophical Transactions* and in that way were made available to a much wider range of readers. Offprints were sent to such figures as the astronomer Pierre-Simon Laplace and the chemist Joseph Louis Gay-Lussac in Paris, and to Ørsted himself in Copenhagen.

On the strength of her accomplishments, Henry Brougham suggested that Mary contribute an account of Newton's *Philosophiae Naturalis Principia Mathematica* and Laplace's famous book on the heavens, *Mécanique céleste*, to the publishing program of the Society for the Diffusion of Useful Knowledge. Brougham—an eccentric Scottish lawyer who was a guiding spirit behind the 1832 Reform Act, and was one of those individuals who had a finger in every pie—had founded the SDUK in 1826 with the aim of spreading knowledge until it "has become as plentiful and as universally diffused as the air we breathe." The SDUK published cheap books in weekly parts, topics ranging from brewing to hydraulics and from insects to Egyptian antiquities. Its most successful venture was the weekly illustrated *Penny Magazine*, which at its peak achieved a circulation of more than 200,000.[8]

The first books that Mary wrote were too detailed and too thorough for a penny magazine readership and so not at all suitable for the SDUK. She told Brougham along the way that her book would have to discuss the calculus, so that was bound to limit its appeal. But John Murray, the London publisher, who was himself a fixture on the intellectual scene in the capital, snapped it up, and so began Mary's successful writing career in science, the second thing that set her apart from other women. In all she wrote five books, *Mechanism of the Heavens* (1831), *On the Connexion of the Physical Sciences* (1834), *Physical Geography* (1848), *On Molecular and Microscopic Science* (1869), and *Personal Recollections* (1874, posthumous).

The book that concerns us is the second one, *On the Connexion of the Physical Sciences*, generally regarded as her most important work. She was preparing it at the peak of her renown, when the Chantrey bust for the Royal Society was being carved and when, a year later—to the envy of many—she was awarded an annual pension of £200 by the government for her services to science. (It was later increased to £300, the same as Michael Faraday and John Dalton received.)

The argument in *Connexion* was sharper then than it might seem now. Its aim was to reveal the common bonds—the links, the convergence—between the physical sciences at a time when they were otherwise being carved up into separate disciplines. Mary was very deliberately her own woman.

"Demonstrating the Unity of the Observable World"

The professed aim of her book, embodied in the title, was to *draw together* a range of subjects in the physical sciences that were undergoing unprecedented change. Secord again: "Through its wide readership, *Connexion* became a key work in transforming the 'natural philosophy' of the seventeenth and eighteenth centuries into the 'physics' of the nineteenth, demonstrating the unity of the observable world."

A key work, indeed, as we shall see, but not quite the first. The desire for an all-embracing vision, even for a cosmic order, is an ancient concern, dating at least from Aristotle, the so-called Ionian Enchantment. The great chain of being, derived from Plato, Aristotle, and others, specified a hierarchy which ranged from nothingness through the inanimate world, into the realm of plants on up through tame and wild animals and then humans, and above that through angels and other "immaterial and intellectual" entities, reaching at the top a superior or supreme being, a terminus or Absolute. In the Middle Ages, in his *Summa Theologica*, Thomas Aquinas had attempted to reconcile Aristotle's science and Christianity. Four hundred years later, Newton brought order to the heavens and other matters, like motion and light. The Enlightenment held to the idea of the unity of all knowledge, Descartes had a vision of knowledge as a system of interconnecting truths that could eventually be abstracted into mathematics, Condorcet pioneered the application of mathematics to social science, while Schelling proposed a "cosmic unity" of all things, even though he thought it was beyond the understanding of man. Linnaeus attempted to order the living world in Latin.

Somerville's approach was much more modern. What she wanted to write about seems to have been clear in her own mind, but the term "the physical sciences," which she used in her title, was only then being formulated. By that time, several philosophers had tried to make physics more unified and coherent, of whom the most successful were John Playfair (1748–1819) and John Herschel (1792–1871). As early as 1812, in the first volume of his *Outlines of Natural Philosophy*, Playfair aimed frankly "to have the elementary truths of Natural Philosophy brought into a small compass, and . . . arranged in the order of their dependence on one another." He distinguished natural philosophy from chemistry. Under natural philosophy he grouped dynamics, mechanics, statics, hydrodynamics, astronomy, optics, electricity, and magnetism. His view—widely shared—was that gravitation was the single principle "which pervades all nature, and connects together the most distant regions of space, as well as the most remote periods of duration." Playfair thought it "probable" that a similar principle obtained for non-gravitational matter.[9]

In his *Preliminary Discourse on the Study of Natural Philosophy* (1831), Herschel used force, motion, and matter as the basis of categorization of the sub-branches of science. He thought there were two great divisions of the science of force—dynamics and statics—and the subdivisions were: mechanics, crystallography, acoustics, light and vision, astronomy and celestial mechanics, geology, mineralogy, chemistry, heat, magnetism, and electricity. As this shows, people were groping for similarities, but not really finding them. It was a time when the differences between substances and processes were still being explored.

Mary Somerville would also have been aware that the French since the late eighteenth century had recognized "*la physique . . .* as a branch of science separate from mathematics on the one hand and chemistry on the other." The properties of matter, heat, light, electricity, and magnetism, plus meteorology, comprised "*la physique.*"

Connexion was therefore a sort of climax to what was in fact a somewhat untidy and unformed nineteenth-century movement to put some unity into natural philosophy. Her book was forceful, and reviewers praised it for exactly that—for bringing the physical sciences together in a new way. The *Mechanics' Magazine* held the book so important that they said it should not be on a bookshelf at all. "Instead of that we say—Read it! Read it!" James Clerk Maxwell, whose great works we shall meet in chapter 1, said *Connexion* was among "those suggestive books, which put into a definite, intelligible, and communicable form, the guiding ideas that are already working in the minds of men of science, so as to lead them to discoveries, but which they cannot yet shape into a definite statement."[10]

Somerville presented mathematics as the most promising

source of ultimate unity, though she accepted that meant it would only ever be available to a very few. With this in mind, she therefore advanced her argument about mathematics without using a single equation.

She wrote most of the book in secret, uncertain of how its female authorship would be received, though she was already celebrated across Europe for her mathematical accomplishments (which is why Brougham had suggested the SDUK project in the first place). And, as Joanna Baillie, the Scottish poet and drama- tist, pointedly remarked, Somerville had "done more to remove the light estimation in which the capacity of women is too often held, than all that has been accomplished by the whole sisterhood of poetic damsels and novel-writing authors." [11]

The first edition of two thousand copies was priced at seven shillings and sixpence and quickly sold out, the book remaining in print for over forty years, in ten editions. It was translated into German, French, and Italian, and publishers in Philadelphia and New York issued pirated editions. The *Athenaeum* conceded that the book was "at the same time a fit companion for the philoso- pher in his study and for the literary lady in her boudoir."

The Search for Meaningful Patterns and "Increasingly Higher Levels of Generalization"

The *Connexion* of the title was further explained in a preface: "The progress of modern science, especially within the last five years, has been remarkable for a tendency to simplify the laws of nature, and to unite detached branches by general principles. In some cases identity has been proved where there appeared to be nothing in common, as in the electric and magnetic influences; in others, as that of light and heat, such analogies have been pointed out as to justify the expectation, that they will ultimately be referred to the same agent; and in all there exists such a bond of union, that proficiency cannot be attained in any one without knowledge of the other." And she concluded: "Innumerable instances might be given in illustration of the immediate connexion of the physical sciences, most of which are united still more closely by the common bond of analysis which is daily extending its empire, and will ultimately embrace almost every subject in nature in its formulae." [12]

Kathryn Neeley reminds us that the aims of science then were not quite the same as they are now. One of the differences was that, since science was not yet professionalized, or as highly specialized as it would become, "omniscience prevailed as an intellectual ideal." She says that early Victorian intellectuals thrived on debate and controversy "but took a unitary approach to intellectual life." They saw culture as a whole and were ambivalent about specialization since it threatened that unity. People should know "something of everything and everything of something." John Herschel, a friend of Somerville, declined the presidency of the British Association for the Advancement of Science (BAAS) because he feared the organization would encourage the compartmentalization of science.

This was an essentially religious view, which held that science advanced by achieving increasingly higher levels of generalization, an approach that was first aired in Germany. "These higher levels of generalisation usually took the form of new laws with greater explanatory power . . . a desire for more and more widely applicable laws to interconnect the diverse phenomena. . . . Increasingly higher levels of generalisation could be achieved reliably only through the accumulation of increasingly large amounts of detailed information and the search for meaningful patterns and analogies." [13] Each early Victorian saw his work as part of an "intellectual totality."

Moreover, this unitary character of intellectual life was regarded as a form of the sublime. "The prevailing belief was that science could not be taught well without reference to the sublime truths of natural theology and that the scientific study of nature revealed God. . . . In the scientific sublime, the reader links with the great in the form of an encounter with the attributes of God revealed in nature by science." Unification was akin to an "enlarged power," a power of intellect embodied or made manifest in science. All this was certainly Mary Somerville's view—that coherence *was* a power.[14]

"United by the Discovery of General Principles"

The book was widely reviewed, almost always favorably.[15] Arguably the most interesting and influential comments were those in the March issue of the *Quarterly Review* by William Whewell, master of Trinity College, Cambridge, and himself the author of several books about the history of science. He acknowledged that "the tendency of the sciences has long been an increasing proclivity to separation and dismemberment. Formerly, the 'learned' embraced in their wide grasp all the branches of the tree of knowledge; the Scaligers and Vossiuses of former days were mathematicians as well as philologers, physical as well as antiquarian speculators.* But those days are past. . . . If a moralist, like Hobbes, ventures into the domain of mathematics, or a poet, like Goethe, wanders into the field of experimental science, he is received with contradictions and contempt . . . the disintegration goes on . . . physical science itself is endlessly subdivided, and subdivisions insulated. . . . The mathematician turns away from the chemist . . . the chemist is perhaps a chemist of electro-chemistry; if so, he leaves common chemical analysis to others. . . . And thus science, even mere physical science, loses all trace of unity."[16]

And then: "A curious illustration of this result may be observed in the want of any name by which we can designate the students of the knowledge of the material world collectively. We are informed that this difficulty was felt very oppressively by the members of the British Association for the Advancement of

---

* Joseph Justus Scaliger (1540–1609) and Gerardus Vossius (1577–1649) both helped extend history and humanist scholarship beyond theology.

Science, at their meetings at York, Oxford, and Cambridge, in the last three summers [i.e., since the very inception of the BAAS]. There was no general term by which these gentlemen could describe themselves with reference to their pursuits. *Philosopher* was felt to be too wide and too lofty a term, and was very properly forbidden them by Mr. [Samuel Taylor] Coleridge, both in his capacity as philologer and metaphysician; *savants* was rather assuming, besides being French instead of English; some ingenious gentleman [in truth, this was Whewell himself, though he didn't say as much in the review] proposed that, by analogy with *artist*, they might form *scientist*, and added that there could be no scruple in making free with this termination when we have such words as *sciolist*, *economist*, and *atheist* but this was not generally palatable; others attempted to translate the term by which the members of similar associations in Germany have described themselves, but it was not found easy to discover an English equivalent for *natur-forscher*. The process of examination which it implies might suggest such undignified compounds as *nature-poker*, or *nature-peeper*, for these *naturæ curiosi*; but these were indignantly rejected."

This was thus the first public airing of the term "scientist," and Whewell, it should be noted, was fond of—and good at—neologisms. Besides "scientist," he is credited with coining the word "physicist" and with suggesting "ion," "anode," and "cathode" to Michael Faraday.

Having coined a word that would in time become commonplace, his review continued, "The inconveniences of this division of the soil of science into infinitely small allotments have been often felt and complained of. It was one object, we believe, of the British Association, to remedy these inconveniences by bringing together the cultivators of different departments. To remove the evil in another way is one object of Mrs. Somerville's book. If we apprehend her purpose rightly, this is to be done by showing how detached branches have, in the history of science, been united by the discovery of general principles."

One reason Whewell was sympathetic to Somerville was because, in his own way, in devising the neologism "scientist," he was engaged in a broadly similar thought process to hers— stressing the similarities between the sciences (including, in his case, their methods), rather than concentrating on the differences. In his 1840 synthesis, *The Philosophy of the Inductive Sciences*, Whewell was the first to use the word "consilience," to mean the "jumping together" of knowledge "by the linking of facts and fact-based theory across disciplines to create a common ground-work of explanation." [17]

But in fact this is as far as the connections of *Connexion* went. The book consisted of thirty-seven sections, over four hundred pages, and covered such topics as "Lunar Theory," "Perturbations to Planetary Orbits," "Tides and Currents," "Laws of Polariza-tion," and "Electricity from Rotation and from Heat." There was no narrative structure, or any large-scale unfolding of logic, and most of the connections listed were those between two of these narrow topics, rather than any deeper underlying totalizing prin-ciples (apart from the mathematical ones which were closest to Mary Somerville's heart). Knowledge of the principle of matter, for example, she said, is needed for predicting its effect on light. She explained why we can look at the sun in the evening, when it is near the horizon, and not look at it at midday, high in the sky. Sound is capable of reflection from surfaces, according to the same laws as light. And so, given what was just around the corner—the great unifying theories of energy conservation and evolution by natural selection, which are the subjects of Part One of this book—the connections in *Connexion* were notable as an early attempt to construct such linkages. In the broader scheme of things, however, they were, in James Secord's verdict, "tame." [18]

Their significance lies in their broad argument, at a time when the sciences were fragmenting, and in their timing.[19] The tenth edition of *Connexion* appeared in 1877, and there were to be no more, for by then the two great unifying theories—arguably of all time—had been announced to the world in the same decade, the 1850s. They had been fleshed out, and, as we shall see, the connections they were about to reveal did indeed make Mary Somerville's argument, for all its originality, seem tame.

A Divinely Inspired Convergence

By the time the last edition of her book appeared in 1877, five years after Somerville's death, it was very much out of date. She had reached the ripe age of nearly ninety-two, and was much missed. And her impact had been such that, two years later, a new all-women college at Oxford was named in her honor, producing its share of no less remarkable figures—Indira Gandhi, Dorothy Hodgkin, and Margaret Thatcher among them.

For its author, the fundamental lesson of *Connexion* was religious. Mary Somerville had rebelled against the strict practices and traditions of the Presbyterian faith into which she had been born, with its belief in original sin, hell, and damnation. Her own most powerful experiences were with nature and, above all, with mathematics. This was where God showed himself, she believed—in the purity of mathematics and in the way a few equations *united* the manifest diversity of the observable world. Such unification was evidence of a divinely inspired convergence that might, one day, reach its end point.

This link with God is no longer so widespread a view, despite the fact that, as this book will argue, the connections between the sciences are much, much stronger than Mary had evidence for. It was by no means her fault that the sciences had not revealed more of their emerging order when she wrote her book. But they began to converge almost immediately after *Connexion* first appeared and well within her lifetime. They have gone on doing so.

The convergence—and perhaps the potential unity—of the sciences is a story all to itself. It is perhaps the most interesting aspect of the history of what can at times seem very different disciplines. As the Stanford-based, Nobel Prize–winning physicist Robert Laughlin has put it, "All of us secretly wish for an ultimate theory, a master set of rules from which all truth would flow." In our secular age this convergence may not point to God—but to what *does* it point?

Unlike Mary Somerville's book, *Convergence* does have a narrative. In fact, it has two narratives. One is a chronological account—switching this way and that—as one breakthrough after another in the various sciences uncovered what were at first quite disparate phenomena, but then, as time went by, began to interconnect. The second narrative is the story of where these interconnections lead us: to a unified—ordered—historical account of the universe we inhabit and our own place within that totality. The piecemeal way this story was unraveled is like a detective story, or a successful effort at code-breaking. And, to repeat what was said earlier, the narrative that has emerged is all the more impressive because, to begin with at least, no one went looking for it.

We shall return to the question of where that narrative is now going at the very end of the text. Meanwhile, what Mary Somerville began we shall endeavor to carry forward.

PART ONE

The Most Important
Unifying Ideas of All Time

The world knows that in 1851 Victorian Britain held a Great Exhibition in London's Hyde Park, under a startling new construction, made almost entirely of glass: the Crystal Palace. Over five months, 6 million people from twenty-eight different countries visited the exhibition, the main theme of which, as one reviewer put it, in the London *Times*, was "The Gifts of Science to Industry." At that time, more people probably knew about science, and its practical possibilities, than ever before. This was no more than fitting because the decade around and following the Crystal Palace exhibition was arguably the most important in the history of science.

Most histories of science begin either with the Ionian Enchantment, in ancient Greece, with the observations and discoveries of Copernicus-Kepler-Galileo-Newtonian astronomy, or with the creation of the Royal Society in London and the Académie des Sciences in Paris in the 1660s, the so-called scientific revolution. Our theme—the coming together of the sciences, the great convergence—starts later, beginning in the 1850s. For, besides the Crystal Palace and all that it represented, that decade saw the emergence of the two most powerful unifying theories of all time.

The idea of the Conservation of Energy and the theory of Evolution by Natural Selection were both introduced to the world in the 1850s. Each was the fruit of the coming together of several sciences: the sciences of heat, optics, electricity, magnetism, food, and blood chemistry in the case of the conservation of energy; deep-space astronomy, deep-time geology, paleontology, anthropology, geography, and biology in the case of evolution. This was the first great coming together, meaning that the 1850s were in many ways the most momentous decade in the history of science, and possibly, as it has turned out, the years which saw the most exciting intellectual breakthrough of all time—the way one science supports and interconnects with another, the beginning of a form of understanding like no other in history. As a result, there was a massive increase in the authority of science, an authority that has gone on expanding as the emerging order of the overlapping and increasingly interlinked sciences has been progressively exposed. These interconnections have been there for all to see, but until now, they have scarcely received the attention they merit.

1

"THE GREATEST OF ALL GENERALIZATIONS"

One morning in late August 1847, James Prescott Joule, a wealthy Manchester brewer but also a distinguished physicist, was walking in Switzerland, near Saint-Martin, beneath the Col de la Forclaz, in the south of the country, not too far from the Italian border. On the road between Saint-Martin and Saint-Gervais he was surprised to meet a colleague, William Thomson, a fellow physicist, later even more distinguished as Lord Kelvin. Thomson noted in a letter the next day to his father—a professor of mathematics—that Joule had with him some very sensitive thermometers and asked if Thomson would assist him in an unusual experiment: he wanted to measure the temperature of the water at the top and bottom of a local waterfall. The request was particularly unusual, Thomson suggested in his letter, because Joule was then on his honeymoon.

The experiment with waterfalls came to nothing. There was so much spray and splash at the foot of the local cataract that neither Joule nor Thomson could get close enough to the main body of water to make measurements. But the idea was ingenious and it was, moreover, very much a child of its time. Joule was homing in on a notion that, it is no exaggeration to say, would prove to be one of the two most important scientific ideas of all time, and a significant new view of nature.

He was not alone. Over the previous few years as many as fifteen scientists, working in Germany, Holland, and France as well as in Britain, were all thinking about the conservation of energy. The historian of science Thomas Kuhn says that there is "no more striking instance of the phenomenon known as simultaneous discovery than conservation of energy." Four of the men—Sadi Carnot in Paris in 1832, Marc Seguin in Lyon in 1839, Carl Holtzmann in Mannheim in 1845, and Gustave-Adolphe Hirn in Mulhouse in 1854—had all recorded their independent convictions that heat and work are quantitatively interchangeable. Between 1837 and 1844, Karl Mohr in Koblenz, William Grove and Michael Faraday in London, and Justus von Liebig in Giessen all described the world of phenomena "as manifesting but a single 'force,' one which could appear in electrical, thermal, dynamical, and many other forms but which could never, in all its transformations, be created or destroyed."[1] And between 1842 and 1847, the hypothesis of energy conservation was publicly announced, says Kuhn, by four "widely scattered" European scientists—Julius von Mayer in Tübingen, James Joule in Manchester, Ludwig Colding in Copenhagen, and Hermann von Helmholtz in Berlin, all but the last working in complete ignorance of the others.

Joule and his waterfalls apart, perhaps the most romantic of the different stories was that of Julius von Mayer. For the whole of 1840, starting in February, Julius Robert von Mayer served as a ship's physician on board a Dutch merchantman to the East Indies. The son of an apothecary from Heilbronn, Württemberg, he was a saturnine, bespectacled man who, in the fashion of his time, wore his beard under—but not actually *on*—his chin. Mayer's life and career interlocked in intellectually productive yet otherwise tragic ways. While a student he was arrested and briefly imprisoned for wearing the colors of a prohibited organization. He was also expelled for a year and spent the time traveling, notably to the Dutch East Indies, a lucky destination for him, as it turned out. Mayer graduated in medicine from the University of Tübingen in 1838, though physics was really his first love, and that was when he enlisted as a ship's doctor with the Dutch East India Company. The return to the East was to have momentous consequences.

On the way there, in the South Atlantic, off South Africa, he observed that the waves that were thrown about during some of the wild storms that the three-master encountered were warmer than the calm seas. That set him thinking about heat and motion. Then, during a stopover in Jakarta in the summer of 1840, he made his most famous observation. As was then common practice, he let the blood of several European sailors who had recently arrived in Java. He was surprised at how red their blood was—he took blood from their veins (blood returning to the heart) and found it was almost as red as arterial blood. Mayer inferred that the sailors' blood was more than usually red owing to the high temperatures in Indonesia, which meant their bodies required a lower rate of metabolic activity to maintain body heat. Their bodies had extracted less oxygen from their arterial blood, making the returning venous blood redder than it would otherwise have been.[2]

Heat and Motion Are the Same

Mayer was struck by this observation because it seemed to him to be self-evident support for the theory of his compatriot, the chemist and agricultural specialist Justus von Liebig, who argued that animal heat is produced by combustion—oxidation—of the chemicals in the food taken in by the body. In effect, Liebig was observing that chemical "force" (as the term was then used), which is latent in food, was being converted into (body) heat. Since the only "force" that enters animals is their food (their fuel) and the only form of force they display is activity and heat, then these two forces must always—by definition—be in balance. There was nowhere else for the force in the food to go.

Mayer originally tried to publish his work in the prestigious *Annalen der Physik und Chemie*. Founded in 1790, the *Annalen der Physik* was itself a symptom of the changes taking place. By the 1840s it was the most important German journal of physics, though many new journals proliferated in that decade. The *Annalen*'s editor since 1824, Johann Christian Poggendorff, a "fact-obsessed experimentalist and scientific biographer," had a very firm idea of what physics was. By the middle of the century, there had emerged "a distinctive science of physics that took quantification and the search for mathematical laws as its universal aims." (This, it will be recalled, is what drew Mary Somerville to the subject.) Poggendorff could make or break scientific careers. All the more so because he edited the *Annalen* for fifty-two years, until he died in 1877.

Owing to a number of basic mistakes, however, due to his poor knowledge of physics, Mayer's paper was rejected by Poggendorff. Disappointed but undeterred, he broached his ideas to the physics professor at Tübingen, his old university, who disagreed with him but suggested some experiments he might do to further develop his ideas. If what Mayer was proposing was true, the professor said, if heat and motion are essentially the same, water should be warmed by vibration, the same thought that had occurred to Joule.

Mayer tried the experiment, and found not only that water *is* warmed by vibration (as he had spotted, months before, aboard the merchantman), but that he was able to measure the different forces—vibration, kinetic energy, and heat. These results, "Remarks on the Forces of Inanimate Nature," were therefore published in the *Annalen der Chemie und Pharmacie* in 1842, and it was here that he argued for a relationship between motion and heat, that "motion and heat are only different manifestations of one and the same force [which must] be able to be converted and transformed into one another." Mayer's ideas did not have much impact at the time, no doubt because he was not a "professional" physicist, though obviously enough the editor of the *Annalen der Chemie und Pharmacie* thought them worth printing. That editor was none other than Justus von Liebig.[3]

"Interwoven into One Great Association"

These experiments, ideas, and observations of Mayer and Joule did not come quite out of the blue. Throughout the early nineteenth century, and apart from Liebig's observations, provocative experimental results had been obtained for some time. In 1799, Alessandro Volta, in Como, north Italy, had stunned the world with his invention of the battery, in which two different metals, laid alternately together in a weak solution of salt, like a multilayered sandwich, generated an electric current. In 1820 Hans Christian Ørsted, in Copenhagen, had noticed that a magnetized compass needle was deflected from magnetic north when an electric current from a battery was switched on and off and passed through a wire near the needle. Five months later, in September that same year, in London, Michael Faraday, working in his basement laboratory in the Royal Institution in Albemarle Street, repeated Ørsted's experiment, and found the same result. Then he moved on to new ground. He brought together a cork, some wire, a glass jar, and a silver cup. He inserted the wire into the cork and put some water in a jar with mercury lying at the bottom. Then he floated the cork in the water, in such a way that the end of the wire in the cork made contact with the mercury. Faraday next fixed the top of the wire into an inverted silver cup with a globule of mercury held under its rim. When connected to a battery, this comprised a circuit that would allow the wire to flex without breaking the flow of electricity. Next, he brought up a magnet near the wire—and it moved. He repeated the action on the other side of the wire, with the same result.

Now came a crucial adjustment. He fixed the magnet in a glass tube and arranged the other contents so that the wire on its cork in the mercury could revolve around it when the current was switched on. Then he joined the circuit and—flick-flick-flick—the wire did a jig around the magnet. Faraday, we are told, did a jig of his own around the workbench.[4]

In Volta's battery, chemical forces produced electricity, Ørsted had demonstrated a link between electricity and magnetism, and in Faraday's experiments, electricity and magnetism together produced movement. On top of this, the new technology of photography, invented in the 1830s, used light to produce chemical reactions. Above all, there was the steam engine, a machine for producing mechanical force from heat. Steam technology would lead to the most productive transformations of all, at least for a time. During the 1830s and 1840s the demand for motive power soared. In an age of colonial expansion, the appetite for railways and steamships was insatiable, and these needed to be made more efficient, with less and less leakage of power, of energy.

But Thomas Kuhn also observed that, of these twelve pioneers in the conservation of energy, five came from Germany itself, and a further two came from Alsace and Denmark—areas of German influence. He put this preponderance of Germans down to the fact that "many of the discoverers of energy conservation were deeply predisposed to see a single indestructible force at the root of all natural phenomena." He suggested that this root idea could be found in the literature of *Naturphilosophie*. "Schelling, for example [and in particular], maintained that magnetic, electrical, chemical and finally even organic phenomena would be interwoven into one great association." Liebig studied for two years with Schelling.[5]

A final factor, according to science historian Crosbie Smith, was the extreme practical-mindedness of physicists and engineers in Scotland and northern England, who were fascinated by the commercial possibilities of new machines. All of this comprised the "deep background" to the ideas of Mayer, Joule, and the others. But the final element, says John Theodore Merz (1840–1922) in his four-volume *History of European Thought in the Nineteenth Century* (1904–12), was that the unification of thought that was brought about by all those experiments and observations "needed a more general term . . . a still higher generalisation, a more complete unification of knowledge . . . this greatest of all exact generalisations [was] the conception of energy."[6]

Nature's Currency System: "Continual Conversion"

The other men who did most, at least to begin with, to explore the conservation of energy—Joule and William Thomson in Britain, Hermann von Helmholtz and Rudolf Clausius in Germany—fared better than Mayer, though there were interminable wrangles in the mid-nineteenth century as to who had discovered what first.

Joule (1818–89), born into a brewing family from Salford, had a Victorian—one might almost say imperial—mane, hair which reached almost as far down his back as his beard did down his front: his head was awash in hair. He is known for just one thing, but it was and is an important thing and was one for which he conducted experiments over a number of years to provide an ever more accurate explanation.

As a young man he had worked in the family's brewery, which may have ignited his interest in heat. This interest was no doubt fanned all the more when he was sent to study chemistry in Manchester with John Dalton. Dalton was famous for his atomic theory—the idea that each chemical element was made up of different kinds of atoms, and that the key difference between different atoms was their weight. Dalton thought that these "elementary elements" could be neither created nor destroyed, based on his observations which showed that different elements combined to produce substances which contained the elements in set proportions, with nothing left over.

With his commercial background, Joule was always interested in the practical end of science—in the possibility of electric motors, for instance, which might take over from steam. That didn't materialize, not then anyway, but his interest in the relation between heat, work, and energy did eventually pay off. "Eventually," because Joule's early reports, on the relationship between electricity and heat, were turned down by the Royal Society—just as Mayer's ideas had been turned down by Poggendorff—and Joule was forced to publish in the less prestigious *Philosophical*

Magazine. But he continued his experiments, which, by stirring a container of water with a paddle wheel, sought to show that work—movement—is converted into heat. Joule wrote that "we consider heat not as *substance* but as a state of *vibration*." (This implicit reference to movement echoes his idea about the different temperatures of water at the top and bottom of waterfalls, and Mayer's observation that storm waves were warmer than calm seas.) Over his lifetime, Joule sought ever more accurate ways to calculate just how much work was needed to raise the temperature of a pound of water by one degree Fahrenheit (the traditional definition of "work"). Accuracy was vital if the conservation of energy was to be proved.[7]

And gradually people *were* won over. For example, Joule addressed several meetings of the British Association for the Advancement of Science, in 1842, and again in 1847. In between these meetings, Mayer published his observations, about body heat and blood color, but Joule had the momentum and, in the BAAS, the stage. The BAAS was well established then, having been founded in 1831, in York, modeled on the German *Gesellschaft Deutscher Naturforscher und Ärzte*. It held annual meetings in different British cities each year. But Joule needed only one individual in his BAAS audience to find what he had to say important, and that moment came in the 1847 meeting, when his ideas were picked up on by a young man of twenty-one. He was then named William Thomson but he would, in time, become better known as Lord Kelvin.

Just as Joule befriended the older Dalton, so he befriended the younger Thomson. In fact, he worked with Thomson on the theory of gases and how they cool and how all that related to Dalton's atomic theory. Joule was in particular interested in nailing the exact average speed at which molecules of gas move (movement that was of course related to their temperature). He focused on hydrogen and treated it as being made up of tiny particles bouncing off one another and off the walls of whatever container they were held in. By manipulating the temperature and the pressure, which affected the volume in predictable ways, he was able to calculate that, at a temperature of sixty degrees Fahrenheit and a pressure of thirty inches of mercury (more or less room temperature and pressure), the particles of gas move at 6,225.54 feet per second. Similarly, with oxygen, the molecules of which weigh sixteen times those of hydrogen, and since the inverse-square law* applies, in ordinary air the oxygen molecules move at a quarter of the speed of hydrogen molecules, or 1,556.39 feet per second. To pin down such infinitesimal activity was an amazing feat, and Joule was invited to address the Royal Society and elected a fellow, more than making up for his earlier rejection.

Joule shared a lot with Thomson, including his religious beliefs, which played an important part in the theory for some people. The principle of continual conversions or exchanges was established and maintained by God, he argued, as a basis for "nature's currency system," guaranteeing a dynamic stability in "nature's economy." "Indeed the phenomena of nature, whether mechanical, chemical, or vital, consist almost entirely in a continual conversion of attraction through space, living force, and heat into one another. Thus it is that order is maintained in the universe—nothing is deranged, nothing ever lost . . . the whole being governed by the sovereign will of God." [8]

Thomson followed on where Joule left off. Born in Belfast in June 1824, he spent almost all his life in university environments. His father was professor of mathematics at the Royal Belfast Academical Institution, a forerunner of Belfast University, and

---

* The inverse-square law applies when some force or energy is radiated evenly from a point source into three-dimensional space, like a lightbulb, say. Since the surface area of a sphere is proportional to the square of the radius, the emitted radiation (light in this example) is spread out over an area that is increasing in proportion to the square of the distance from the source. When you sit next to a light to read by, the phenomenon shows itself. If you move your chair so that you are twice as distant from the source of light as you were initially, the light diminishes by the square of that—i.e., 2^2: it is four times as dim.

William and his brother were educated at home by their father (his brother James also became a physicist). Their mother died when William was six, and in 1832 their father moved to Glasgow, where again he became professor of mathematics. As a special dispensation, both his sons were allowed to attend lectures there, matriculating in 1834, when William was ten. After Glasgow, William was due to go to Cambridge, but there were concerns that graduating in Glasgow might "disadvantage" his prospects down south, so although he passed his finals and the MA exams a year later, he did not formally graduate. At the time, he therefore signed himself as William Thomson BATAIAP (Bachelor of Arts To All Intents And Purposes).

William transferred to Cambridge in 1841, graduating four years later, having won a number of prizes and publishing several papers in the *Cambridge Mathematical Journal.* He then worked for a while in Paris, familiarizing himself with the work of the brilliant French physicist Sadi Carnot (who had died tragically young), and then joined his father in Glasgow, as professor of natural philosophy. James Thomson Senior, who had worked tirelessly to bring his son to Glasgow, died shortly afterwards from cholera. But William remained at Glasgow from when he was appointed professor (in his mid-twenties) until he retired at seventy-five, when, "to keep his hand in," he enrolled as a student all over again. This, as historian John Gribbin rejoices in saying, made him "possibly both the youngest student and the oldest student ever to attend the University of Glasgow."[9]

Thomson was much more than a scientist. He had a hand in the first working transatlantic telegraph, between Great Britain and the USA (after other attempts had failed), which transformed communication almost as much as, and maybe more than, the Internet of today. He made money from his scientific and industrial patents, to such an extent that he was, first, knighted in 1866 and then made Baron Kelvin of Largs in 1892 (the River Largs runs through the campus of Glasgow University).

"A Principle Pervading All Nature"

Thomson echoed Joule in his theology as well as his science. "The fact is," he wrote, "it may I believe be demonstrated that work is *lost to man* irrecoverably [when conduction occurs] but not lost in the material world." Employing the word "energy" for the first time since 1849, says Crosbie Smith, Thomson expressed his analysis in theological and cosmological terms. "Although no destruction of energy can take place in the material world without an act of power possessed only by the supreme ruler, yet transformations take place which remove irrecoverably from the control of man sources of power which, if the opportunity of turning them to his own account had been made use of, might have been rendered available."[10] God, as "supreme ruler," had established this law of "energy conservation," but nonetheless there were sources of energy in nature (such as waterfalls) that could be made use of—in fact, it was a mistake for Thomson if they were *not* made use of, because that implied waste, the Presbyterian's abiding sin. Finally, nature's transformations had a *direction* which only God could reverse: "The material world could not come back to any previous state without a violation of the laws which have been manifested to man."

In purely scientific terms, however, Kelvin's most important contribution was to make thermodynamics (as the conservation of energy became more formally known) a consolidated scientific discipline by the middle of the century. Together with Peter Guthrie Tait, another Scot, their joint work, *Treatise on Natural Philosophy* (1867), was both an attempt to rewrite Newton and to place thermodynamics and the conservation of energy at the core of a new science—nineteenth-century physics. Kelvin may even have been the first person to use the word "energy" in this new sense. In 1881 he said, "The very name energy, though first used in the present sense by Dr. Thomas Young about the beginning of this century, has only come into use practically after the doctrine which defines it had . . . been raised from a mere formula of mathematical dynamics to the position it now holds of a principle pervading all nature and guiding the investigator in every field of science."[11] Tait and Kelvin planned a second volume of their book, never written, which would have included "a great section on 'the *one* law of the Universe', the Conservation of Energy'."

On top of all this, Kelvin established the absolute scale of temperature, which also stems from the idea that heat is equivalent to work (as Joule had spent his lifetime demonstrating) and that a particular change in temperature is equivalent to a particular amount of work. This carries the implication that there is in fact an absolute minimum possible temperature: $-273°$ Fahrenheit, now written as $0°K$ (for Kelvin), when no more work can be done and no more heat can be extracted from a system.

"The Human Engine Is Little Different from the Steam Engine"

Thomson's ideas were being more or less paralleled in Germany by the work of Hermann von Helmholtz and Rudolf Clausius. With hindsight, everything can be seen as pointing toward the theory of the conservation of energy, but it still required someone to formulate these ideas clearly, and that occurred in the seminal memoir of 1847 by von Helmholtz (1821–94). In *On the Conservation of Force* he provided the requisite mathematical formulation, linking heat, light, electricity, and magnetism by treating these phenomena as different manifestations of "energy."

Like Kelvin, von Helmholtz had many fingers in many pies. He was born in Potsdam when it was "a one-class" garrison town, and von Helmholtz's parents were part of the intellectual middle class (his father was a high school teacher) and no fewer than twenty-three godparents graced Hermann's baptism. His early studies were funded by a Prussian Army scholarship in the course of which he studied physiology. In return for his education being paid for, von Helmholtz served as a medical officer before becoming, in 1849, associate professor of physiology at

the University of Königsberg. In 1850 he invented the ophthal-
moscope, which allows the far wall of the eye to be inspected,
and contributed many papers on optics and the physiology of
stereoscopic perception, as well as such subjects as fermentation.
But von Helmholtz fits in here because of his 1847 pamphlet,
"On the Conservation of Force." [12]

Like Mayer, he had sent his paper to Poggendorff at the
Annalen der Physik but was rebuffed, and he chose to publish his
pamphlet privately. And, like Mayer, von Helmholtz approached
the problem of energy from a medical perspective. His previous
physiological publications had all been designed to show how
the heat of animal bodies and their muscular activity could be
traced to the oxidation of food—that the human engine was little
different from the steam engine. He did not think there were
forces entirely peculiar to living things but insisted instead that
organic life was the result of forces that were "modifications" of
those operating in the inorganic realm. He had parallel ideas not
just with Mayer and Kelvin, but with Liebig too.

In the purely mechanical universe envisaged by von Helmholtz
there was an obvious connection between human and machine
work. For him, *Lebenskraft*, as the Germans called the life force,
was no more than an expression of "organisation" among related
parts which carried no implication of a vital force. [13] "The idea
of work is evidently transferred to machines from comparing
their performances with those of men and animals, to replace
which they were applied. We still reckon the work of steam
engines according to horse power." Which led him to the principle of the conservation of force: "We cannot create mechanical
force, but we may help ourselves from the general storehouse
of Nature. . . . The possessor of a mill claims the gravity of the
descending rivulet, or the living force of the moving wind, as his
possession. These portions of the store of Nature are what give
his property its chief value." His idea of the "store" of nature
complemented Joule's notion of the "currency" of nature.

In making his case without any experimental evidence (which

the members of the Berlin Academy noticed, while being impressed by his presentation), von Helmholtz "first established a clear distinction between theoretical and experimental physics."

The Tendency Toward Increasing Disorder

While Mayer and von Helmholtz, being doctors, came to the science of work through physiology, von Helmholtz's fellow Prussian Rudolf Clausius approached the phenomenon, like his British and French contemporaries, via the ubiquitous steam engine.

In later life Clausius had a rather forbidding appearance: a very high forehead, rather hard, piercing eyes, a thin, stern mouth, and a white beard fringing his cheeks and chin. In fairness to him, this sternness may have reflected nothing more than the pain he was in continuously after suffering a wound in the Franco-Prussian War of 1870–71. At the same time he was a fervent nationalist and that may also have been a factor.

He was born in January 1822, in Köslin, Prussia (now Koszalin, Poland), where his father was a pastor with his own private school. The sixth of his father's sons, Rudolf attended the family school for a few years, before transferring to the gymnasium at Stettin (now Szczecin, Poland) and then going on to the University of Berlin in 1840. To begin with he was drawn to history and studied under the great Leopold von Ranke, which may have had something to do with his subsequent nationalism. But Clausius switched to math and physics. In 1846, two years after graduating from Berlin, he entered August Böckh's seminar at Halle, and worked on explaining the blue color of the sky. The theory Clausius came up with about the blue of the sky, and its redness at night and morning, was based on faulty physics. He thought it was caused by reflection and refraction of light, whereas John Strutt, later Lord Rayleigh, was able to show it was due to the scattering of light.[14]

But Clausius's special contribution was to apply mathematics far more deeply than any of his predecessors, and his work was an important stage in the establishment of thermodynamics and theoretical physics. His first paper on the mechanical theory of heat was published in 1850. This was his most famous work and we shall return to it in just a moment. He advanced rapidly in his career, at least to begin with, being invited to the post of professor at the Royal Artillery and Engineering School at Berlin in September 1850 on the strength of his paper, then moving on to the Polytechnikum in Zurich, where he remained for some time despite being invited back to Germany more than once. He eventually accepted a chair at Würzburg in 1869, moving on to Bonn after only a year, when the Franco-Prussian War intervened. A "burning nationalist," as someone described him, Clausius volunteered, despite being just short of his fiftieth birthday, and was allowed to assume the leadership of an ambulance corps, which he formed from Bonn students, helping to carry the wounded at the great battles of Vionville and Gravelotte—the Germans suffered twenty thousand casualties at the latter battle. During the hostilities, Clausius was wounded in the leg, which caused him severe pain and disability for the rest of his life.[15] He was awarded the Iron Cross in 1871.

Unlike Mayer and von Helmholtz, Clausius did succeed in having his first important paper, "On the Moving Force of Heat, and the Laws Regarding the Nature of Heat That Are Deducible Therefrom," accepted by the *Annalen*. It appeared in 1850 and its importance was immediately recognized. In it he argued that the production of work resulted not only from a change in the *distribution* of heat, as Sadi Carnot—the French physicist and military engineer—had argued, but also from the *consumption* of heat: heat could be produced by the "expenditure" of work. "It is quite possible," he wrote, "that in the production of work . . . a certain portion of heat may be consumed, and a further portion transmitted from a warm body to a cold one: and both portions may stand in a certain definite relation to the quantity of work produced." In doing this he stated two fundamental principles, which would become known as the first and second laws of thermodynamics.

The first law may be illustrated by how it was later taught to Max Planck, the man who, at the turn of the twentieth century, would build on Clausius's work. Imagine a worker lifting a heavy stone onto the roof of a house. The stone will remain in position long after it has been left there, storing energy until at some point in the future it falls back to earth. Energy, says the first law, can be neither created nor destroyed. Clausius, however, pointed out in his second law that the first law does not give the total picture. In the example given, energy is expended by the worker as he lifts the stone into place, and is dissipated in the effort as heat, which among other things causes the worker to sweat. This *dissipation*, which Clausius was to term "entropy," was of fundamental importance, he said, because although it did not disappear from the universe, this energy could never be recovered in its original form. Clausius therefore concluded that the world (and the universe) must always tend towards increasing disorder, must always add to its entropy.[16]

Clausius never stopped refining his theories of heat, becoming in the process interested in the kinetic theory of gases, in particular the notion that the large-scale properties of gases were a function of the small-scale movements of the particles, or molecules, which comprised the gas. Heat, he came to think, was a function of the motion of such particles—hot gases were made up of fast-moving particles, colder gases of slower particles. Work was understood as "the alteration in some way or another of the arrangement of the constituent molecules of a body."

This idea that heat was a form of motion was not new. In addition to the ideas of Joule and Mayer, the American Benjamin Thompson had observed that heat was produced when a cannon barrel was bored, and in Britain Sir Humphry Davy had likewise noted that ice could be melted by friction. What attracted Clausius's interest was the exact form of motion that comprised heat. Was it the vibration of the internal particles, was it their "translational" motion as they moved from one position to another, or was it because they rotated on their own axes?

Clausius's second seminal paper, "On the Kind of Motion That We Call Heat," was published in the *Annalen* in 1857. He argued that the heat of a gas must be made up of all three types of movement and that therefore its total heat ought to be proportional to the sum of these motions. He assumed that the volume occupied by the particles themselves was vanishingly small and that all the particles moved with the same average velocity, which he calculated as being hundreds—if not thousands—of meters per second (building on Joule). This prompted the objection from several others that his assumptions and calculations could not be right, since otherwise gases would diffuse far more quickly than they were known to do. He therefore abandoned that approach, introducing instead the concept of the "mean free path"—the average distance that a particle could travel in a straight line before colliding with another one.[17]

The Unification of Electricity, Magnetism, and Light

Clausius was elected a fellow of the Royal Society in 1868, and awarded its Copley Medal in 1879. Others were attracted by his efforts, in particular James Clerk Maxwell in Britain, who published "Illustrations of the Dynamical Theory of Gases" in the *Philosophical Magazine* in 1860, making use of Clausius's idea of the mean free path.

According to one of his biographers, James Clerk Maxwell had a scientific idea "that was as profound as any work of philosophy, as beautiful as any painting, and more powerful than any act of politics or war. Nothing would be the same again." These are big things to say, but, in a nutshell, Maxwell conceived four equations that, at a stroke, united electricity, magnetism, and light and in so doing showed that visible light was only a small band in a vast range of possible waves, "which all travelled at the same speed but vibrated at different frequencies."[18] Physicists, says the same biographer, honor Maxwell alongside Newton and Einstein, yet among the general public "for some reason he is much less well known." This was all carried through while Mary Somerville was still alive and, in effect, helped make *Connexions* out of date.

Maxwell was brought up for the first eight years of his life on his father's estate at Glenlair, in the Galloway region of southwest Scotland. His family were well connected—his grandfather was a composer, as well as having various official municipal jobs, and a fellow of the Royal Society. An uncle was a friend of James Hutton and had illustrated Hutton's seminal work, *Theory of the Earth* (see chapter 2). Maxwell's parents had married late; their first child had died in infancy, and James's mother was almost forty when he was born.

His late arrival made his parents indulgent. It became plain soon enough that he was an exceptional child, intent on finding out how everything worked and having an explanation for everything. As a result, as a boy he learned how to knit, bake, and weave baskets. Like Humphry Davy and Michael Faraday he shared the nineteenth-century scientist's fascination with writing poetry, though none was published in his lifetime, and it is not hard to see why. One read:

> Then V_n/V_1 the tangent will equal
> Of the angle of starting worked out in the sequel.

Another poem actually had a graph in it.

The Vale of Urr, where Glenlair was situated, was known to its residents as Happy Valley, but, when she was forty-seven, Maxwell's mother, Frances, contracted abdominal cancer and died soon after undergoing an operation (performed without anesthetic). The loss brought father and son together, but there was a problem with James's education. It had been planned that he would be educated at home until he was thirteen, but now his father had too many calls on his time. An aunt who lived in the capital came to the rescue and took him in, which enabled him to attend Edinburgh Academy, one of the best schools in Scotland.[19]

It was not, at first, a success. Because the school was almost full, James was obliged to enter a class of boys a year older than he was, who had all been at the school for months, and had established their own conventions and cliques. On top of which they mostly came from well-heeled Edinburgh families—when they saw his rough-hewn country clothes and heard his rural accent, they picked on him mercilessly. He wore (to begin with) a loose tweed tunic, with a frilly collar, and square-toed shoes with brass buckles. No one had ever seen clothing like this in Edinburgh, where the pupils wore close-fitting tunics and slim shoes. The boys nicknamed him "Dafty."

School continued difficult for a while, not helped by a hesitancy of speech verging on, but not quite, a stutter. And it contrasted strongly with his aunt's house, which he loved. It was full of books, drawings, and paintings, and his cousin—his aunt's daughter—was a rising artist, who even Landseer had complimented.

Then things started to improve. In his second year, the speed with which he mastered geometry impressed his teachers and, no less, his classmates. In the academy at that time, boys sat in order of ability, so he was now moved forward to sit with more congenial company, and began to make friends. Among them was Peter Guthrie Tait—P. G. Tait was to become one of Scotland's finest physicists and, as we have seen, coauthor, with William Thomson, of the *Treatise on Natural Philosophy*.[20]

At the age of fourteen Maxwell published his first paper. It was about how to draw an oval. Everyone knows that if you attach string to a pin and a pencil to the other end, you can draw a perfect circle. Maxwell observed that if you have two pins, with one piece of string attached to each, and then push a pencil against the string, so that it remains taut, you can draw a perfect oval. Then he undid one end of the string and looped it around the free pin, and got another oval, egg-shaped. He played with more curves and studied their mathematical relationships, coming up with some formulas to describe what he had found. Some of this had been worked out earlier by no less a figure than René Descartes, but Maxwell's system was simpler and was judged good enough to be read before the Royal Society of Edinburgh. Because he was so young, the paper had to be read for him.

He was a devout Christian, of the austere Presbyterian kind, something that paid off when he visited other Presbyterian relatives in Glasgow. One of his cousins, Jemima, had married Hugh Blackburn, professor of mathematics at Glasgow and a great friend of William Thomson, the new professor of natural philosophy there. Maxwell and Thomson struck up a friendship that would continue for years.

As mentioned in the Introduction, in mid-nineteenth-century Britain the word "scientist" had not yet come into common use. Physicists and chemists called themselves "natural philosophers" and biologists called themselves "natural historians." Maxwell decided to enroll at Edinburgh University, to study mathematics, natural philosophy, and logic. He matriculated at sixteen.

This was when Maxwell himself began to experiment, aided by the practice of the Scottish universities of closing from late April to early November to allow students home to help with the farming. He read and read and read and carried out his first experiments at Glenlair, developing an interest in electromagnetism and polarized light. These DIY adventures did more than develop his experimental skill, though that was important. They helped give him a deep feeling for nature's materials and processes that later pervaded his theoretical work. While in Edinburgh he produced two more papers for the Royal Society there. So, when he left for Cambridge at the age of nineteen, he had a solid body of knowledge, a handful of publications to his name, and some valuable and potentially influential friends in the world of academia and science.

He started at Peterhouse but found it dull and moved to Trinity, which was more congenial and much more mathematically minded (the master at the time being William Whewell). In Cambridge Maxwell joined the class of the famous (in mathematical circles) "senior wrangler maker," William Hopkins—wranglers being those who gained first-class degrees in the mathematical tripos, which all had to take. The reward for wranglers was lifelong recognition in whatever field they chose. The tripos was an arduous seven-day affair, six hours a day, and James came second, after E. J. Routh, who went on to be a remarkable mathematician, with a function named after him, the routhian. (P. G. Tait, Maxwell's erstwhile Edinburgh Academy friend, had been senior wrangler two years before.)[21]

With the tripos out of the way, Maxwell was now free to give rein to the ideas that had been brewing in his mind over his two stints as an undergraduate. There were two aspects of the physical world he wanted to explore. One was the process of vision, particularly the way we see colors, and the other was electricity and magnetism.

In his color research he had an early breakthrough, finding that there is a fundamental difference between mixing pigments, as one does with paints or dyes, and mixing lights, as one does when spinning a multicolored disc. Pigments act as extractors of color, so that the light you see after mixing two paints is whatever color the paints have failed to absorb. In other words, mixing pigments is a subtractive process, whereas mixing lights is additive—so that, for instance, blue and yellow do not make green, as they do with pigments, but *pink*. And by experiment he was able to show that there are, in light terms, three primary colors—red, blue, and green—and that it is possible to mix them in different proportions to obtain all the colors of the rainbow. This was a major advance and is the theory behind the colors in color television, for example.

At the same time, he was getting to grips with electricity and magnetism, and in 1855 the first of his three great papers appeared. Michael Faraday had thought of lines of force as discrete tentacles (analogous to the lines of iron filings that form around a magnet). Maxwell now conceived them as merged into one continuous essence, which he called "flux"—the higher the density of flux at any particular location, the stronger the electrical or magnetic force there. And he grasped moreover that the electric and magnetic forces between bodies vary inversely as the square of their distance apart—much as Newton had said of gravity.[22]

In this way, lines of force became the "field," and *this* was the concept that set Maxwell apart and put him on a par with Newton and Einstein. More than that, he would build on it six years later with his concept of *electromagnetic waves.*

In between times, his father fell ill, and James was forced to spend time nursing him. But it wasn't enough: he needed a post nearer home. This cropped up when he was offered the position as professor of natural philosophy at Marischal College in Aberdeen, one of the colleges that would, not much later, become Aberdeen University. The post buoyed both father and son, but it had its drawbacks. James later wrote to a friend, "No jokes of any kind are understood here. I have not made one for 2 months, and if I feel one coming on I shall bite my tongue." But it wasn't all hopeless, as James found the daughter of the college principal exactly to his taste, proposed, and was accepted.[23]

In June 1858 he and Katherine were married and then, a few months later, he read the paper by Clausius about the diffusion of gases. The problem, which several people had pointed out, was that, to explain the pressure of gases at normal temperatures, the molecules would have to move very fast—several hundred meters a second, as Joule had calculated. Why then do smells—of perfume, say—spread relatively slowly about a room? Clausius proposed that each molecule undergoes an enormous number of collisions, so that it is forever changing direction—to carry a smell across a room the molecule(s) would actually have to travel several kilometers.

Clausius had assumed that, at any given moment, all the molecules would travel at the same speed. He knew that couldn't be the correct answer, but he couldn't think of anything better. Maxwell was also stymied at first, but then he had a brain wave. At a stroke, says Basil Mahon, it "opened the way to huge advances in our understanding of how the world works."

Maxwell saw that what was needed was a way of representing many motions in a single equation, a *statistical* law. He devised one that said nothing about individual molecules but accounted for the *proportion* that had the velocities within any given range. This was the first-ever statistical law in physics, and the distribution of velocities turned out to be bell-shaped, the familiar normal distribution of populations about a mean. But its shape varied with the temperature—the hotter the gas, the flatter the curve and the wider the bell.

This was a discovery of the first magnitude, which would in time lead to statistical mechanics, a proper understanding of thermodynamics, and to the use of probability distributions in quantum mechanics. This alone was enough to put Maxwell in the first rank of scientists. The Royal Society certainly thought so, awarding him the Rumford Medal, its highest award for physics. No less important in the long run, King's College London was looking for a professor of natural philosophy—James entered his name and was appointed. And he still had more than one breakthrough in him.

King's, in the Strand, just north of the Thames, had been founded in 1829 as an Anglican alternative to the nonsectarian University College, a mile further north, which was itself intended as an alternative to the strictly Church of England Oxbridge universities. Unlike the traditional courses, to be found at Aberdeen and Cambridge, King's' courses were much more modern.

Being in London meant that Maxwell could attend the meetings of the Royal Society, and the Royal Institution, where he was able to cement his friendship with Faraday. They had corresponded a great deal, but now at last met and struck up a genial friendship. And Maxwell homed in on his final great insight.

In his paper, "On Faraday's Lines of Force," he showed how he had found a way of representing the lines of force mathematically as continuous fields, and had made a start towards forming a set of equations governing the way electrical and magnetic fields interact with one another. But that was still only part of the picture. Picture is in fact the wrong word here, because it is at this point that physics began to enter a world where the familiar visual analogies break down. The image of a "field" is easy enough to imagine in itself, but what Maxwell was struggling to explain, in his equations, could only be explained with difficulty in ordinary language, and this came home to him—and then to everyone—in his 1862 paper, where he concluded, dramatically, and using the mathematics that he had himself created, that light is also a form of electromagnetic disturbance and, moreover, could be understood as both a wave and a beam of particles. This was unheard of, inexplicable when put into language, but made sense in mathematics.[24]

In fact, Maxwell derived four equations that between them "summed up everything that it is possible to say about classical electricity and magnetism." Which is why, among physicists, if not yet the general public, Maxwell is placed on a par with Newton. "Between them, Newton's laws and his theory of gravity and Maxwell's equations explained everything known to physics at the end of the 1860s." Maxwell's achievement was the greatest breakthrough since Newton's *Principia Mathematica* in 1687.[25]

As if all this were not enough, Maxwell's equations contained within them the implication that there must be other forms of electromagnetic waves with much longer wavelengths than those of visible light. Their discovery would not be long in coming.

The final chapter in Maxwell's extraordinary career arrived when he was invited to accept an important new professorship at Cambridge. The duke of Devonshire, who was chancellor of the University, had offered a large sum of money to build a new laboratory for teaching and research, which was to compete with the best of what then existed on the Continent, especially in Germany. Cambridge was being left behind in experimental science, not just in France and Germany, but by many of the new British universities as well.

Maxwell was not over keen to accept Cambridge's offer. His theories were so new that not everyone understood them and he couldn't be certain of his reception more generally. But many of the younger physicists at Cambridge, who *had* kept up with his work, implored him to come, and that settled it.[26]

Time Becomes a Property of Matter

Clausius had assumed that every particle in a gas traveled at the same average velocity. Maxwell relied on the new science of statistics to calculate a *random distribution* of particle velocities, arguing that the collisions between particles would result in a distribution of velocities about a mean rather than an equalization. (Just what these particles *were* was never settled, not then, though Maxwell was convinced they were "proof of the existence of a divine manufacturer.")

The statistical—probabilistic—element introduced into physics in this way was a very controversial and yet fundamental advance. In his 1850 paper, Clausius had drawn attention in the second law of thermodynamics to the "directionality" of the heat flow—heat tends to pass from a hotter to a colder body. He had not at first bothered with the implications of the irreversibility or otherwise of processes, but in 1854 he argued that the transformation of heat into work and the transformation of heat from a higher temperature into heat of a lower temperature were in effect equivalent and that in some circumstances they could be counteracted—reversed—by the conversion of work into heat, where heat would flow from a colder to a warmer body. This, for Clausius, only emphasized the difference between reversible (man-made) and irreversible (natural) processes: a decayed house never puts itself back together, a broken bottle never spontaneously reassembles.[27]

It was only later, in 1865, that Clausius proposed the term

"entropy" (from the Greek word for "transformation") for the irreversible processes. The tendency for heat to pass from warmer to colder bodies could now be described as an instance of the increase in entropy. In doing this, Clausius emphasised the *directionality* of physical processes. Entropy was a counterpart of energy "because the two concepts had analogous physical significance." Clausius set out the two laws of thermodynamics as follows: "The energy of the universe is constant" and "The entropy of the universe tends to a maximum." Time, in some mysterious way, had become a property of matter.

For some people, the second law had a much greater significance than even Clausius understood. William Thomson thought that the irreversibility that was such a feature of the second law—the dissipation of energy—also implied a "progressivist cosmogony," one that moreover underlined the biblical view about the transitory character of the universe. In particular, Thomson drew the implication from the second law that the universe, known by then to be cooling, would "in a finite time" run down and become uninhabitable. Von Helmholtz had also noticed this implication of the second law. It was only in 1867 that Clausius himself, who had by then moved back to Germany from Zurich, acknowledged the "heat death" of the universe.[28]

The Marriage of Mathematics and Physics

The statistical notions aired by Clausius and Maxwell attracted the attention of the Austrian physicist Ludwig Boltzmann. Boltzmann (1844–1906) was born in Vienna during the night of the Mardi Gras Ball, between Shrove Tuesday and Ash Wednesday, a coincidence which, he half-jokingly complained, helped to explain the frequent and rapid mood swings that tossed him between unalloyed happiness and deep depression. The son of a tax official, he was short and stout, with curly hair that made him look younger than he was. His fiancée called him her "sweet fat darling."

He took his doctorate at the University of Vienna, then taught at Graz before moving to Heidelberg and afterward Berlin, where he studied under Bunsen, Kirchhoff, and von Helmholtz. In 1869, at the age of twenty-five, he was appointed to the chair of theoretical physics in Graz. After that, Boltzmann had a very unsettled career—he changed professorships numerous times, more than once because he couldn't get on with colleagues. The constant arguments depressed Boltzmann and he attempted suicide for the first time.

Finally, in 1901, after all this chopping and changing, Boltzmann returned to Vienna to the chair he had vacated in one of his arguments with colleagues and which had not been filled in the meantime. In addition he was given a course to teach on the philosophy of science, which quickly became very popular, so much so that he was invited to the palace of Emperor Franz Joseph.

This was impressive, but Boltzmann's main achievement lay in two famous papers that described in mathematical terms the velocities, spatial distribution, and collision probabilities of molecules in a gas, all of which determined its temperature (heat and motion again). The mathematics were statistical, showing that— whatever the initial state of a gas—Maxwell's velocity distribution law would describe its equilibrium state. This became known as the Maxwell-Boltzmann distribution. Boltzmann also produced a statistical description of entropy.[29]

In 1904 Boltzmann went to the United States and visited the World's Fair at St. Louis, where he gave some lectures before going on to visit Berkeley and Stanford. While there he behaved oddly—people couldn't make out whether his elevated manner was an illness or pretentiousness. He returned home and went on vacation with his family to Duino, near Trieste. While his wife and daughter were swimming he hanged himself. No one can be certain whether his general instability was the cause of his suicide, or the continual attacks on his ideas. What is certain, unfortunately, is that he couldn't have been aware, at the time of his death, that his ideas would very soon receive experimental confirmation.

What is important about the work of Mayer, Joule, and von Helmholtz, and in particular Clausius, Maxwell, and Boltzmann, is that—whether one can follow the mathematics or not—they brought *probability* into physics. How can that be? Matter definitely exists, transformations (as when water freezes, say) obey invariant laws. What can probability have to do with it? This was the first appearance of "strangeness" in physics, heralding the increasingly weird twentieth-century quantum world. These early physicists also made "particles" (atoms, molecules, or something else, not yet clearly understood) integral to the behavior of substances.[30]

The understanding of thermodynamics was the high point of nineteenth-century physics, and of the early marriage between physics and mathematics, building richly on Mary Somerville's previous ideas. It signaled an end to the strictly mechanical Newtonian view of nature. It would prove decisive in leading to a spectacular new form of energy: nuclear power. This all stemmed, ultimately, from the concept of the conservation of energy.

2

"A SINGLE STROKE UNIFIES LIFE, MEANING, PURPOSE, AND PHYSICAL LAW"

I t may be difficult for us to understand now but, in the late eighteenth and early nineteenth centuries, when the philologists were attacking the very basics of Christianity—seeking to pillory the absurdities and inconsistencies of the Bible, for instance—the men of science did not for the most part join in. For the most part, biologists, chemists, and physiologists remained devoutly religious. That even applied, again for the most part, to the practitioners of the two sciences that were to produce the most convincing evidence that the biblical chronology had to be wrong: astronomy and geology.

Astronomy underwent its greatest change since Copernicus and Newton thanks to an unlikely couple who might never have achieved what they did had not the British powers that be decided to invite a German to be their king.

The way Richard Holmes tells the story, Joseph Banks—botanist and explorer, who accompanied James Cook on his first great voyage—shortly after he was elected president of the Royal Society in 1778, began to hear stories about a gifted amateur astronomer "working away on his own" in the West Country. This news reached Banks via the secretary to the Royal Society,

Sir William Watson, whose son lived in Somerset and was a central member of the Bath Philosophical Society. According to these accounts, this maverick was a German who built his own (very powerful) telescopes and was making extravagant claims about the moon.

The man's name was Wilhelm, or William, Herschel. Though tall and well dressed, "and wearing his hair powdered," he spoke with a thick German accent (he was from Hanover) and had no manservant with him when Watson's son encountered him, in a cobbled backstreet of Bath, looking at the moon.[1] Watson's son had asked if he might look through the telescope, which he was smart enough to note was a reflecting instrument, not the usual refracting type used by amateurs. And he found that though the whole seven-feet-long contraption was evidently homemade, it nonetheless offered a better resolution than he had ever seen and he observed the moon more clearly than ever before.

Watson's son formed a friendship with Herschel, finding him to be in fact the organist at Bath's Octagon Chapel, who made ends meet by giving music lessons. He also composed, had a house full of astronomical and other books, and lived with his sister, who looked after him but whom he also described as his "astronomical assistant."

On the strength of this, Watson's son invited Herschel to join the Bath Philosophical Society, where he began sending papers. These were so unconventional, but so striking, that Watson sent them on to his father, and some of the more surprising were published in the Royal Society's *Philosophical Transactions*. The first of them, "Observations on the Mountains of the Moon," claimed that, with his homemade telescope, he had observed "forests" on the surface of the moon and concluded that it was, "in all probability," inhabited. This outraged the more established types in the Royal Society, some of whom decided to visit Herschel in Bath. Nothing much came out of this meeting, other than the fact that they were impressed by his telescopes and intrigued by his diminutive sister, Caroline, who seemed as mad about astronomy as he was.[2]

All that changed a year later when Herschel announced that he had discovered a new planet, something that hadn't happened since the days of Pythagoras. Moreover, Herschel's new planet had important implications for the nature of the solar system.

But first some background. Herschel had been born in Hanover on November 15, 1738, twelve years before his sister. They had been deeply attached since childhood and what we know about their life is drawn from the journal Caroline kept. William and Caroline's parents produced ten children—one every two years—of whom six survived.[3] William's father made him a tiny violin, which he learned to play as soon as he could hold it—this seems to be where he got his facility for working with wood, and making instruments. On winter nights, the children were taken outside to view the stars. In those days before widespread light pollution, the night skies were much more vivid than now.

But not everything was rosy. The eldest child in the family, Jacob, the apple of his mother's eye, soon became spoiled and a bully, who whipped Caroline and teased Wilhelm when he did well at school, as he often did. Jacob was a virtuoso musician and thought nothing else mattered in life. When he was fourteen, William joined the regimental band, alongside his father and brother, and learned in time to play the oboe, the violin, the harpsichord, the guitar, and, eventually, the organ.

In the spring of 1756, when William was seventeen and Caroline six, the Hanover Foot Guard was posted to England, to serve under their ally, the Hanoverian King George II. The three men of the family were stationed in Maidstone, Kent, returning home a year later. Richard Holmes tells us that Jacob took with him "a beautifully tailored English suit," while Wilhelm took a copy of John Locke's *An Essay Concerning Human Understanding*. But the family now became embroiled in the French-German wars, even to the extent of having French troops billeted in their home. Jacob and Wilhelm escaped, fleeing back to England, where they arrived together in London, penniless.

They obtained employment as musicians—playing or

teaching. Eventually, when events in Germany quietened down, Jacob decided to return home. William, however, was happier in England—he liked the freer way of life, without the petty restrictions of Germany's principalities, and he liked English culture: novels, theater, politics. He obtained a position as music master to the Durham militia, stationed at Richmond in Yorkshire, then a most civilized regional center. At night he went out onto the Yorkshire moors, laid a blanket on the grass, and studied the layout of the stars that he would come to know so well.

Aware that Jacob was still bullying his sister, and that she had suffered a bout of typhus, which she survived but which appeared to have stunted her growth, he spirited her away to England. He took her to Bath, where he had been appointed organist at the Octagon Chapel. Caroline spoke almost no English, her face was badly marked with smallpox scars, and she was less than five feet tall. Her appearance was boyish, though her hair was full of curls bubbling about her face, which had a "neat, very determined little chin."[4] On the journey to England, Wilhelm made her sit on top of the coach, so she could study the stars. He treated her affectionately but sternly. However, she was no maidservant—he made it clear that she was now to act as the hostess. He was thirty-four, she was twenty-two. She gradually became his assistant.

In February 1766 he started his first astronomical observations journal and began collecting and building telescopes, moving on in time from refractor scopes to Newtonian reflectors. And it was now that he had his first big idea. At that time most astronomers considered the night sky as so many stars set out like inlaid diamonds on a giant glass dome. It was realized, of course, that some stars and planets were closer to earth than others, but it was Herschel who seems to have first conceived the idea of *deep space*.

In 1774 William built his first five-foot reflector telescope, and it was immediately apparent that he had created an instrument of unparalleled light-gathering power and clarity.[5] He could see, for example, that the Pole Star—which had been the key to navigation, and the poet's traditional emblem of steadiness and singularity for centuries—was not one star *but two*. He also focused on the nebulae, or "star clouds," which were a mystery.

The recognition of nebulae was relatively new then. About thirty nebulae were known in the 1740s, though by the time Herschel came on the scene, according to Charles Messier, France's most notable astronomer and star cataloguer, that had swelled to about a hundred. Within ten years, Herschel's observations swelled it still further, to about a thousand. No one knew their composition, origins, or distance. The most popular understanding was that they were loose agglomerations of gas, hanging about in the Milky Way, "some loose flotsam of God's creation."[6]

With the family more or less settled (he had also brought a younger brother over from Hanover), from about 1779 Herschel began to expand and deepen his astronomical inquiries in earnest. His first idea was a catalogue of double stars, several of which were known, and some of which had been catalogued by John Flamsteed (1646–1719), the first astronomer royal. The point about double stars was that, via parallax, they might offer a more exact clue as to how far away they were. At that point there was no real understanding of astronomical distance, or how big the Milky Way was. Kant thought that, because of its brightness, Sirius, the Dog Star, was probably the center of the Milky Way and even of the entire universe.[7] A lot of astronomers assumed that other stars were inhabited.

A New Planet

All the while, William continued his interest in cataloguing double stars. And it was as he was doing this that, on Tuesday, March 13, 1781, slightly before midnight, Herschel spotted a new and unidentified "disc-like object moving through the constellation of Gemini." From his observation journal it seems that he thought at first that he had found a comet, but after three or four days he began to think again, noting that the object was well defined and had no trail. That could only mean a new "wanderer"—a new planet—and indeed that is what he had found, the seventh planet in the solar system, beyond Jupiter and Saturn: the first new planet to be discovered since the days of Ptolemy (c. AD 90–c. AD 168). Herschel at first named it for the Hanoverian king, "Georgium Sidus" (George's Star), but it eventually became known as Uranus, "Urania" being the goddess of astronomy.[8]

For some time, however, no one could agree on what, exactly, Herschel had found. He eventually communicated his observations to Watson, who conveyed them to the Royal Society, who asked Messier in Paris for an opinion. Given Herschel's earlier faux pas over life on the moon, not everyone was immediately convinced.

Nevil Maskelyne, the astronomer royal at the time, was in an especial quandary. There were dangers to his credibility in acceding too quickly to what Herschel was claiming. On the other hand, it had the potential to be a feather in the cap of British science (albeit one produced by a German), which otherwise the predatory French might appropriate for their own, by recognizing Herschel first (and maybe even naming the star). Moreover, Banks was pressing: the Royal Society needed to cement its relations with the new king, who was known to be keen on stars. Maskelyne observed the object himself, and confirmed its existence, though he refrained from saying at that moment whether it was a planet or a comet. Later he changed his mind, and opted to support Herschel. Messier agreed, writing from Paris that, having himself discovered no fewer than eighteen comets, Herschel's discovery resembled none of them. The result was that Herschel gave a paper before the Royal Society in late April. The paper, entitled "An Account of a Comet," said plainly that Herschel had discovered a new planet. He was elected a fellow of the Royal Society immediately.[9]

The Order of the Heavens

The discovery began a revolution in the popular conception of cosmology. Astronomers from all over Europe wrote to Herschel, asking for details of his equipment, though there were still skeptics in the Royal Society who doubted what they could actually achieve. One who wasn't skeptical was the king, George III, who was fascinated by the heavens himself, and invited Herschel to court at Windsor to congratulate him. When they met, in May 1782, with both Hanoverians speaking English, the encounter was a great success.

This rubbed off on Banks, ever mindful to promote the interests of science, and he now sought to obtain for Herschel a salary and a better place to live. The post of astronomer royal was filled, so Banks convinced the king to create a new position, the king's personal astronomer, on a salary of £200 per annum, with a house near Windsor, at Datchet, thrown in.[10] Caroline continued to keep her journal so that the chronology of their joint careers is well attested.

Between 1784 and 1785, Herschel began to draw together his new and very radical ideas about the cosmos, which were published in two "revolutionary papers" in the Royal Society's *Philosophical Transactions.* In "An Investigation of the Construction of the Heavens," published in June 1784, Herschel identified 466 new nebulae (four times the number identified by Messier) and for the first time raised the possibility that many of them, if not all, must be huge, independent star clusters or galaxies that were *outside* the Milky Way. This led him to propose that the Milky Way wasn't flat but three-dimensional, that we are in effect *inside* it, part of it, and that it is discus-shaped with arms extending out into deep space.

In his second paper, a year later, and headed "On the Construction of the Heavens," he began by saying that astronomy needed a "delicate balance" of observation and speculation if it were to proceed by induction—mere observation was not enough. And he went on to observe that the universe was not static, that the heavens far away were constantly changing, that all gaseous nebulae were "resolvable" into stars, and were in reality enormous star clusters far beyond the Milky Way. In so doing he immeasurably increased the size of the cosmos—by this time his nebulae count had risen to well over nine hundred, many of which, he insisted, were larger than the Milky Way.[11] And he estimated that deep space was "not less than 6 or 8 thousand times the distance of Sirius." He conceded that these were coarse estimates, and they are, certainly, much less than we now know, though thoroughly outlandish for their time.

In this paper, incidentally, he credited his sister with discovering one of the nebulae clusters. This mention, though brief, probably did wonders for Caroline's self-confidence, for she went on to make a name for herself as an astronomer in her own right, discovering no fewer than eight comets.

For his part, William also observed that the many nebulae he had identified varied in systematic ways—some were more "compressed" than others; others appeared to be "condensing." He advanced the idea that some nebulae were older than others, more *evolved*—that nebulae aged, matured, and climaxed. The fundamental force was gravity, gradually compressing nebulous gas into huge, bright galactic systems, which eventually condensed into individual stars.

It was in this paper that astronomy changed its character, fundamentally, from a mathematical science concerned primarily with navigation, to a cosmological science concerned with the evolution of stars and the origins of the universe.[12]

Although the implications of this were slow to be absorbed (perhaps because people didn't *want* to absorb them), one of the first to follow up Herschel's ideas was the French mathematician and astronomer (and, perhaps significantly, atheist) Pierre-Simon Laplace, who published a paper on "the nebular hypothesis" in 1796.

Laplace drew on Herschel's ideas and applied them to the

formation of the solar system. He is sometimes known as the French Newton, being as much a mathematician as a physicist and astronomer. Many details of his early life went up in smoke, literally, when the family château burned down in 1925. But we know he was born in Beaumont-en-Auge, in Normandy, in 1749, into an agricultural background. We know too that Laplace was schooled in a Benedictine priory, his father intending him for the church, before he was sent to the University of Caen to read theology.

Laplace's adaptation of Herschel's ideas to the solar system was published in his *Exposition du système du monde* (1796) and the first volume of his classic *Mécanique céleste* (1799), in which he argued that the sun had slowly condensed out of a nebulous cloud of stardust and then spun off our entire planetary system, in exactly the same way as other planetary systems across the universe.* He also came close to predicting the existence of black holes when he pointed out that there could be massive stars in the universe where gravity was so great that not even light could escape from their surface. The significance of all this, of course, was as much theological as scientific. He was saying there had been no special Creation, that instead there had been a purely material origin of the earth with no divine intention needed, no Genesis. Nor was divine creation visible in any other part of the universe.

Herschel was made a baronet in May 1838, in time to attend Queen Victoria's coronation in Westminster Abbey, and was elected president of the Royal Society in the same year. By the 1850s he was the leading public scientist of Victorian England and was photographed by Julia Margaret Cameron, using a process that he himself had partly invented. But, without belittling all this in any way, his real achievement was to reveal the philosophical significance of astronomy. He calculated that the light rays that

---

* It was this title that Mary Somerville wrote about in her first effort for the Society for the Dissemination of Useful Knowledge, and which John Murray eventually published (see the Introduction).

reached his telescopes from faraway nebulae must have been, in some cases, "two million years on their way."[13] In other words, the universe was almost unimaginably bigger and older than anyone had previously thought. Without Herschel, Charles Darwin would not have been plausible.

The First Geological Synthesis

Alongside cosmology another discipline matured that would put prehistory onto a different footing and further prepare the way for Darwin. Geology differed fundamentally from all the other sciences, and from philosophy. It was, as Charles Gillispie has said, the first science to be concerned with the history of nature rather than its order.

In the seventeenth century Descartes had been the first to link the new astronomy and the new physics to form a coherent view of the universe, in which even the sun—let alone the earth—was just another star. He speculated that the earth might have formed from a ball of cooling ash and become trapped in the sun's "vortex." The idea that physics operated on the same principles throughout the universe was a major change in thinking that could not have occurred to the medieval mind. The basic ideas of heaven and earth, as understood in the West at least, were Aristotelian and were held to be fundamentally different: one could not give rise to the other. Eventually, Descartes's physics were replaced by Newton's, and the "vortex" by gravity. In 1691 Thomas Burnet published his *Sacred Theory of the Earth*, in which he argued that various materials had coalesced to form the earth, with dense rock at the center, then less dense water, then a light crust, on which we live. A few years later, in 1696, William Whiston, Newton's successor at Cambridge, proposed that the earth could have been formed from the cloud of dust left by a comet, which coalesced to form a solid body, and was then deluged with water from a second passing comet. This idea, that the earth was once covered by a vast ocean, which then retreated, proved enduring. Leibniz added the thought that the earth had once been much hotter than it is now, and that earthquakes would therefore have been much more violent in the past.[14]

Slowly, then, a view was forming that the earth itself had changed over time. Nonetheless, however the earth had formed, the central problem faced by the early geologists was to explain how sedimentary rocks, formed by deposition from water, could now stand on dry land. As Peter Bowler has pointed out, there can be only two answers—either the sea levels have subsided, or the land has been raised. The belief that all sedimentary rocks were deposited on the floor of a vast ocean that has since disappeared became known as the Neptunist theory, after the Roman god of the sea. The alternative became known as Vulcanism, after the god of fire.[15]

By far the most influential Neptunist in the eighteenth century, in fact the most influential geologist of any kind, was Abraham Gottlob Werner, a teacher at the mining school in Freiberg, Germany. He proposed that, once one assumed that the earth, when it cooled, had an uneven surface, and that the waters retreated at a different rate in different areas, the formation of rocks could be explained. Primary rocks would be exposed first. Then, assuming the retreat of the waters was slow enough, there would be erosion of the primary rocks, which would drain into the great ocean, and then these sediments would be revealed as the waters retreated still further, to create secondary rocks, a process that could be repeated and repeated. In such a way the different types of rock had been formed in a succession. The first phase produced the "primitive" rocks—granite, gneiss, porphyry—which had crystallized out of the original chemical solution during the flood. The last, which was not formed until all the flood waters had receded, was generated by volcanic activity—accounting for how lavas and tuff, for example, had been produced. According to Werner, volcanoes around the earth were caused by the ignition of coal deposits. Though Werner was himself in no way interested in religion, his Neptunist theory fitted very well with the biblical account of the Flood, which is one reason why it was so popular and gave rise to the phrase "scriptural geology."

This theory had tidiness to recommend it. But it did not even begin to explain why some types of rock, which according to Werner were more recent than other types, were often found situated *below* them. Still more problematical was the sheer amount of water that would have been needed to hold all the land of the earth in solution. It would have to have been a flood many miles deep: what had happened to all that water when it had receded?[16]

The chief rival to Werner, though nowhere near as influential to begin with, was a Scot, from the Edinburgh Enlightenment—none other than Mary Somerville's friend James Hutton. From the middle of the eighteenth century, some naturalists began to suspect that volcanic activity *had* produced some effect on the earth. It was noticed, for instance, that certain mountains in central France had the form of volcanoes, though there was no record of such activity in history. Others pointed to the Giant's Causeway in Ireland, which appeared to consist of columns of basalt that had solidified from a molten state and were therefore of volcanic origin. Hutton did not begin with the origins of the earth, but instead confined himself to observation rather than speculation.

Best known as a geologist, Hutton was in fact a trained chemist and a qualified doctor, who may have completed his medical studies simply because that was the way to learn most chemistry. His father William was a wealthy Edinburgh merchant, who died when his son was two.[17] Even so, his mother brought up James and his three sisters in relative comfort.

In 1736, when he was ten, James entered Edinburgh High School, where he studied the classics and mathematics, going on to university at the age of fourteen. (This was not then as surprising as it sounds now, for at the time Scottish universities competed with schools to educate the brightest pupils.) At university he studied mathematics, logic, and metaphysics and graduated when he was only seventeen. This smooth progression was interrupted in 1745 when Bonnie Prince Charlie's rebellion broke out and, soon after, Hutton found he had fathered an illegitimate child, no small thing in Presbyterian Edinburgh. He fled Scotland, continuing his studies in Paris, Leiden, and London, and did not return to Edinburgh until 1767. But when he did go back, he returned to live in the family home with his three sisters.

And what a homecoming it was. The Edinburgh Enlightenment was in full flood and he quickly formed lasting friendships with Joseph Black, James Watt, and Adam Smith. Alongside the presence of three sisters, the Edinburgh house soon became Hutton's laboratory as well as his home. One visitor wrote, "His study is so full of fossils and chemical apparatus of various kinds that there is barely room to sit down."[18]

Fossils formed part of the picture for Hutton, but not the main part. He looked around him at the geological changes he could see occurring in his own day and adopted the view that these processes had always been going on. In this way he observed that the crust of the earth, its outermost, most accessible layer, is formed by two types of rock, one of igneous origin (formed by heat), and the other of aqueous origin. He further observed that the main igneous rocks (granite, porphyry, basalt) usually lie beneath the aqueous ones, except where subterranean upheavals have thrust the igneous rocks upward. He also observed what anyone else could see, that weathering and erosion are even today laying down a fine silt of sandstone, limestone, clay, and pebbles on the bed of the ocean near river estuaries. He then asked what could have transformed these silts into the solid rock that is everywhere about us. He concluded that it could only have been heat. Water was ruled out—an important breakthrough—because so many of these rocks are clearly insoluble. And so where did this heat come from? Hutton concluded that it came from inside the earth, and that it was expressed by volcanic action. This would explain the convoluted and angled strata that could be observed at many places all over the world. He pointed out that volcanic action was still occurring, and that the rivers—again as anyone could see—were still carrying silt to the sea.

Hutton first published his theories in the *Transactions* of the Royal Society of Edinburgh in 1788. (The Royal Society of Edinburgh was created in 1783 but grew out of the earlier Edinburgh Philosophical Society. It was and is more broadly based than the Royal Society in London, including literary figures and historians among its interests and fellows.) This first Hutton publication was followed by the two-volume *Theory of the Earth* in 1795, "the earliest treatise which can be considered a geological synthesis rather than an imaginative exercise." [19]

At the time Hutton's book appeared, the historical reality of the Flood was beyond question. Just as the Flood was undisputed, so the biblical narrative of the Creation of the world, as revealed in Genesis, was also beyond question. On this account, the length of time since Creation was still believed to be about six thousand years (based on the wording in Genesis that it took God six days to create the world and, elsewhere, that to God one day is like a thousand years). And though some people were beginning to wonder whether this was long enough, hardly anyone thought the earth *very* much older.

There was no question but that Hutton's Vulcanism fitted many of the facts better than Werner's Neptunism. Many critics resisted it, however, because Vulcanism implied vast tracts of geological time, "inconceivable ages that went far beyond what anyone had envisaged before." And so there were many eminent men of science in the early nineteenth century who, despite Hutton's theories, still subscribed to Neptunism: Sir Joseph Banks, Humphry Davy, not to mention Hutton's friend James Watt. Hutton's theory did not really begin to catch on—his books weren't very well written—until John Playfair published a popular version in 1802 (this was the same John Playfair who had tried to bring "the elementary truths of Natural Philosophy into a small compass"—see the Introduction).[20]

But Hutton (a deist) was not alone in believing that the observation of processes still going on would triumph. In 1815, William Smith, a canal builder often called the "father" of British geology, pointed out that similar forms of rock, scattered across the globe, contained similar fossils. Many of these species no longer existed. This, in itself, implied that species came into existence, flourished, and then became extinct, over the vast periods of time that it took the rocks to be laid down and harden. This was significant in two ways. In the first place, it supported the idea that successive layers of rock were not formed all at once, but over time. And second, it reinforced the notion that there had been separate and numerous creations and extinctions, quite at variance with what it said in the Bible.[21]

The Biological Order in the Rocks

Objections to the biblical account were growing. Nevertheless, it was still the case that hardly anyone at the beginning of the nineteenth century questioned the Flood. Peter Bowler says that at this time geological texts sometimes outsold popular novels, but that science "was respectable only so long as it did not appear to disturb religious and social conventions of the day." Neptunism did, however, receive a significant twist in 1811 when Georges Cuvier published his *Recherches sur les ossemens fossiles* ("Researches on Fossil Bones"). With his book going through four editions in ten years, Cuvier argued that there had been not one but several cataclysms—including floods—in the history of the earth. Looking about him, in the Huttonian manner, he concluded that, because entire mammoths and other sizeable vertebrates had been "encased whole" in the ice in mountain regions, these cataclysms must have been very sudden indeed. He also argued that if whole mountains had been lifted high above the seas, these cataclysms could only have been—by definition—unimaginably violent, so violent that entire species had been exterminated and, conceivably, earlier forms of humanity.[22]

Cuvier also observed, and this was important, that in the rocks the deeper the fossils, the more different they were from life forms in existence today and that, moreover, fossils occur in a consistent order everywhere in the world: fish, amphibia, reptilia, mammalia. He therefore concluded that the older the strata of rock, the higher the proportion of extinct species. Since, at that time, no human fossils had turned up anywhere, he concluded that "mankind must have been created at some time between the last catastrophe and the one preceding it." He also observed that Napoleon's expedition to Egypt (1798–1801) had brought back mummified animals thousands of years old, which were identical to those now living, which confirmed the stability of species. Fossil species must therefore have lived for a long time too, before dying out.[23]

There was in fact one other reason why many geologists subscribed to the great flood theory. This was the existence of huge rocks of a completely different type from the land surrounding them. These would later be shown to have been deposited by the ice sheets during the ice ages, but to begin with their distribution was attributed to the great deluge.

The man who played the first major role in marrying fossils and geology was Professor William Buckland, Oxford's first professor of geology. Born in Axminster, Devon, Buckland was the son of a rector, whose interest in road improvements led to his collecting fossil shells, which interest he passed on to his son. Ordained as a priest, Buckland nonetheless kept up his interest in geology, making many excursions on horseback. In 1818 he was elected a fellow of the Royal Society and in the same year persuaded the prince regent to endow an additional readership in geology and paleontology at Oxford University, and became the first holder of the new position.

Before he had been at Oxford very long, in 1821 some miners stumbled upon a cave at Kirkdale in the Vale of Pickering in Yorkshire, where they discovered a huge deposit of "assorted bones." Buckland saw his chance. Hurrying to Yorkshire, he quickly established that while most of the bones belonged to hyenas, there were also many birds and other species, including animals no longer found in Britain—lions, tigers, elephants, rhinoceroses, and hippopotamuses. Moreover, each of the bones and skulls was deformed or broken in much the same way, and he concluded that what the miners had found was a den of hyenas. He wrote up the discovery first as an academic paper, which won the Royal Society's Copley Medal, and then followed it with a more popular account.

His thesis was nothing if not neat. Most of the bones in Kirkdale belonged to species now extinct in Europe. Such bones are never found in alluvial (riverine) deposits of sand or silt; there is no evidence that these animals have lived in Europe since the Flood. It therefore followed, said Buckland, that the animals whose remains the miners had found must have been interred *prior* to Noah's time. He argued finally that the top layer of remains was so beautifully preserved in mud and silt "that they must have been buried suddenly and, judging by the layer of postdiluvial stalactite covering the mud, not much more than five or six thousand years ago." He also noticed a number of very small balls of "calcareous excrement" from an animal that had fed on the bones. This was subsequently recognized as resembling the feces of the Cape hyena, which is how fecal remains began to be used as part of paleontology. Buckland coined the term "coprolite" for fossil fecal matter.[24]

Buckland was as devoted to geology as he was to God. When he married, his idea of a honeymoon (in which his wife seems to have happily cooperated) was a grand European tour of all the famous geological sites, not unlike Joule taking *his* wife to sites around Europe that he found interesting as a physicist. Whether Buckland's wife enjoyed her husband's other exotic habit is less certain. This was zoophagy: Buckland claimed to have eaten his way through as much of the entire animal kingdom as he could lay hands on. The most distasteful items, he said, were mole and bluebottle. Guests at his house record his dining on panther, crocodile, and mouse.[25]

The Ordering of Life Forms: Lyell's Synthesis

However, there were still problems with the flood theory, not least the fact that, as even Buckland acknowledged, the various pieces of evidence around the world placed the flood at widely varying epochs. In addition, by the 1830s the cooling earth theory was gaining coherence as an explanation as to why geological activity was greater in the past than now, further fueling the view that the earth developed, and that life forms had been very different in years gone by. This helped give rise to the idea, in 1841, of John Phillips, professor of geology at Dublin, who identified order in the great sequence of geological formations: the Paleozoic, the age of fishes and invertebrates; the Mesozoic, the age of reptiles; and the Cenozoic, the age of mammals.

This was based in part on the work of Adam Sedgwick and Sir Roderick Murchison, in Wales, which began the decoding of the Paleozoic system. The Paleozoic period would eventually be shown to have extended from roughly 550 million years ago to 250 million years ago, and during that time plant life had moved out of the oceans onto land, fish appeared, and then amphibians, and subsequently reptiles had reached land. These new forms of life were all wiped out, about 250 million years ago, for reasons that are still hard to fathom. But it was clear from the analyses of Sedgwick and Murchison that early forms of life on earth were very old, that life had begun in the sea, and then climbed ashore. Deluge or no deluge, all this was again a dramatic contradiction of the biblical account.[26]

The importance of the discoveries of Cuvier, Buckland, Sedgwick, and Murchison—over and above their intrinsic merit—was that they brought about a decisive change of mind on the part of Charles Lyell. In 1830 he published the first volume of what would turn into his three-volume *Principles of Geology*.

Lyell, another Scot, was born into a well-educated and sophisticated family—his father was known both as a botanist and as the translator of several works of Dante. Charles attended William

Buckland's celebrated lectures at Oxford, but after graduation took up law at Lincoln's Inn. He had private means, so an income was not his main concern, and when he was sent out on the western circuit he was able to make geological observations as he traveled. But he never had good eyesight, and as it began to deteriorate further, he abandoned the law for geology full-time, to begin what would be his life's work—developing James Hutton's ideas on how the processes we see around us today have always existed and have shaped the surface of the earth as we know it.

Lyell's argument in the *Principles* was contained in the (fashionably) long subtitle: *Being an Attempt to Explain the Former Changes of the Earth's Surface, by Reference to Causes Now in Operation.* Essentially, *Principles* was a work of synthesis, rather than of original research, in which Lyell clarified and *interpreted* already published material to support two conclusions. The first, obviously enough, was to show that the main geological features of the earth could be explained as the result of actions in history that were exactly the same as those that could be observed in the present. In a review of his book, the term "uniformitarianism" was used and caught on. Lyell's second aim, despite his being a devout Christian, was to resist the idea that a great flood, or series of floods, had produced the features of the earth that we see around us.[27]

On the religious front, Lyell took the common-sense view, arguing that it was unlikely God would keep interfering in the laws of nature by provoking a series of major cataclysms. Instead, he said, *provided that one assumed that the past extended back far enough*, the geological action that could be observed as still in operation today was enough to explain "the record in the rocks." And (his most intellectually original contribution) he compared the findings of stratigraphy, paleontology, and physical geography to identify three separate eras with three distinct forms of life. These became known as the Pliocene, Miocene, and Eocene epochs, the last of which went back 55 million years. Yet again this was a much longer time frame than anything in the Old Testament.

Volume one of the *Principles* took issue with the Flood, and began the process whereby the idea would be killed off. In volume two, Lyell demolished the biblical version of Creation. Inspecting the fossils as revealed in the order of the rocks, he showed that there had been a continuous stream of creation, and extinction, involving literally countless species. He thought that man had been created relatively recently, but by a process that was just the same as for other animals.[28]

The Ascending Order of Life in the Rocks

More radical still than Lyell was Robert Chambers, yet another Edinburgh figure, whose *Vestiges of the Natural History of Creation*, published in 1844, was so contentious that Chambers published the book anonymously. This work also promoted the basic idea of evolution, though again without in any way anticipating Darwinian natural selection. But Chambers described the progress of life as a purely natural process, his main contribution being to order the paleontological record in an ascending system and to argue that man did not stand out in any way from other organisms in the natural world. Though he had no grasp of natural selection, or indeed of how evolution might actually work, he did introduce people to the *idea* of evolution fifteen years before Darwin.

James Secord, in his recent book *Victorian Sensation* (2000) has explored the full impact of *Vestiges*. He goes so far as to say that Darwin was, in a sense, "scooped" by Chambers, that wide and varied sections of (British) society discussed *Vestiges*—at the British Association, in fashionable intellectual salons and societies, in London, Cambridge, Liverpool, and Edinburgh, among feminists and freethinkers but also among "lower" social groups. Moreover, the ideas the book promoted passed into general discussion, being referred to in paintings, exhibitions, and cartoons in the new, mass-circulation newspapers. Chambers believed his book would create a sensation: one reason he published it anonymously was in case it didn't do well; another reason was in case it *did* do well. Secord's especially important point is that it was

Vestiges that introduced evolution to a huge range of people (there were fourteen editions) and that, viewed in such a light, Darwin's *Origin of Species* did not so much create a crisis as resolve one.[29]

The Link Between Ice and Agriculture

One other set of events also paved the way for Darwin. This was the discovery of the great Ice Age, by Louis Agassiz and others. Born in Môtier, in the Swiss canton of Fribourg, Agassiz obtained a PhD in botany from Erlangen and a medical degree in Munich. He was appointed professor of natural history at Neuchâtel, where his work on glaciation was to earn him enduring fame.

By observing the behaviors of present-day glaciers (of which there was no shortage in the Swiss Alps), Agassiz came to the conclusion that much of northern Europe had once been buried by a covering of ice, in places up to three kilometers thick. This conclusion was based chiefly on three types of evidence found at the edges of glaciers even today—"erratics," moraines, and till. Erratics are large boulders—like those near Geneva—whose constitution is quite different from the rock all around them. They are pushed by the edges of glaciers, as the ice expands, and then left in a "foreign" environment when the earth warms up again and the ice retreats. Thus geologists suddenly find a massive boulder of, say, granite, in an area otherwise made up of limestone. Till is a form of gravel formed by the ice as it expands over the earth and acts, in J. D. Macdougall's words, like a giant sheet of sandpaper. (Till provides a lot of gravel resources for modern construction industries.) Moraines are mounds of till that build up at the edges of glaciers and can be quite large: most of Long Island, in New York State, is a moraine more than 110 miles from end to end. Agassiz and others concluded that the most recent great Ice Age began about 130,000 years ago, peaked at 20,000 years ago, and ended quickly at 12,000 to 10,000 years ago. In time this would prove extremely significant, in that it tallied with the emerging evidence for the beginnings of agriculture.[30]

Order Without Design

The term "evolution" was originally used in biology exclusively for the growth of the embryo. In the original Latin it means "to unfold." Outside that usage, terms like "progressionism," or development, were used to convey the cohering notion that simpler organisms had, in some way that was yet to be discovered, given rise to more complex ones. Evolution was next used in a cultural sense, following the observations of Giambattista Vico, Johann Gottfried Herder, and others, who saw in the developing order of human societies a progression from primitive to more advanced forms of civilization.[31]

Because of these factors, and others, it has been said that there was something "in the air" in the middle of the nineteenth century, which helped give rise to what Darwin would call natural selection. A struggle for existence had been implied as long ago as 1797 by Thomas Robert Malthus, who had argued that while population increases exponentially, food supply increases only arithmetically, meaning that population must in time outgrow food supply, with disastrous consequences. Each tribe in history would have completed for resources, he said, with the less successful becoming extinct.

And it was by now difficult to contradict the evidence of the rocks, where the basic order was clear. "The earliest rocks [600 million years ago] yielded only the remains of invertebrates, with the first fish appearing only in the Silurian [440–410 million years ago]. The Mesozoic [250–265 million years ago] was dominated by the reptiles, including the dinosaurs. Although present in small numbers in the Mesozoic, the mammals only became dominant in the Cenozoic [65 million years ago to the present], gradually progressing to the more advanced creatures of today, including the human species." (The dates in square brackets were not, of course, accepted in the nineteenth century.) It was hard for people not to read some sort of "end" in this ordering, "leading," via stages, to humans, "and thus revealing a divine plan with a symbolic purpose."[32]

A final element in this "climate of opinion," this "something in the air," as regards "progressionism," and how it was achieved, was the work of Alfred Russel Wallace. Wallace's reputation and role in the discovery of natural selection have gone through their own evolution in recent times. Wallace was probably Welsh, of Scottish extraction. "Probably" Welsh because he was born in Llanbadoc, near Usk in Monmouthshire, but spent little time there, his father claiming to be descended (as so many Wallaces do) from William Wallace, one of the Scots' leaders in the Wars of Scottish Independence. Wallace (the son) considered himself English. The seventh of nine children, he always had a wanderlust and he seems always to have had money troubles. He was far more than a scientist, being very interested in politics and social change.

He read avidly—everything from insects to engineering to architecture, Malthus, Humboldt's adventures, Darwin's voyages, Lyell's *Principles*—and this augmented his urge to travel. On his first expedition, to Brazil, the ship's cargo caught fire on the voyage home and all the specimens were lost, save for Wallace's diary and a few sketches, and he spent ten days in an open boat. A subsequent expedition to the Far East was more successful, with Alfred Russel collecting 126,000 specimens, including 80,000 beetles and 80 bird skeletons. His book on this set of travels was published in 1869 as *The Malay Archipelago* and has never been out of print (Joseph Conrad called it his "favourite bedside reading" and apparently used it as a source for some of his novels, especially *Lord Jim*).[33]

It was on the Far East trip that Wallace began to have his ideas about evolution and natural selection, and in 1858 he sent an article outlining his theory to Darwin (asking him to send it on to Lyell, whom Darwin knew and Wallace didn't). For many years it was accepted that this paper sent to Darwin, "On the Tendencies of Varieties to Depart Indefinitely from the Original Type," contained a clear exposition of natural selection, such that Darwin was forced to begin a move towards publication of his own book, *On the Origin of Species.* (Wallace's article was published in that year by the Linnean Society of London, together with a paper by Darwin.) As a result, some scholars have argued that Wallace was never given the recognition he deserves.

More recently, however, a closer reading of Wallace's paper has shown that his idea about natural selection was *not* the same as Darwin's, and that it was much less powerful as an explanatory device. Wallace did not stress competition between individuals, but between individuals and the environment. For Wallace, the less fit individuals, those less well adapted to their environment, will be eliminated, especially when there are major changes in that environment. Under this system, each individual struggles against the environment, and the fate of any one individual is independent of others. This difference, which is fundamental, may explain why Wallace appears to have shown no resentment when Darwin's book was published the year after he had sent him his paper. Indeed, he dedicated *The Malay Archipelago* to Darwin, visited him at Down House (Darwin's home) in 1862, and became friendly with Lyell and Herbert Spencer. And he certainly did not mope about, but instead pursued other—very different—interests, such as spiritualism and land reform.[34]

None of the foregoing, however, should cloud the fact that when *On the Origin of Species* did appear, in 1859, it introduced "an entirely new and—to Darwin's contemporaries—an entirely unexpected approach to the question of biological evolution." Darwin's theory explained, as no one else had done, a new mechanism of change in the biological world. It showed how one species gave rise to another and, in Harvard historian Ernst Mayr's words, "represented not merely the replacement of one scientific theory ('immutable species') by a new one, but demanded a complete rethinking of man's concept of the world and of himself." For Peter Bowler, "The historian of ideas sees the revolution in biology as symptomatic of a deeper change in the values of Western society, as the Christian view of man and nature was replaced by a materialistic one."[35]

The most notable flash of insight by Darwin was his theory of natural selection, the backbone of the book (its full title was *On the Origin of Species by Means of Natural Selection, or the Preservation of Favoured Races in the Struggle for Life*). Individuals of any species show variations, and those better suited were more likely to reproduce and give rise to a new generation. In this way, accidental variations that "fitted" better than others were encouraged. No "design" was necessary in this theory, which could be observed on all sides. Although Darwin had been stimulated to publish the *Origin* after being contacted by Wallace, he had been germinating his ideas since the late 1830s, after his now-famous voyage on the *Beagle*.

Unfashionable Subjects

Charles Robert Darwin was born on February 12, 1809 in Shrewsbury. His father, Robert Waring Darwin, was a successful and wealthy doctor, so successful that in his later years he grew corpulent, as Charles recalled, "the largest man whom I ever saw."[36] Charles was the grandson of the physician Erasmus Darwin (1731–1802), and his mother was Susannah Wedgwood, daughter of the potter Josiah Wedgwood (1730–95), both men members of the celebrated Lunar Society of Birmingham, which had done so much to promote science in the eighteenth century. The men in Darwin's family were liberals and Whigs, freethinkers on matters of religion, "having no interest in the platitudes of those who used Christianity to bolster the status quo." However, the women in the family (Darwin had three sisters and one brother) were inclined to take the Bible seriously. The Darwins were supporters of free enterprise who, having gained their wealth by their own efforts, expected to hold on to it: they had as little patience with aristocratic privilege as they had with radicals who pushed the merits of socialism.

Charles was sent as a boarder to Shrewsbury, where the head-master, Dr. Samuel Butler (grandfather of the novelist of the same name), provided a classical education that didn't appeal to Darwin. He did become interested in minerals and birdwatching and chemistry, but was cautioned by the headmaster for taking an interest in "so unfashionable a subject."[37] He then went on to Edinburgh to study medicine at the early age of sixteen. This choice shows the family's unorthodox style. Erasmus's son had gone there, and Charles's father had gotten his degree at Leiden— not for them the orthodox route of Oxbridge. However, after one or two unpleasant experiences in the operating theater in Edinburgh, Charles decided that he wasn't cut out for medicine.

His father therefore suggested the church, and as a require-ment to take holy orders, he needed a degree from an English university, which is how he came to go to Cambridge. At that time, he said later in his *Autobiography*, he did not in the least doubt the literal truth of every word in the scriptures.

At Cambridge he was surrounded by other students studying for the church, and the natural sciences did not form part of his curriculum.[38] The professor of botany, John Henslow, gave pub-lic lectures on his subject and Darwin enjoyed them enormously. But he steered clear of Adam Sedgwick's geology course and only showed an interest toward the end of his time there. Yet his interest in natural history was growing, encouraged by Henslow, who recognized his seriousness and started inviting him to his "open houses," held every week.

After his degree, Darwin stayed on in Cambridge for two more terms and now did attend Sedgwick's lectures. In the summer of 1831, Charles accompanied him on a field trip to Wales, so it seems the geologist was taking Darwin as seriously as Henslow was.

In fact, it was Henslow who was responsible for Darwin going on the *Beagle*. The captain was looking for a naturalist to accompany him on a voyage to chart the coast of South America and the South Seas, and didn't want a doctor doubling up as a naturalist, as was usual, because he felt that would mean neither job would be done properly. The hydrographer of the navy had a friend in Cambridge, who was a friend of Henslow, and Henslow recommended Darwin.[39]

Darwin moved aboard ship at Plymouth on October 24, 1831, though because of bad weather they didn't leave until December 27. He took guns, hand lenses, microscopes, equipment for geological and chemical analysis, and books, including the first volume of Lyell's *Principles of Geology* which had just appeared, "a work that Henslow advised him to read but on no account to believe."[40]

Patterns in Populations

The voyage of the *Beagle* was a watershed in Darwin's career but not in the way it is usually presented. He did not conceive natural selection on board the five-year circumnavigation, but he did return to England determined to become a leading scientist rather than a churchman. The American psychologist Frank Sulloway's computerized study of Darwin's letters from the time has shown that, to begin with at least, he concentrated far more on geology than on biology, gradually starting out as a critic of Lyell, but, as time went by, and as he studied rock formations in South America, he came round to Lyell's uniformitarian views.

On several of the stops he journeyed inland on horseback, and it was in this way that, after meeting some gauchos, he became aware of the two forms of rhea—the South American ostrich. He saw that each of the two forms had its own territory but that there was an overlap area, where both could be found competing for supremacy.

Early in 1834 the *Beagle* reached the west coast of the continent, and while he was ashore in Valdivia, the largest city in the region, Darwin witnessed an earthquake severe enough for its rolling motion to make him giddy.[41] The fact that parts of the coast were raised by the earthquake, between two and ten feet, also made an impression on him, further underlining Lyell's uniformitarian argument.

And then it was on to the Galápagos Islands, in the Pacific Ocean, part of Ecuador, which the *Beagle* reached in September 1835. These islands, several hundred miles out into the Pacific from Ecuador, have since become famous as the location where Darwin realized that the finches had adapted to each island, as revealed by the differing shapes of their beaks. Except that it didn't work like that. Darwin did spend time on Galápagos, alone at one stage, while the *Beagle* went in search of water. But, as Sulloway and Bowler make clear, there was no "eureka" moment for Darwin on Galápagos—he left the islands without realizing the full significance of the finches and only pieced together the full picture after he returned to Britain, when he had to borrow specimens from other collections than his own to test his conclusions. It was not until the long journey home (via Tahiti, which he liked, and New Zealand, which he didn't, and Bahia in Brazil for the second time) that he began to suspect that the evidence from the island groups "might upset the accepted view on the stability of species."[42]

His time in South America, in particular the Galápagos Islands, had taught him to think in terms of *populations* rather than individuals, as he studied variation from island to island. And he began to wonder why there were related species on different islands and continents—would the Creator have visited each location and made these fine adjustments? From a study of barnacles he noted how much variety was possible in a single species, and all these observations and inferences gradually came together in the two decades after he returned.

The Explanation of Design

When the book was published, on November 24, 1859, 1,250 copies were snapped up on the first day. Darwin himself took the waters at Ilkley, in Yorkshire, waiting for the storm to break. It did not take long and it is not hard to see why. Ernst Mayr, the historian of biology, concluded that there were six major philosophical implications of Darwin's theories: (1) the replacement of a static by an evolving world; (2) the demonstration of the implausibility of creationism; (3) the refutation of cosmic teleology (the idea that there was a purpose in the universe); (4) the abolition of any justification for absolute anthropocentrism (that the purpose of the world is the production of man); (5) the explanation of "design" in the world by purely materialistic processes; (6) the replacement of essentialism by population thinking.[43]

We must be clear about the impact of the *Origin*. It owed something to Darwin's solid reputation and because his book was packed with supporting details—it was not produced by a nobody. Yet its impact also had something to do with the fact that, as James Secord has pointed out, the book *resolved*—or appeared to resolve—a crisis, not because it sparked one. Natural selection was, essentially, the last plank in the evolutionary argument, not the first one, the final filling in of the theory, providing the mechanism by which one species gives rise to another. The "non-revolutionary" nature of the *Origin*, to use Peter Bowler's term, is underlined by Secord's chart in his book, which records that the *Origin* did not decisively outsell *Vestiges* until the twentieth century.

That said, the *Origin* did promote enormous opposition. John F. W. Herschel, a philosopher, polymath, and photographer whom Darwin admired (and the son of William Herschel), called natural selection the "law of higgledy-piggledy," while Sedgwick (who, remember, was both a divine and a scientist) said that when he read the book parts of it made him laugh "till my sides were almost sore," while other parts he read with "absolute sorrow" because he thought them false and "grievously mischievous." To another he wrote: "It repudiates all reasoning from final causes; and seems to shut the door on any view (however feeble) of the God of Nature as manifested in His works." Despite this the two men remained friendly until Sedgwick's death. Agassiz, a staunch creationist, was physically repulsed by the idea that all humans were equal, and he never accepted natural selection.[44]

Many of the favorable reviews of the *Origin* were lukewarm about natural selection. Lyell, for example, never accepted it fully, and described it as "distasteful," while T. H. Huxley did not think it could be proved or falsified. In the late nineteenth century, while the theory of evolution was widely accepted, natural selection was ignored, and this was important because it allowed people to assume that evolution was "intended to develop toward a particular goal, just as embryos grew to maturity." Viewed in this way, evolution was not the threat to religion it is sometimes made to appear. Ernst Mayr says the selection aspect of Darwin's theory was not finally accepted until the evolutionary synthesis of the 1930s and 1940s (chapter 9).

Many people simply thought that Darwinism was selfish and wasteful. What was the Darwinian purpose of musical ability, or the ability to perform abstract mathematical calculations? Darwin, it should be said, was never entirely happy with the word "selection."

Darwin's theory certainly had a major weakness. There was no account of the actual mechanism by which inherited characteristics were passed on ("hard heredity"). These were discovered by the monk Gregor Mendel in Moravia in 1865, but Darwin and everyone else missed their significance and they were not rediscovered and given general circulation until 1900.

The Overlap Between Apes and Humans

Darwin didn't stop with the *Origin*. No account of Darwinism can afford to neglect *The Descent of Man*. For many people, the crucial issue underlying the debate as to whether man was evolved from apes revolved around the question of the soul. If man was, in effect, little more than an ape, did that mean that the very idea of a soul—the traditional all-important difference between animals and men—would have to be rejected? Darwin's

The Descent of Man, published in 1871, tried to do two things at once: convince skeptics that man really was descended from the animals and yet also explain what exactly it meant to be human— how humans had acquired their unique qualities. In the *Descent*, he knew that, above all, he had to explain the very great—the enormous—increase in mental power from apes to humans. If evolution was a slow, gradual process, why did such a large gap exist?[45]

Darwin's answer came in chapter four. There, he advanced the proposition that man possesses a unique *physical* attribute, namely, an upright posture. Darwin argued that this upright posture, and the bipedal mode of locomotion, would have freed the human's hands which, as a result, eventually developed the capacity to use tools—the stone handaxes then being made so much of. And it was this, he said, which would have sparked the rapid growth in intelligence among this one form of great ape.

Darwin did not offer any cogent reason as to why ancient man had started to walk upright, and it was not until 1889 that Wallace suggested it could well have been an adaptation to a new environment. He speculated that early man was forced out of the trees onto the open savannah plains, perhaps as a result of climate change, which shrank the forests. On the savannah, he suggested, bipedalism was a more suitable mode of locomotion.[46]

The Significance of the New Synthesis

The legacy of Darwinism is complex. The way it drew on—and synthesized—geology, anthropology, animal behavior; the way it fitted in with recent advances in astronomy and cosmology, was for many people both fascinating and naturally convincing. Furthermore, the bringing together of different disciplines, and its timing, played a major role in the secularization of European thought. Darwinism forced people to a new view of history: that it occurred by accident, and that there was no goal, no ultimate end point. As well as killing the need for God, it challenged the idea of wisdom as some definite attainable state, however far off. It was Darwinism's model of societal change that led Marx to his view of the inevitability of revolution, and it was Darwin's biology that led Freud to conceive the "prehuman" nature of subconscious mental activity. Since the rediscovery of the gene, in 1900, and the flowering of the technology based on it, Darwinism has triumphed and is beginning to make a practical difference to our lives.

It is important to point out one big difference between the concept of the conservation of energy and evolution by natural selection. As John Theodore Merz made clear in his *History of European Thought in the Nineteenth Century*, the concept of "energy" was rapidly accepted by the scientists of the time. He himself observed that the unification of thought that had been brought about by so many observations and experiments in optics, electricity, magnetism, and biochemistry "needed a more general term . . . a still higher generalisation, a more complete unification of knowledge . . . this greatest of all exact generalizations, the conception of energy."

This was not at all the case with evolution by natural selection. Almost a century later, in 1995, the American philosopher Daniel Dennett stated flatly that "if I were to give an award for the single best idea anyone ever had, I'd give it to Darwin. . . . In a single stroke, the idea of evolution by natural selection unifies the realm of life, meaning and purpose with the realm of space, time, cause and effect, mechanism and physical law. . . . Darwin's dangerous idea is reductionism incarnate."[47] But note that even then Darwinism was a "dangerous" idea—it still worried some people and was by no means universally accepted. In some backwaters, it still isn't.

Why the difference? Possibly it is because physics is more abstruse than biology, at a greater distance from everyday life. Its mathematical content puts it beyond the reach of many, and it has little direct impact on our understanding of ourselves, our motivations or what it means to be human. We respect physics and its achievements, but we don't, so to speak, want—or feel the need—to put our heads under the hood.

In one sense, though, that difference serves our purpose. As the decades passed after their discoveries, the concepts of energy and evolution would themselves come together to underline that a basic, wide-ranging, and deep coherence is at the very root of reality and is arguably its most significant feature. The converging coherence of the sciences—underpinned by energy and evolution—helps us grasp everything that we now know. At the same time, it also embodies—as was introduced in the Preface—what we might call the final mystery. This process—and this era of history—began in the 1850s.

PART TWO

The Long Arm of the Laws of Physics

The two great unifying themes of Part One—the conservation of energy and evolution by natural selection—set the scene for this book. However, they embody a contradiction. The thermodynamicists said plainly that the world was running down, its useful energy leaking out, that organization was being lost, that "entropy tends to a maximum." But evolution showed that biological systems at least were running up, becoming more—not less—complex and organized.

This contradiction is a secondary theme of the book. The synthesis, the unification between energy and evolution, would be some time coming—not, in fact, until the final decades of the twentieth century. To an extent, this was due to the fact that, as this section will show, physics up until World War II became more unified and more diversified—more ordered and more complex—all at the same time. The emerging order, the coherence of the sciences, was at that stage by no means obvious everywhere.

In this part, evolution takes a backseat as we follow the impressive progress of physics as it exposes order among new forms of energy—radioactivity—and elementary particles, links the first elementary particles to atomic structure, explains the relationship between electrons and quantum states, and shows above all how physics underpins chemistry in a marriage of extraordinary intimacy. It was then, also, that the first steps were taken to show how physics would eventually join forces with astronomy, to create an entirely new cosmology that transformed our ideas of the universe and its evolution. It was the beginning of another marriage: that between the breathtakingly small and the unimaginably large.

3

BENEATH THE PATTERN
OF THE ELEMENTS

Just as more than a dozen individuals had the same idea—at more or less the same time—about the conservation of energy, much the same thing happened with the discovery of the periodic table. Different individuals, in different countries, were struck by a pattern among the elements, which perhaps pointed to an underlying order.

The same idea, more or less, occurred to the Italian Stanislao Cannizzaro, a passionate nationalist who fought under Garibaldi and became a politician as well as a scientist; to John Newlands, a British industrial chemist; to the French mineralogist Alexandre-Émile Béguyer de Chancourtois; to the German physicist and chemist Lothar Meyer; and, perhaps most famously, to the Russian Dmitri Mendeleyev. Here again there is a raft of priority disputes, centered around the realization, on the part of all these men, that if the elements are arranged in order of their atomic weight, there is a repeating pattern in which at regular intervals elements have similar properties to one another. These intervals, to begin with at least, appeared to recur at multiples of eight of hydrogen (atomic weight 1).*

---

* "Atomic weight" is now a historical term. Relative atomic mass is preferred these days. While the atomic number of hydrogen is 1, its relative atomic mass

82

CONVERGENCE

Mendeleyev, whose version was in the end the most widely adopted (though the Royal Society awarded its Davy Medal* to Mendeleyev *and* Meyer), completed his work in ignorance of the others because he was a Russian and at that stage Russia was intellectually cut off from Western Europe. Russia then was still feudal land, most people being serfs owned by the landowners.[1]

In fact, Mendeleyev didn't come from European Russia at all but from Tobolsk in Siberia, thousands of miles to the east. He was the youngest of fourteen or seventeen children (someone seems to have lost count along the way), and Dmitri's early life was bleak. His father, the headmaster of the local school, became blind the year Dmitri was born and his wife had to look after the family. Then, a worse catastrophe struck. Dmitri's father died, and the following year the glass factory that his mother's family owned burned down. And so, in 1849 Maria Dmitrievna, already fifty-seven, and her two children who were still dependent on her, plus Dmitri, set out for Moscow, 1,300 miles away. She left behind a number of workers who depended on her, but it seems she appreciated how brilliant her son was and made the move for his sake.

Success took time. Moscow University operated a quota system for students from the provinces, but Siberia was so distant that there was no quota and his qualifications were not recognized. The Mendeleyevs therefore moved on to St. Petersburg. Here it was the same story, except that Dmitri's mother found that she knew the head of the Central Pedagogical Institute, who had been a friend and colleague of her dead husband. Mendeleyev was admitted to study mathematics and science and even given a small government scholarship.

---

is 1.008. This is a relative number, relative to the atoms of other elements, and takes into account the slightly different isotopes (such as deuterium in the case of hydrogen), which were not known about in Mendeleyev's time.

* The Davy Medal is awarded for outstanding work in chemistry and is named in honor of Humphry Davy (1778–1829).

Things got worse before they got better. Before he could shine, his mother died. Then his sister died. Then he suffered a throat hemorrhage and was placed in the institute hospital. TB was diagnosed and he was given only a few months to live.

Paul Strathern describes this as Mendeleyev's Dickensian nadir. But then he seems to have enjoyed an equally Dickensian reversal of fortune. With his deep-set blue eyes, flowing hair, and luxuriant beard, he was adopted as the institute mascot. He spent long periods in bed, but gradually recovered, being allowed up for spells in the laboratories, where he began doing original research.[2]

Similarities Between Heaven and Earth

In 1855 he qualified as a teacher, his brilliance showing through as he took the gold medal for the best student of his year. On the advice of the chemist and composer Alexander Borodin he managed to secure a government grant to study in Paris, where he was placed under Henri Regnault, the man who first established that absolute zero was –273°C.[3] After Paris, Mendeleyev transferred to Heidelberg, where he worked with Robert Bunsen, of Bunsen burner fame, and Gustav Kirchhoff, who between them developed spectroscopy, which uses a prism to refract light. As Newton had shown, when white light passes through a prism it breaks up into a rainbow of colors as different hues are refracted differently. Kirchhoff and Bunsen found that, when an element was heated, the light it emitted produced its own characteristic spectrum of colors.

Just before Mendeleyev arrived in Heidelberg, Kirchhoff and Bunsen had observed that light which passed through gas produced lines that matched the substance of the element the gas was made of. This enabled them to produce the remarkable result that the light from the sun, which passed through the atmosphere, produced bands which corresponded to sodium vapor, meaning that one of the sun's constituents was sodium. This was a hint that the heavens were made of the same components as the earth, a result as important theologically as it was scientifically.

Mendeleyev returned to St. Petersburg in 1861 to find that he was virtually alone in Russia in being up to date on the latest discoveries in chemistry. His lectures at last began to attract attention, on the strength of which he became a full professor and turned his hand to some very successful textbooks.[4]

Chemical Patience

During the writing of one of his textbooks, however, he had an epiphany about the way the elements might be related. He made some notes on the back of an envelope (we know this because the letter is preserved, along with the rings from a coffee mug that he also laid on the envelope), writing down the atomic weights of the halogens, which he knew had similar properties but very different atomic weights:*

$$F = 19 \qquad Cl = 35 \qquad Br = 80 \qquad I = 127$$

And it was the same with the oxygen group:

$$O = 16 \qquad S = 32 \qquad Se = 79 \qquad Te = 128$$

And again with the nitrogen group:

$$N = 14 \qquad P = 41 \qquad As = 75 \qquad Sb = 122$$

Yet when he put them all together, a pattern did begin to emerge:

$$F = 19 \qquad Cl = 35 \qquad Br = 80 \qquad I = 127$$
$$O = 16 \qquad S = 32 \qquad Se = 79 \qquad Te = 128$$
$$N = 14 \qquad P = 41 \qquad As = 75 \qquad Sb = 122$$

---

* Halogens are produced from minerals or salts, and form acids binding with hydrogen.

The pattern—if it *was* a pattern—was that, beginning with the element at the left-hand foot of the table (N = 14), and reading up the vertical columns, and from left to right, most of the atomic weights were in order—save for P and Te.[5]

According to a colleague who visited him that weekend, Mendeleyev had been working on his problem for three days and three nights without rest. But only now did an analogy occur to him.[6] His favorite card game was patience and he realized that what he was looking at was not unlike a card game, with families of chemicals rather like suits of cards (he actually called it "chemical patience"). Tired, he rested his head on his arm, fell asleep, and, sitting there in his study, had a dream. In the dream, he later said, all the elements fell into place in what he called the Periodic Table of Elements—periodic because elements with similar properties occurred at periodic intervals.

Mendeleyev was so struck by the periodicity that he made two bold claims. One, that in those cases where the atomic weight of elements did not place them where their properties indicated they should go, the atomic weights had been calculated wrongly. And second, he left gaps where no atomic weights were known, predicting that the gaps would be filled in later by subsequent discovery.

An added factor was that the way he grouped the elements, those in the same group all had the same valency, the same affinity for other elements (measured by the number of hydrogen atoms they typically combined with). This tended to confirm that valency was related to chemical properties, and his bold predictions—both borne out by later experimentation—confirmed Mendeleyev as the man who really first understood the layout of the periodic table *and its significance.*

With the periodic table, as Paul Strathern puts it, chemistry came of age. Mendeleyev's discovery gave chemistry a central idea to put alongside Newton's in physics and Darwin's in biology. "Mendeleyev had classified the building blocks of the universe."[7] In 1955, element 101 was identified and, in honor of the Russian's achievement, was named mendelevium.

At that time chemistry was chemistry and physics was physics (and, for that matter, biology was biology). But, as the periodic table implied, matter was both discrete (it existed in discrete forms chemically) and yet patterned. The years around the turn of the twentieth century were to show that the discreteness of nature was fundamental.

Energy, Not Matter, Is the Essence

At the time, most physicists still clung to a mechanical view of the universe, even James Clerk Maxwell, whose field theory found many supporters. This was accompanied by the rise to prominence of the idea of a universal ether as a "quasi-hypothetical," continuous, and all-pervading medium through which forces were propagated at a finite speed. This helped people consider the possibility that the foundation of these forces was electromagnetic, rather than mechanical. In this environment, new ideas began to proliferate—most importantly a new field of "energetics," put forward by the German physicist Georg Helm and his chemist colleague Wilhelm Ostwald. On this view, energy not matter was the essence "of a reality that could be understood only as processes of actions."[8]

One man who was impressed by this argument was Heinrich Hertz. Born in Hamburg in 1857, the son of a Jewish lawyer who converted to Christianity, Heinrich attended university in Munich and afterward at Berlin, where he studied under Gustav Kirchhoff and Hermann von Helmholtz. Hertz's PhD thesis in 1880 was so well received that he became von Helmholtz's assistant, after which he was appointed a lecturer in theoretical physics at Kiel.

At Kiel, Hertz produced his first important contribution when he derived Maxwell's equations but in a way different from Maxwell's own, and which did not involve the assumption of an ether. On the strength of this, Hertz was appointed the following year—at the early age of twenty-eight—to the chair of physics at Karlsruhe, a much bigger, better-equipped university. There, his first significant discovery was the photoelectric effect, whereby ultraviolet radiation releases electrons from the surface of a metal.

In 1888 Hertz built on all this, producing his own important breakthrough by devising a most innovative apparatus. The central element here was a metal rod in the shape of a hoop with a tiny (3mm) gap at its midpoint, not unlike a large key ring. When a strong enough current was passed through the hoop, sparks were generated across the gap. At the same time, violent oscillations were set up in the rod forming the hoop. Hertz's crucial observation was that these oscillations sent out waves through the nearby air, a phenomenon he was able to prove because a similar circuit some way off could detect them. In later experiments Hertz showed that these waves could be reflected and refracted—like light waves—and that they traveled at the speed of light but have much longer wavelengths than light. Later still, he observed that a concave reflector could focus the waves and that they passed through non-conducting substances unchanged. These were originally called Hertzian waves, and their initial importance lay in the fact that they confirmed Maxwell's prediction (see chapter 1) that electromagnetic waves could exist in more than one form—light. Later, Hertz's waves were called radio.[9]

Asked by a student what use might be made of his discovery, Hertz famously replied, "It's of no use whatsoever. This is just an experiment that proves Maestro Maxwell was right—we just have these mysterious electromagnetic waves that we cannot see with the naked eye." Asked "So what's next?" he shrugged. "Nothing, I guess." A young Italian, on holiday in the Alps, read Hertz's article about his discovery and immediately wondered whether the waves set off by Hertz's spark oscillator might be used for signaling. Guglielmo Marconi rushed back home to see whether his idea might work.[10]

From Karlsruhe, Hertz transferred to Bonn to follow Rudolf Clausius as director of the Physics Institute. His abandonment of exciting experimental work in favor of theory has always baffled his fellow physicists. Had he lived (he died from bone disease at thirty-seven) he would have been as surprised as anyone at the direction physics was about to take. Rollo Appleyard, the historian of electrical engineering, says Hertz was in all respects "a Newtonian."[11]

A New Form of Energy

A new electromagnetic wave was one thing, one very useful thing as it turned out before too long. But it didn't help anyone understand what electromagnetism *was*. And how did it relate to the statistical properties of matter also being unearthed by Maxwell and Boltzmann?

An advance was made when it was noticed that flashes of "light" sometimes occurred in the partial vacuums that existed in barometers. What, exactly, was going on? This brought about the invention of a new—and as it turned out all-important—instrument: glass vessels with metal electrodes at either end. Air was pumped out of these vessels, creating a more complete vacuum than existed in barometers, before gases were introduced, and an electrical current passed through the electrodes to see how the gases might be affected. In the course of these experiments, it was noticed that if an electric current was passed through a vacuum, a strange glow could be observed, reminiscent of—but not the same as—the flashes that had been observed in barometers. The exact nature of this glow was not understood at first, but because the rays emanated from the cathode end of the electrical circuit, and were absorbed into the anode, Eugen Goldstein, a Polish physicist who had studied under von Helmholtz in Berlin, christened them *Kathodenstrahlen*, or cathode rays. Goldstein and almost all German physicists at the time were convinced that these "rays" were waves, whereas almost all British physicists were equally convinced they were particles.[12] It was not until the 1890s that three experiments stemming from cathode-ray tubes finally made everything clear and set modern physics on its triumphant course.

The first was carried out by the professor of physics at the University of Würzburg, Wilhelm Röntgen. Röntgen—"darkly handsome," according to his biographer—grew up in Holland before studying in Zurich under Clausius. At Würzburg, towards the end of 1895, when he was already fifty, he started investigating cathode rays—in particular their penetrating power. It was by then common to use a barium platinocyanide screen to detect any fluorescence caused by cathode rays. This screen was not really part of the experiment, more a fail-safe device should there be any anomalies. In Röntgen's case, the screen was some way from the cathode-ray tube, which was in fact covered with black cardboard and operated only in a darkened room. On November 8, 1895, now a famous date in the history of science, Röntgen noticed—out of the corner of his eye—that the screen, though a good distance from the tube, also fluoresced, or phosphoresced, with a greenish-yellow glow. This could not possibly have been caused by the cathode rays. But did that mean the apparatus was giving off *other* rays, invisible to the naked eye? He confirmed his results, noting also that the paper screen covered with barium platinocyanide fluoresced "whether the treated side or the other be turned towards the discharging tube." So excited was Röntgen that he remained in his lab for seven straight weeks, speaking to no one other than this wife. He had all his meals brought to him at his workbench and even slept in the lab.[13]

Then, totally by chance, he made another discovery, when he accidentally put his hand between the cathode-ray tube and a screen. "And there on the screen was an image of the bones in his hand"—the rays could penetrate flesh. This discovery would revolutionize medicine very quickly. Just over two weeks after he gave his first public lecture on the subject, the rays were used in the United States to locate a bullet in a patient's leg and to set the broken arm of a boy in Hanover, New Hampshire. For as much as a decade, physicists worked fruitfully with X-rays (X for unknown, though they are called *Röntgenstrahlen* in Germany), without understanding exactly what they were.

Before Röntgen went fully public with his discoveries, he had written to several eminent scientists and one, Henri Poincaré, had shared the news with some of the seventy-eight members of the French Académie des Sciences at their weekly meeting in Paris on January 24, 1896. One of those present at Poincaré's talk was a fellow physicist, Henri Becquerel. Intrigued by the fluorescing that Röntgen had observed, Becquerel decided to see whether naturally fluorescing elements had the same effect. In a famous but accidental experiment, he put some uranium salt on a number of photoelectric plates, and left them in a closed (light-tight) drawer. Four days later, he found images on the plates, given off by what we now know was a radioactive source. Becquerel had discovered that "fluorescing" was naturally occurring radioactivity.[14]

In some ways, Becquerel's findings were more important than Röntgen's, which is why a more detailed investigation of the phenomenon was taken up by his fellow Frenchman Pierre Curie and his wife, Marie.

The Discovery of Radioactivity

Born in Warsaw on November 7, 1867, Marie Curie was originally named Maria Skodowska. Though precociously bright at school, she had no hope of ever attending university in what was then the Russian sector of a divided Poland, so she scraped together enough funds to move to Paris in 1891. She traveled the four-day, 1,200-mile journey in a bare, unheated, "ladies only" fourth-class carriage. In Paris she attended the Sorbonne, where, John Gribbin tells us, she "almost literally starved in a garret," subsisting at times on a cup of chocolate and a slice of buttered bread a day. These dismal conditions were somewhat relieved by her friendship with the pianist Ignacy Paderewski, later to be prime minister of a free Poland.[15]

At the Sorbonne, however, she met and married Pierre Curie, the son and grandson of doctors and already established as an expert on the properties of magnetic materials. The marriage soon led to a pregnancy, meaning that Marie could only settle down to her PhD work on "uranium rays" in the autumn of 1897. At that time no woman had yet completed a PhD at a European university, and for her work Marie was given the use of a leaky shed, once used as a morgue, "a cross between a stable and a potato cellar." She was banned from the main laboratories "for fear that the sexual excitement of her presence" might distract the other students.

Despite this, Marie produced her first breakthrough within six months, in February 1898. While most physicists focused only on uranium as the source of the unique energy that Becquerel had identified, Marie undertook the immense task of looking for it in all the metals, metallic compounds, salts, and minerals she could get her hands on. In this way she found that thorium was even more radioactive (a term she coined) than uranium and that cerium, niobium, and tantalum were also slightly radioactive. But then, testing pitchblende, the ore from which uranium is extracted, she found it was actually *four times* more radioactive than it should have been, judging by its uranium content. Acting on the advice of a colleague, Marie ground pitchblende to a powder, an arduous, heroic task that took weeks and weeks, dissolved the residue in acid, and then went through the seemingly endless task of separating the elements. And what she found was a substance that, according to her notebooks, handwritten in this way, was: "**150 times more active than uranium**."[16]

Working beside her, Pierre was heating a solution of bismuth sulphide when the test tube cracked. However, a thin black powder remained attached to the slivers of glass which, he found, was *330 times* more active than uranium. What they had found consisted in fact of two new elements, not one. The first they named "polonium," in a gesture to Marie's homeland, which at that time did not officially exist, and the second they named radium.

Before the Curies could take their discovery much further, disaster struck. On April 19, 1916 Pierre was killed when he slipped on the Rue Dauphine in Paris in the rain, and his skull was crushed under the rear wheels of a horse-drawn wagon loaded with military uniforms. He had by some accounts been suffering from dizzy spells just prior to the accident and had been put on a special diet of eggs and vegetables for what was thought to be "rheumatism." The feeling now is that he was the first victim that we know about to have suffered radiation sickness.

Marie lived on, serving with distinction in World War I, where she developed "Little Curies"—radiological ambulances X-raying wounded soldiers—for which she was awarded the Military Medal. In July 1934, she died of pernicious anemia, also a victim of radiation sickness, the other symptom of which, in her case, was sleepwalking.* Her laboratory notebooks are still so radioactive that they are locked in a lead-lined safe, and only allowed out now and then under strictly controlled circumstances.[17]

Radon had been discovered in the same year, 1898, by Ernest Rutherford, who encountered it while experimenting at Cambridge with Becquerel rays. This underlies the equally important discovery that radioactivity is not a permanent phenomenon, but decreases over time—this would have consequences, later, for fields such as archaeology and paleontology (see chapter 12). There were several spurious discoveries of "rays" around this time (N-rays, blacklight, etherion), but it was J. J. Thomson's 1897 discovery that capped everything: it produced the first of the Cavendish Laboratory's great successes and set modern physics on its way.

The Discovery of the Electron

Established in 1871, the Cavendish Laboratory opened its doors three years later. The new laboratory became a success only after a few false starts. Having tried—and failed—to attract first William Thomson from Glasgow, and second none other than Hermann

---

* In 1995, *Nature* reported that X-rays, not radium, might have killed her.

von Helmholtz, Cambridge finally offered the directorship to James Clerk Maxwell.

Maxwell died in 1879, when the laboratory was only five years old, and was succeeded by Lord Rayleigh, who built on his work, but retired after another five years to his estates in Essex. The directorship then passed, somewhat unexpectedly, to a twenty-eight-year-old, Joseph John Thomson. Despite his youth, Thomson had already made a reputation in Cambridge as a man who, like Mary Somerville before him, could use mathematics to bring order to physics. Universally known as "JJ," Thomson, it can be said, helped the Cavendish play a leading role in the new physics, which came into view one step at a time.

JJ was born in 1856, the son of a bookseller in Manchester at a time when "What Manchester thinks today the world thinks tomorrow." His father was a member of the "Lit. and Phil. Society" and introduced his young son to Joule, telling him that someday he would be proud to have met that gentleman. At seventeen, JJ secured a certificate of engineering and published his first paper at Owens College, the forerunner of Manchester University. Partly on the strength of that he went up to Trinity College, Cambridge, where he became second wrangler. A slight, bespectacled man with a bushy mustache, he steered the Cavendish towards more directly practical work, though he did have his lapses. The story is told that one day he bought a pair of new trousers on his way home for lunch, having been convinced by a colleague that his old pants were too baggy and worn. At home he changed into his new trousers and returned to the lab. His wife, arriving home from a shopping trip, found the worn-out pair on the bed. Alarmed, she hurriedly telephoned the Cavendish, convinced that her somewhat absentminded husband had gone back to work without any trousers on.[18]

In a series of experiments, JJ pumped different gases into glass vacuum tubes, passed an electric current, and then surrounded them either with electrical fields or with magnets. As a result of this systematic manipulation of conditions, he convincingly demonstrated that cathode "rays" were in fact infinitesimally minute *particles* erupting from the cathode and drawn to the anode. He further found that the particles' trajectory could be altered by an electric field and that a magnetic field shaped them into a curve. More important still, he found that the particles were lighter than hydrogen atoms—the smallest known unit of matter—and exactly the same *whatever* the gas used. Thomson had clearly identified something fundamental. This was in fact the first experimental establishment of the particulate theory of matter, one of the discoveries that explains the discrete nature of nature.

The "corpuscles," as Thomson called these particles at first (others termed them "protyles"), are today known as electrons, a name first proposed by the Anglo-Irish physicist and astronomer George Johnstone Stoney.[19] But the excitement of this was soon matched by that of the quantum.

The Discovery of the Quantum

In 1900 Max Planck was forty-two, two years younger than JJ A diffident, soft-spoken man, Planck was the sixth child of a professor of jurisprudence at Kiel University and the daughter of a pastor. As this sounds, his was a very religious, rather academic family, on top of which Max became an excellent musician (he had a harmonium specially built for him). His letters of this time speak of summers spent in the Baltic resort of Eldena, croquet on the lawn, evenings spent reading Sir Walter Scott, and musicals put on by the family. But science was Planck's calling, and by the turn of the century he was near the top of his profession, a member of the Prussian Academy and a full professor at the University of Berlin.

In 1897, the year Thomson discovered electrons, Planck began work on the project that was to make his name. It had been known since antiquity that as a substance (iron, say) is heated, it first glows dull red, then bright red, then white. This is because longer wavelengths (of light) appear at moderate temperatures, and as temperatures rise, shorter wavelengths appear. When the material becomes white-hot, all wavelengths are given off. Studies of even hotter bodies—stars, for example—show that in the next stage the longer wavelengths drop out, so that the color gradually moves to the blue part of the spectrum. Planck was fascinated by this and by its link to a second mystery, the so-called black body problem. A perfectly formed black body is one that absorbs every wavelength of electromagnetic radiation equally well. Such bodies do not exist in nature, though lampblack, for instance, comes close, absorbing 98 percent of all radiation. According to classical physics, a black body should only emit radiation according to its temperature, and then such radiation should be emitted at every wavelength—it should only ever glow white. But studies of the black bodies available to Planck—made of porcelain and platinum and located at the Bureau of Standards in Charlottenburg—showed that when heated, they behaved more or less like a lump of iron, giving off first dull red, then bright red-orange, then white light. Why?

Planck's revolutionary idea first occurred to him around October 7, 1900. On that day he sent a postcard to his colleague Heinrich Rubens on which he sketched an equation to explain the behavior of radiation in a black body. The essence of Planck's idea—mathematical only to begin with—was that electromagnetic radiation was not continuous, as classical physics had claimed, but could only be emitted in packets of a definite size. It was, he said, as if a hosepipe could spurt water only in "packets" of liquid. Planck knew how important his idea was the moment he had it. That very afternoon he had taken his young son for a walk and confided to him: "I have had a conception today as great as the kind of thought that Newton had."[20]

By December 14 that year, when Planck addressed the Berlin Physics Society, he had worked out his full theory. Part of this was the calculation of the dimensions of this small packet of energy, which Planck called h and which later became known

as Planck's constant. This, he calculated, had the value of 6.55 x
1027 ergs each second (an erg is a small unit of energy). Planck
had identified this very small packet as a basic indivisible building
block of the universe, an "atom" of radiation, which he called a
"quantum" (after the Latin word for "how much"). As with the
electron, the quantum also helps explain the discrete nature of
nature.

The Organization of the Atom

Physics was transformed again a few years later, on the evening
of Tuesday, March 7, 1911, this time in Manchester. We know
about this event thanks to James Chadwick, who was a student
then but later became a famous physicist. A meeting was held at
the Manchester Literary and Philosophical Society (the "Lit. and
Phil."), where the audience was made up mainly of municipal
worthies—intelligent people but scarcely specialists. A local fruit
importer spoke first, giving an account of how he had been surprised to discover a rare snake mixed in with a load of Jamaican
bananas.

The next talk was delivered by Ernest Rutherford, who introduced those present to what is certainly one of the most influential ideas of the modern world—the basic structure of the
atom. He told the audience that the atom was made up of "a
central electrical charge concentrated at a point and surrounded
by a uniform spherical distribution of opposite electricity equal
in amount." It sounds dry, but James Chadwick said later that he
remembered the meeting all his life. It was, he wrote, "a most
shattering performance to us, young boys that we were. . . . We
realised that this was obviously the truth, this was it." [21]

When he was in a good mood, Ernest Rutherford would
march around the Cavendish Laboratory in Cambridge singing
"Onward Christian Soldiers" at the top of his voice. He grew up
in an isolated rural community in New Zealand, one of twelve
children, a stocky character with a weather-beaten face who

learned engineering from his father, a builder of bridges and railway lines. At his school, Nelson College, where the head-master was a famous cricketer, he became head boy—known in the jargon as "Dux," which gave him the nickname he always answered to: "Quacks."

Despite earning three degrees by the time he was twenty-three, Rutherford failed three times to find employment as a teacher in New Zealand and therefore applied for (and obtained) a scholar-ship to study at any university in the British Empire. His choice was the Cavendish Laboratory in Cambridge.

To begin with, Rutherford was regarded by some of the other physicists as a rough "colonial," but JJ quickly realized Ruther-ford's skills. As a result, he obtained a professorship at McGill in Canada. Once there, however, he found it an intellectual and scientific backwater, so as soon as he could he returned to Britain, as professor at Manchester. Rutherford's final move was back to Cambridge as director of the Cavendish.

After he arrived there the first time, in October 1895, he quickly began a series of experiments designed to elaborate on the work of Röntgen and Becquerel. As Richard P. Brennan puts it, "Rutherford was well known for his deeply held belief that swearing at an experiment made it work better, and considering his results he might have been right." As Mark Twain said, "In times of stress, swearing affords a relief denied even to prayer."[22]

There were three naturally radioactive substances—uranium, radium, and thorium—and Rutherford and his assistant Fred-erick Soddy focused their attentions on thorium, which gave off a radioactive gas. When they analyzed the gas, however, Rutherford and Soddy were shocked to discover that it was completely inert—in other words, *it wasn't thorium*. How could that be? Soddy later described the excitement of those times in a memoir. He and Rutherford gradually realized that their results "conveyed the tremendous and inevitable conclusion that the element thorium was spontaneously transmuting itself into [the chemically inert] argon gas!" This was the first of Rutherford's many important experiments. What he and Soddy had discovered was the spontaneous decomposition of the radioactive elements, a modern form of alchemy.[23]

An Analogy with the Heavens

This wasn't all. Rutherford also observed that when uranium or thorium decayed, they gave off two types of radiation. Some of the radiation could be stopped by a sheet of paper, but a proportion could only be stopped by a sheet of aluminum one-five-hundredth of a centimeter thick. The weaker of the two forms Rutherford called "alpha" radiation; later experiments showed that "alpha particles" were in fact helium atoms and therefore positively charged. The stronger "beta radiation," on the other hand, consisted of electrons with a negative charge.* The electrons, Rutherford said, were "similar in all respects to cathode rays." So exciting were these results that in 1908 Rutherford was awarded the Nobel Prize (then worth £7,000) at the age of thirty-seven.[24]

By now he was devoting all his energies to the alpha particle. He reasoned that because it was so much larger than the beta electron (the electron had almost no mass), it was far more likely to interact with matter, and that interaction would obviously be crucial to further understanding. If only he could think up the right experiments, the alpha might even tell him something about the *structure* of the atom. "I was brought up to look at the atom as a nice hard fellow, red or grey in colour, according to taste," he said. That view had begun to change while he was in Canada, where he had shown that alpha particles sprayed through a narrow slit and projected in a beam could be deflected by a magnetic field. All these experiments were carried out with very

---

* A third type of radiation, similar to X-rays, was discovered in 1900 by the French physicist Paul Ulrich Villard. This was the most penetrating of all, being stopped only by lead. This was called gamma radiation.

basic equipment—that was the beauty of Rutherford's approach. And it was a refinement of this equipment that produced the next breakthrough.

In one of the many experiments he tried, he covered the slit with a very thin sheet of mica, a mineral that splits fairly naturally into slivers. The piece Rutherford placed over the slit in this experiment was so thin—about three-thousandths of an inch—that, in theory at least, alpha particles should have passed through it. They did, but not in quite the way Rutherford had expected. When the results of the spraying were "collected" on photographic paper, the edges of the image appeared fuzzy. Rutherford could think of only one explanation for that: some of the particles were being deflected. That much was clear, but it was the *size* of the deflection that excited Rutherford. From his experiments with magnetic fields, he knew that powerful forces were needed to induce even small deflections. Yet his photographic paper showed that some alpha particles were being knocked off course by as much as two degrees. And only one thing could explain *that*. As Rutherford himself was to put it, "the atoms of matter must be the seat of very intense electrical forces."

This idea of Rutherford's, though surprising, did not automatically lead to further insights, and for a time, he and his new assistant, Ernest Marsden, went doggedly on, studying the behavior of alpha particles, spraying them onto foils of different material— gold, silver, aluminum. Nothing notable was observed. But then Rutherford arrived at the laboratory one morning and "wondered aloud" to Marsden whether (with the deflection result still in his mind) it might be an idea to bombard the metal foils with particles sprayed *at an angle*.

The most obvious angle to start with was forty-five degrees, which is what Marsden did, using foil made of gold. And this simple experiment, it was said later, "shook physics to its foundations." Sprayed at an angle of forty-five degrees, the alpha particles did not pass *through* the gold foil—instead they were bounced back by ninety degrees onto the zinc sulphide screen.

"I remember well reporting the result to Rutherford," Marsden wrote in a memoir, "when I met him on the steps leading to his private room, and the joy with which I told him." Rutherford was quick to grasp what Marsden had already worked out: for such a deflection to occur, a massive amount of energy must be locked up somewhere in the equipment used in their simple experiment.

But for a while Rutherford remained mystified. "It was quite the most incredible event that has ever happened to me in my life," he wrote in his autobiography. "It was almost as incredible as if you fired a 15-inch shell at a piece of tissue paper and it came back and hit you. On consideration I realised that this scattering backwards must be the result of a single collision, and when I made calculations I saw that it was impossible to get anything of that order of magnitude unless you took a system in which the greatest part of the mass of the atom was concentrated in a minute nucleus." In fact, he brooded for months before he could feel confident he was right. (He didn't call the nucleus a nucleus to begin with but "a central charge"; "nucleus" was first proposed by John Nicholson, a mathematical physicist also at the Cavendish.)[25] One reason for the delay was because Rutherford was slowly coming to terms with the fact that the idea of the atom he had grown up with—the notion that it was a miniature plum pudding, with electrons dotted about like raisins—would no longer do. Another model entirely was far more likely. He made an analogy with the heavens. The nucleus of the atom, made up of positively charged particles that he named "protons" (Greek for "first things"), was orbited by negatively charged electrons, just as planets went around the stars.

As a theory, the planetary model was elegant, much more so than the "plum pudding" version. But was it correct? To test his theory, Rutherford suspended a large magnet from the ceiling of his laboratory. Directly underneath, on a table, he fixed another magnet. When the pendulum magnet was swung over the table at a forty-five-degree angle, and when the magnets were matched in polarity, the swinging magnet bounded through ninety degrees, just as the alpha particles did when they hit the gold foil. His theory had passed the first test, and atomic physics had become nuclear physics.* It was "a new view of nature . . . the discovery of a new layer of reality, a new dimension of the universe."[26]

The Unification of JJ, Planck, and Rutherford

Niels Bohr could not have been more different from Rutherford. He was a Dane and an exceptional athlete. He played soccer for Copenhagen University, he loved skiing and sailing, and was "unbeatable" at table tennis. Undoubtedly one of the most brilliant men of the twentieth century, C. P. Snow described him as a tall man, with "an enormous, domed head," who spoke with a soft voice, "not much above a whisper." Snow also found him to be "a talker as hard to get to the point as Henry James in his later years."[27]

This extraordinary man came from a civilized, scientific family—his father was a professor of physiology, his brother was a mathematician, and all were widely read in four languages. Bohr's early work was on the surface tension of water, but he then switched to radioactivity, which was the main reason that drew him to Rutherford, and England, in 1911. He studied first in Cambridge, but as we have seen, he didn't get along with JJ, and after he heard Rutherford speak at a dinner at the Cavendish Laboratory, he moved to Manchester.

At that time, although Rutherford's theory of the atom was widely accepted by physicists, there were serious problems with it, the most worrying of which was its stability—no one could see

---

* Just how small, exactly, were atoms? Jean Baptiste Perrin (1870–1942) used the work of Maxwell and Einstein's mathematics to estimate the size of water molecules and the atoms of which they were comprised. He concluded, in 1913, that atoms were roughly one-hundred-millionth of a centimeter: 250 million of them, laid out in a line, would be an inch long. Nowadays (i.e., the twenty-first century), they can be photographed and magnified to where they can be seen by the naked eye.

why the orbiting electrons didn't just collapse in on the nucleus.

Shortly after Bohr arrived to work with Rutherford, he had a series of brilliant intuitions, the most important of which was that although the radioactive properties of matter originate in the atomic nucleus, *chemical* properties reflect primarily the distribution of electrons. At a stroke he had explained the link between physics and chemistry.[28]

As we have seen, Rutherford's model of the atom was essentially unstable. According to "classical" theory, if an electron did not move in a straight line, it lost energy through radiation. But electrons went round the nucleus of the atom in orbits—such atoms should therefore either fly apart in all directions or collapse in on themselves in an explosion of light. Clearly, this did not happen: matter, made of atoms, is by and large very stable.

Bohr's contribution was to put together a proposition and an observation. He proposed "stationary" states in the atom. Rutherford found this difficult to accept at first, but Bohr insisted, flatly, against classical theory, that there must be certain orbits electrons can occupy without flying off or collapsing into the nucleus and without radiating light. He immeasurably strengthened this idea by adding to it an observation that had been known for years— that when light passes through a substance, each element gives off a characteristic spectrum of color and, moreover, one that is stable and discontinuous. Bohr's brilliance was to realize that this spectroscopic effect existed because electrons going round the nucleus cannot occupy "any old orbit" but only certain "permissible" orbits. These orbits meant that the atom was stable.[29]

But the real importance of Bohr's breakthrough was in his unification of Rutherford, Planck, and JJ: confirming the quantum—discrete—nature of nature, the stability of the atom (half-quantum states were inadmissible), and the nature of the link between physics and chemistry. At first many of the old guard were skeptical of Bohr's claims—JJ himself was one: he didn't publicly voice his support of Bohr until 1936, when he himself was eighty. But Rutherford was a strong supporter, and when Albert Einstein was told of how the Dane's theories matched the spectroscopes so clearly, he remarked, "Then this is one of the greatest discoveries."

In short order, "On the Constitution of Atoms and Molecules," the collective title of Bohr's three papers on the subject ("The Trilogy"), became a classic, and after nearly three years in Manchester, he was offered a professorship in his home city of Copenhagen. Soon afterward he was given his own Institute of Theoretical Physics, also in Copenhagen, which became a major center of the subject in the years between the wars. Bohr's quiet, agreeable, reflective personality—when speaking he often paused for minutes on end while he sought the appropriate word—was an important factor in this process. But also relevant to the rise of the Copenhagen institute was Denmark's position as a small, neutral country where, in the dark years of the twentieth century, physicists could meet away from the frenetic spotlight of the major European and North American centers. That, in its way, had its own kind of order.

In going from Mendeleyev's periodic table to Bohr's "Trilogy," we have gone from one order of chemistry to another via a major detour through fundamental physics—new forms of energy (radioactivity), new entities (particles), new concepts (the quantum), and new structures (orbits around a nucleus). At this new level of reality, the distinction between traditional disciplines began to break down, as the way nature was organized became clearer. For the time being, it all seemed fairly simple: the nucleus, the electron in its orbit, and the discipline of the quantum fitted together in a comfortable structure, a neat explanatory mechanism that was easy to grasp. It was as tidy and profound a piece of unification as one could wish for. But the tidiness wouldn't last.

4

THE UNIFICATION OF SPACE AND TIME, AND OF MASS AND ENERGY

G ermany, as we have seen, led the way in the tradition of theoretical physics—Clausius, Boltzmann, Hertz, Planck. But the most famous theoretical physicist in history, by far, was Albert Einstein, and he arrived on the intellectual stage with a bang. Of all the scientific journals in the world, the single most sought-after collector's item is the *Annalen der Physik*, volume xvii, for 1905. In that year Einstein published not one but three papers in the journal, causing 1905 to be dubbed the *annus mirabilis* of science. It was the greatest triumph since Newton had holed up in his mother's house at Woolsthorpe-by-Colsterworth in the plague year of 1666 and developed another three seminal breakthroughs—calculus, his analysis of the light spectrum, and the laws of gravity.

And, just as Newton had been born in the year Galileo died, Einstein was born in the year that James Clerk Maxwell died. He saw it as part of his task to extend the ideas of the Scotsman.[1]

Einstein was born in Ulm, between Stuttgart and Munich, on March 14, 1879. Hermann, his father, was an electrical engineer. Einstein was not an only child (he had a sister, Maria, always called Maja), but he was fairly solitary by nature and independent, a trait that contributed to his unhappiness at school. He hated the autocratic atmosphere just as he hated the crude nationalism and vicious anti-Semitism that he saw all around him. He discovered Hume, Kant, and Darwin for himself, but argued incessantly with his fellow pupils and teachers, to the point where, although he did well academically (he always came top, or next to top, in mathematics and Latin), he was expelled. There is little doubt that his reaction to the coercive teaching standards of the time brought on his later independent and questioning attitude. His Greek teacher once told him that whatever field in life he chose, he would fail at it.

Albert was very intense in his teens, never mixed with others his own age, and never read "light" literature. Aside from science his main interest was music. His mother was an accomplished pianist, and he taught himself both the piano and violin and played them throughout his life. He adored Mozart and Bach above all, but felt uncomfortable listening to Beethoven: "I think he is too personal, almost naked."[2]

Aged sixteen, he moved with his parents to Milan, and attended the polytechnic in Zurich at nineteen (because it did not require a diploma for admission). Even then, he failed the admission examination and needed to enroll in a prep school for a year. The Zurich polytechnic, the ETH, was less prestigious than the neighboring universities of Zurich, Basel, or Geneva, but was solid enough, being partly funded by Werner von Siemens, the German electrical engineering magnate. Einstein, as he said himself later, never took mathematics seriously enough at the ETH, nor was he especially assiduous in attending lectures, making opportunistic use of meticulous notes kept by a friend.

In any case, he came to feel that many of the physics lectures at the ETH were out of date, and he and a handful of other students started to read the works of Gustav Kirchhoff, Hermann von Helmholtz, and Heinrich Hertz in private, since they were not included in the course. They also read the work of a less well-known theorist, August Föppl, whose *Introduction to Maxwell's Theory of Electricity* was filled with concepts that would soon echo in Einstein's work, especially one that called into question the concept of "absolute motion." And he read the treatises of the French mathematician Henri Poincaré, in which he called into question absolute time and space.[3]

On graduation Einstein spent a period of time jobless (he was not offered a post as teaching assistant, it was said, because he was "irreverent" to his teachers). But then he found a job as a patent officer ("technical expert, third class") in Bern. Half-educated and half-in and half-out of academic life, he began in 1901 to publish scientific papers.

The first were unremarkable. Einstein did not, after all, have access to the latest scientific literature and either repeated or misunderstood other people's work. One of his specialities, however, was statistical techniques—Boltzmann's methods—and this stood him in good stead. More important still, perhaps, the fact that he was out of the mainstream of science may have helped his originality, which flourished suddenly in 1905.

Einstein's three great papers were published in March (on quantum theory), in May (on Brownian motion), and in June (on the special theory of relativity). Though Planck's original paper had caused little stir when it was read to the Berlin Physics Society in December 1900, other scientists soon realized that Planck must be right. His idea explained so much, including the observation that the chemical world is made up of discrete units—the elements. Discrete elements implied fundamental units of matter that were themselves discrete (as Dalton had said, years before). At the same time, for years experiments had shown that light behaved as a wave.[4]

In the first part of his first paper, Einstein, showing early the openness of mind for which physics would become celebrated, therefore made the hitherto unthinkable suggestion that light was *both*: a wave at some times and a particle at others. This idea took some time to be accepted, or even understood, except among physicists, who realized that Einstein's insight fitted the available facts. In time the wave-particle duality, as it became known, formed the basis of quantum mechanics in the 1920s.*

Two months after this paper, Einstein published his second great work, on Brownian motion. When suspended in water and inspected under the microscope, small grains of pollen, no more than a hundredth of a millimeter in size, jerk or zigzag backward and forward. Einstein's idea was that this "dance" was due to the pollen being bombarded by molecules of water hitting the grains at random. Here his knowledge of statistics paid off, for his complex calculations were borne out by experiment. This is generally regarded as the first proof that molecules exist.

It was Einstein's third paper, on the special theory of relativity, that would make him famous. (Though not immediately. He received only one note after publication, from Max Planck, requesting more details.) It was this theory which led to his conclusion that $E=mc^2$, or, in Einstein's own words, "The mass of a body is a measure of its energy content." The few scholars who understood what Einstein was getting at were skeptical, not seeing how his ideas could be tested experimentally. He continued to work at the patent office until 1909.[5]

The Conception of Spacetime

Relativity didn't come out of nowhere. Both Ernst Mach and Henri Poincaré had wondered whether the concepts of absolute space and time could be justified and whether, given the finite speed of light, local time and general time could be the same and whether simultaneity and uniform motion needed to be rethought (though quite how was a different matter).[6]

In fact, if one looked hard enough there were all sorts of

---

*If you have difficulty visualizing something that is both a particle and a wave, you are in good company. We are dealing here with qualities that are essentially mathematical, and all visual analogies and ordinary language will be inadequate. Niels Bohr said that anyone who wasn't made "dizzy" by the very idea of what later physicists called "quantum weirdness" had lost the plot.

anomalies that needed to be accounted for. For example, there were the anomalies of motion. A person who feels he or she is "at rest" in an armchair on earth is actually spinning with the planet's rotation at 1,040 miles per hour, and orbiting with the earth around the sun at 67,000 miles per hour.[7] Then there are the anomalies of light. Newton had conceived of light as essentially a stream of emitted particles, but by Einstein's day most scientists accepted the rival theory, put forward by Newton's Dutch contemporary Christiaan Huygens, that light should be understood as a wave. Numerous experiments had confirmed this view, including the most famous, by Thomas Young, showing how light passing through two slits produces an interference pattern that resembles that of water waves sluicing through two slits. In each case, the peaks and troughs sometimes reinforce each other and sometimes cancel each other out, producing striking patterns. Moreover, Maxwell's famous equations linking electricity, light, and magnetism, about which we have heard so much, also predicted electromagnetic waves that had to travel at the speed of light—186,282.4 miles per second in a vacuum. Maxwell confirmed his equations with his discovery that this is the speed of electricity through a wire.

Through all this, it became clear that light was a visible part of a whole spectrum of electromagnetic waves, which include what we now call AM radio signals (with a wavelength of three hundred yards), FM signals (three yards), and microwaves (three inches). As the wavelengths get shorter (and the frequency of the cycles increases), they produce the visible spectrum of light, ranging from red (twenty-five-millionths of an inch) to violet (fourteen-millionths of an inch). Even shorter wavelengths produce ultraviolet rays, X-rays, and gamma rays.[8]

This was all clear enough, but in turn it posed two fundamental questions. What was the medium through which these rays were traveling? And their speed—of ~186,000 miles per second—was *relative to what*? To begin with, scientists thought we must be surrounded by a difficult-to-detect, and presumably inert, medium; an "ether" filling up universal space, as Einstein himself observed. This led, as Walter Isaacson puts it, to the great ether hunt of the late nineteenth century.

A famous experiment, carried out in 1887 at Cleveland, Ohio, by Albert Michelson and Edward Morley, was part of the hunt. Their apparatus split a light beam, sending one part back and forth to a mirror at the end of an arm that faced the direction of the earth's movement. The second part went back and forth along an arm that was at a ninety-degree angle. The two parts of the beam were then rejoined, the aim being to see if the beam going against the supposed ether would take longer. They found that it made no difference. No one was ever able to detect the elusive ether—whichever way the light was moving, its speed was observed to be exactly the same.

People played around with all sorts of ideas as to why the ether was so elusive, but Einstein was more interested in the astronomical data. No one had been able to find any evidence that the speed of light depended on its source—light coming from any star, however far away, seemed to arrive at the same speed.[9] He came to the conclusion that "all light should be defined by frequency and intensity alone, completely independently of whether it comes from a moving or a stationary light source." This is what it said in Maxwell's equations, but for Einstein it was troublesome. As he wrote at the time, "In view of this dilemma, there appears to be nothing else to do than to abandon either the principle of relativity or the simple law of the propagation of light."

But it was now, in May 1905—while pondering the great dilemma—that he had his first major breakthrough. As he said later, "I suddenly understood the key to the problem." Five weeks later he sent off to the *Annalen* his most famous paper of all, "On the Electrodynamics of Moving Bodies." The key insight here is that two events which appear to be simultaneous to one observer will *not* appear as simultaneous to another observer who is moving very quickly. Moreover, there is no way to say which one of the observers is correct. Put more starkly, there is no way to declare that the two events are truly simultaneous.

He sought to explain this later in a thought experiment that he dreamed up and is encapsulated in the diagram below, involving a train running alongside an embankment:

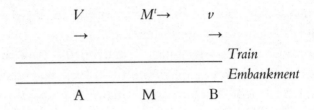

Einstein asks us to imagine two bolts of lightning striking at the same time at points A and B. They do so at exactly the moment an observer on the train is at point M', midway between A and B. The train is traveling to the right, in the direction of the arrows, and there is a second observer, on the embankment, at point M, also midway between the two lightning strikes.

Now, if the train is motionless, the observer on the train would see the two lightning strikes at exactly the same time, as does the observer on the bank. But, with the train traveling to the right—very fast—the observer on the moving train will have moved slightly to the right, closer to the source of light at point B, in the time it takes to reach him. Thus he will see the light from point B *before* he will see the light from point A: the two events are not simultaneous.

"We thus arrive at an important result," said Einstein. "Events that are simultaneous with respect to the embankment are not simultaneous with respect to the train." Furthermore, the principle of relativity says that there is no way to decree that the embankment is "at rest" and the train "in motion." We can only say they are in motion relative to each other, and there is no other right or wrong answer. There is no way to say that two events are "absolutely" or "really" simultaneous. Further, it means that *there is no absolute time.* "Instead, all moving reference frames have their own relative time." [10]

Given that there is no such thing as absolute simultaneity, that therefore time is relative, Einstein showed that space and distance are too. If the man in the train carriage covers a distance in a unit of time, then the distance for the person on the embankment is not necessarily equal to that. A moving measuring device will record different dimensions from a stationary one.

Consider it this way. Imagine that on the train there is a "clock" in the form of a mirror on the floor and another on the ceiling. A beam of light traveling from the floor to the ceiling and back again will be seen by the person on the train as going straight up and then straight down. For a person on the embankment, however, the light beam will travel up in a slightly diagonal direction, because the train has hurried forward during that time. And then, on the way down, the light beam moves in another diagonal, because again the train has hurried forward. For both observers, the speed of light is the same, but the person on the embankment sees the light beam as traveling further than does the person on the train. It follows that from the perspective of the observer on the embankment, time is going more slowly than for the passenger on the train.

How much more slowly? The difference is in fact tiny, minuscule. If the train is moving at three hundred kilometers per hour, for example, v^2/c^2 $(300^2/1,072,000,000^2) = 0.000000000000077$. To get the "time stretching" factor, we need $1/\sqrt{000000000000077} = 1.0000000000039$. Which means that, traveling for an entire lifetime—say, one hundred years—a train would extend that lifetime by 0.0000000000039 years, relative to someone on the platform. That comes out at slightly more than one-tenth of a millisecond. Nearer the speed of light, however, say 90 percent, the time-stretching factor would be bigger than two, meaning that the moving clock would be ticking at less than half the rate of that on the embankment. Put more dramatically, if one of a pair of twins—a space-age twin—was to go into space, traveling at 1g, and if then they undergo acceleration for ten years, followed by ten years of deceleration, before turning round and returning to earth in the same fashion, arriving home after traveling, as they think, for forty years, time on earth will actually have moved on by no fewer than 59,000 years.[11]

And so time and space are inextricably linked in mysterious ways. Time dilation, as it is called, has been tested using very sensitive clocks on commercial flights, and has proved that Einstein was correct. But the effect, in our ordinary lives, is minuscule. In fact, it has been calculated that if one twin had the misfortune of spending his or her whole life on an airplane, when he or she returned, he or she would have aged 0.00005 seconds or so less than the twin on earth.[12]

This was Einstein's "special theory" of relativity: that all inertial frames of reference are equally valid, that observers in different states of motion will measure things differently, that there are no privileged states of movement. This is not the same as saying that everything is subjective, because in fact there are strict rules as to how dimensions vary. Special relativity means that measurements of time, including duration and simultaneity, can be relative, depending on the motion of the observer, and so can measurements of space, such as distance and length. But the union of the two, now called spacetime, remains invariant in all inertial frames. Einstein briefly considered calling his discovery/creation "Invariance Theory," but the term never caught on, and when Max Planck referred to it as *Relativtheorie* in 1906, that was that.

Einstein incubated the *general* theory of relativity for ten years. In the interim came $E = mc^2$.

The Great Unifying Equation

This idea didn't come out of nowhere either. The concept of the conservation of mass had been known since at least Lavoisier, in the late eighteenth century. He was just one who observed that, for example, when iron rusts it actually gains weight—oxygen has, in effect, "stuck" to the metal. So far as light was concerned, people had been trying to measure its speed since Galileo put lanterns on different Tuscan hills and tried (and failed) to measure how long it took for the light from one lantern to reach the other. This situation was eventually clarified and the problem solved in the late seventeenth century, when a young Danish astronomer named Ole Rømer, still only twenty-one, was called in to assist the famous Italo-French scientist Giovanni Domenico (or Jean-Dominique) Cassini over a problem with the planet Jupiter. Cassini was then the world authority on Jupiter, which was especially fascinating because, like the earth, it has a moon (in fact it has three). The problem with Jupiter that had plagued astronomers for decades was that its innermost moon, Io, was *supposed* to orbit Jupiter every forty-two and a half hours. However, for some unknown reason it never stuck to this schedule. At some times its orbit was a little shorter, at others a little longer.

No one could explain these discrepancies. Was there a hidden body somewhere, perhaps tugging at it? When Rømer arrived in Paris, he went through Cassini's meticulous records. Being young and therefore not embroiled totally in the scientific orthodoxies of the day (a little like Einstein himself), he had an idea that had not occurred to anyone else. This was that the earth, in its orbit around the sun, was nearer Jupiter at certain times of the year and this affected the amount of time it took for the light from Io to reach earth. Cassini and the older astronomers of the time didn't accept this argument, pointing to the fact that, among other things, Galileo's experiment had failed. Surely light traveled instantaneously.

Rømer demurred. And, at a public meeting in Paris in August 1671, he challenged Cassini directly, arguing that when Io next appeared—and it was due on November 9—it would not be at 5:27, as the orthodoxy had it, but nearer 5:37. Which is exactly what happened—at 5:37 and 49 seconds to be precise. But this did more than embarrass Cassini. It enabled Rømer and others to calculate that the speed of light was, approximately, 670 million miles per hour. To emphasize how fast this is, it means that you could travel from London to Los Angeles in less than one-twentieth of a second. The speed of sound is Mach 1. The speed of light, c, is Mach 900,000. (C stands for *celeritas*, Latin for "swiftness." Light is made up of photons, which have no mass, so far as anyone can tell, and this is why they travel at the speed of light.)

The notion of "squared," as in X^2 (referred to earlier, in the footnote on p. 25), is more a feature of our world than we tend to think. It is a fundamental feature of space. When you sit next to a reading lamp, say, and then move closer, this phenomenon shows itself. If you move twice as close to the lamp, the light's intensity increases four times. If you roll a ball down a slope into sand, it will penetrate the sand to a certain extent: X. If you double the speed of the ball it will penetrate four times as deeply as it did before. If you roll it at three times the speed it will penetrate nine times as deeply as it did to begin with. If you increase your speed in a car from 20 mph to 80 mph you quadruple the speed, but it will take you sixteen times as long to stop. So far as momentum is concerned, the "square" value is central.

The speed of light is 670 million miles per hour, so c^2 is 448,900,000,000,000,000 mph, or 718,240,000,000,000,000 kph.[13]

The Interchangeability of Mass and Energy

So much for deep background. Starting in the 1890s, a raft of investigators had observed certain materials behaving in ways that had got Einstein thinking. Among these was the fact that a number of "metal-streaked ores" had been discovered in the Congo, and, in what was then Czechoslovakia, rocks had been found which sprayed out mysterious "energy beams." Marie Curie had coined the term "radioactivity" for these beams (chapter 3), though even she—as we have seen—didn't fully appreciate the force of what she had named, radioactivity quite possibly being responsible for her husband's and her own death. However, these ores could spray out many trillions of high-speed alpha particles every second without—so it seemed—any appreciable loss of weight.

Einstein put all this together: Maxwell's equations, with their dependence on the speed of light; his own idea of relativity; the constancy—and importance—of the speed of light; the discovery of radioactivity. He conjoined this with the very great numbers involved (trillions of alpha particles, 670 million miles per hour for the speed of light) to come up with what was a surprisingly short paper, just three pages long. It was entitled "Does the Inertia of a Body Depend on Its Energy-Content?" [14]

This paper also included a thought experiment in which this time he calculated the properties of two light-pulses emitted in opposite directions by a body at rest. Then he worked out the properties of these same light-pulses when observed from a moving frame of reference. It was this that produced equations regarding the relationship between speed and mass: mass and energy are different manifestations of the same thing. As Walter Isaacson puts it in his biography of Einstein, "There is a fundamental interchangeability between the two." Or, as Einstein himself said, "The mass of a body is a measure of its energy content."

Many examples have been given about the enormous amounts of energy the equation yields. Perhaps the most vivid is that which says that the mass of just one raisin, suitably treated, could supply most of New York City's energy needs for one day. We need to destroy only one microgram of matter every second to power a city. Chemical energy has it roots in the structure of atoms, and as the bonds between molecules snap and re-form in reactions, energy is released and mass reduces. At its simplest, a hydrogen atom contains a single electron in orbit around a single proton. Importantly, the mass of this atom is less than the combined mass of an electron and proton that are not related *in* an atom, albeit by a tiny amount—0.00000000000000000000000000000002 kilograms. This difference, however small, has massive consequences when converted into energy. This difference—negative energy in effect—is known to physicists as "binding energy." It is this difference, in uranium, that we know as nuclear energy, and, occurring naturally on earth, it is the energy that drives tectonic plates and pushes up mountains (chapter 12).[15]

Einstein ended his paper (which was not immediately recognized as the revolutionary idea that it became) by suggesting an experiment that could be done to verify his notion. "Perhaps it will prove possible to test this theory using bodies whose energy content is variable to a high degree, e.g., salts of radium."

"The Greatest Discovery Ever Made . . . The Most Amazing Combination of Philosophical Penetration, Physical Intuition, and Mathematical Skill"

Einstein sat on the general theory of relativity for ten years before going public. In fact, he had the basic idea as early as November 1907. As he recalled, "I was sitting in a chair in the patent office at Bern when all of a sudden a thought occurred to me. If a person falls freely, he will not feel his own weight." That realization "startled" him, he said, and he embarked on his epic journey, which "impelled me toward a theory of gravitation." Later he would describe the moment in the patent office as "the happiest thought in my life."* The story of the falling man has become famous, and even been embellished to the extent that there may have been a real person, a painter, who fell from the roof of a nearby building. This is probably true in the same sense that Newton got the idea of gravity when an apple fell on his head.

The thought experiment that Einstein had this time was of a man in an enclosed chamber, such as an elevator, in free fall above the earth, or in deep space "far removed from stars and other appreciable masses." Such a person would not feel he was moving, and would not feel his weight; anything he took out

---

* This is the usual translation, but the original German phrase was *die glück-lichste Gedanke*, which is better rendered as "luckiest."

of his pockets would float alongside him. The only thing he would feel would be acceleration or deceleration, and since, as every schoolchild knows, gravitation exerts its influence in an accelerative way, Einstein concluded that "the local effects of gravity and acceleration are equivalent." This was the basic idea he now played with to extend relativity beyond a theory that was restricted to systems that moved with a uniform velocity.

Over the next years he derived certain mathematical calculations that led to various predictions. One was that clocks would run more slowly in a more intense gravitational field; another was that light should be bent by gravity and that a wavelength of light emitted from a source with a large mass, such as the sun, should increase slightly in what he termed the "gravitational red shift."

The way that gravity bends light can be shown by a return to the thought experiment of the elevator. Imagine a light particle entering the elevator through a tiny hole on the right side as it is moving upward. An observer inside the elevator will see the light beam hit the far wall (the left wall) at the same height as it entered. But an observer outside the elevator, and relatively stationary with respect to it, will see the beam bend downward because, by the time the beam has crossed the elevator, from right to left, the elevator has risen by a small amount. Since gravity and acceleration are equivalent, it follows that a gravitational field (like a heavenly body) should also bend light. Einstein worked out that "a ray of light going past the sun would undergo a deflection of 0.83 second of arc."*

The idea that light beams could be bent had some interesting implications. In everyday life, light travels in straight lines, but if it can be bent by changing gravitational fields, how can a straight line be determined? Einstein's answer was that spacetime is curved. This is, as Isaacson concedes, a challenging notion for most of us. One analogy is with the surface of the earth. We all

---

* Einstein later doubled his calculation. A second of arc is an angle of 1/3,600th of a degree.

know that the earth is a globe, but in everyday life it feels flat. This has consequences. For example, we can see that parallel lines, lines that never meet in a Euclidean world, do in fact meet at the poles. Furthermore, the three angles of a triangle formed with its base at the equator, and with its two other sides joining at the North Pole, say, would each be 90 degrees, whereas in the Euclidean world the three corners of a triangle add up to 180 degrees. Einstein asks us to go one step further, and accept that spacetime is curved and that the shortest path through a region of space that is curved by gravity might seem quite different from the straight lines of Euclidean geometry.

The central idea of general relativity therefore can be put this way: "gravity is geometry." [16] For a time very few physicists took to Einstein's new theory; many even denounced it. He wasn't too concerned, and he looked forward to an eclipse of the sun, which would be total in Crimea, and was due on August 21, 1914. Some colleagues were going there to test his prediction that light beams would be bent in the region of the sun. Unfortunately, twenty days before the eclipse, World War I broke out and Germany declared war on Russia. The German scientists who were already in Crimea were captured and their equipment confiscated.

Paul Dirac, one of the creators of quantum mechanics (chapter 5), said general relativity was "probably the greatest scientific discovery ever made." Max Born called it "the greatest feat of human thinking about nature, the most amazing combination of philosophical penetration, physical intuition and mathematical skill." [17]

Eddington's Eclipse

Owing to the war, between 1914 and 1918 all direct links between Great Britain and Germany were cut off. But Holland, like Switzerland, remained neutral. At the University of Leiden, Willem de Sitter, director of the observatory, and the man who would coin the term "dark matter," was a friend and collaborator of

Einstein. In 1915 he was sent Einstein's paper on the general theory of relativity. An accomplished mathematician and physicist, de Sitter was well connected and realized that as a Dutch neutral he was an important go-between. He therefore passed on a copy of Einstein's paper to Arthur Eddington in London.

Eddington was by then a central figure in the British scientific establishment, despite having a "mystical bent," according to one of his biographers. Born in Kendal in the Lake District in 1882, into a Quaker family of farmers, he was educated first at home, then at Owens College in Manchester, from which he went up to Trinity College, Cambridge, as JJ had done. At Trinity he became the first ever person to become senior wrangler in his second (rather than his third) year. And it was at Cambridge that he came into contact with JJ and Rutherford. Fascinated by astronomy since he was a boy, he subsequently took up an appointment at the Royal Observatory in Greenwich from 1906, and in 1912 became secretary of the Royal Astronomical Society.

His first important work was a massive and ambitious survey of the structure of the universe. Its main discovery, made in 1912, was that the brightness of the so-called Cepheid stars (stars with variable—"pulsated"—luminosity) pulsated in a regular way associated with their sizes. This helped establish real distances in the heavens and showed that our own galaxy has a diameter of about 100,000 light-years and that the sun, which had been thought to be at its center, is in fact about 30,000 light-years eccentric.* The second important result of Cepheid research was the discovery that the spiral nebulae were in fact extragalactic objects, entire galaxies themselves, and very far away (the nearest, the Great Nebula in Andromeda, being 750,000 light-years

---

* The light-year—the distance light travels in a year (about 6 trillion miles)—was first proposed by Friedrich Bessel (1784–1846), director of the Königsberg Observatory, who realized that a large unit of length was needed to make the huge interstellar distances comprehensible. Until that time the astronomical unit (AU) was used. This was equivalent to the radius of the earth's orbit about the sun, roughly 93 million miles.

away). This eventually provided a figure for the distance of the farthest objects, 500 million light-years away, and an age for the universe of between 10 and 20 billion years.[18]

But Eddington was also a keen traveler and had visited Brazil and Malta to study eclipses. His work and his academic standing thus made him the obvious choice when the Physical Society of London, during wartime, wanted someone to prepare a *Report on the Relativity Theory of Gravitation*. This, which appeared in 1918, was the first complete account of general relativity to be published in English. Eddington, as we have seen, had already received a copy of Einstein's 1915 paper from Holland, so he was well prepared, and his report attracted widespread attention, so much so that Sir Frank Dyson, the astronomer royal, offered an unusual opportunity to test Einstein's theory. On May 29, 1919, there was to be a total eclipse of the sun. This offered the chance to assess if, as Einstein predicted, light rays were bent as they passed near a large body. It says something for the astronomer royal's influence that, during the last full year of World War I, Dyson obtained from the government a grant of £1,000 to mount not one but two expeditions, to Príncipe off the coast of West Africa and to Sobral, across the Atlantic, in Brazil.[19]

Eddington was given Príncipe, together with E. T. Cottingham, a Northamptonshire clock maker by trade and an expert on astronomical timekeeping. In the astronomer royal's study on the night before they left, Eddington, Cottingham, and Dyson sat up late calculating how far light would have to be deflected for Einstein's theory to be confirmed. At one point, Cottingham asked rhetorically what would happen if they found twice the expected value. Drily, Dyson replied, "Then Eddington will go mad and you will have to come home alone!"

Eddington's own notebooks continue the account: "We sailed early in March to Lisbon. At Funchal we saw [the other two astronomers] off to Brazil on March 16, but we had to remain until April 9 . . . and got our first sight of Principe in the morning of April 23. . . . About May 16 we had no difficulty in getting the check photographs on three different nights." Then the weather changed. On the morning of May 29, the day of the eclipse, the heavens opened, and Eddington began to fear that their arduous journey was a waste of time.

At one thirty in the afternoon, however, by which time the partial phase of the eclipse had already begun, the clouds began to clear. "I did not see the eclipse," Eddington wrote later, "being too busy changing plates, except for one glance to make sure it had begun and another half-way through to see how much cloud there was. We took sixteen photographs. . . . The last six show a few images which I hope will give us what we need. . . . June 3: We developed the photographs, 2 each night for 6 nights after the eclipse and I spent the whole day measuring. The cloudy weather upset my plans. . . . But the one plate that I measured gave a result agreeing with Einstein." Eddington turned to his companion. "Cottingham," he said, "you won't have to go home alone." [20]

The publicity given to Eddington's confirmation of relativity made Einstein the most famous scientist in the world. Relativity theory had not found universal acceptance when he had first proposed it. Eddington's Príncipe observations were therefore the point at which many scientists were forced to concede that this exceedingly uncommon idea about the physical world was, in fact, true.

The Siren Song of Unified Theory

Given that Einstein created this huge synthesis of some of our more basic—and seemingly so variable—concepts, it is worth adding a coda: he failed, later in life, despite years of trying, to come up with what it is that brings together electricity, magnetism, gravity, and quantum mechanics (which latter we shall come to). He fervently hoped that he would be able to extend the gravitational field equations of general relativity so that they would describe the electromagnetic field as well. As he put it in his Nobel Prize acceptance lecture in 1921, "The mind striving after unification cannot be satisfied that two fields should exist which, by their nature, are quite independent. We seek a mathematically unified field theory in which the gravitational field and the electromagnetic field are interpreted only as different components or manifestations of the same uniform field." Such a unified theory, he hoped, would make quantum mechanics compatible with relativity.

Hitherto, Einstein's genius had been in finding missing links between different theories, but, says Walter Isaacson, "the siren song of unified theory had come to mesmerize Einstein. . . . Over it lies the marble smile of nature," as Einstein himself conceded at one point. Across the next two decades nothing Einstein came up with ever resulted in a successful unified field theory. In fact, in some ways, with the discovery of more and more particles, physics for the moment was becoming *less* unified.[21]

Even on his deathbed, in April 1955, Einstein was still working on equations that might create a unified field theory, and in this he was not entirely alone. In a recent book, Paul Halpern has shown how both Einstein and Erwin Schrödinger (who we shall come to) spent far longer on the convergence—the hunt for a unification of quantum physics and relativity—than they did on the breakthroughs for which they are known.[22] The fact that the universe is comprehensible at all is what Einstein found most awesome, most worthy of reverence. For a scientist, therefore, worship is in effect the act of seeking out the ultimate order that explains everything. Einstein has probably come closer than anyone to achieving that aim.

Over the decades since the 1850s, more and more order has revealed itself, as Einstein had cause to know better than most. But his fate in his later years—and the same would subsequently be true of Schrödinger—shows that order is a little bit like happiness. The more you go deliberately looking for it, the harder it is to pin down. That Einstein, of all people, should be mesmerized in this way is a most poignant irony. And, perhaps, a warning.

5

THE "CONSUMMATED MARRIAGE" OF PHYSICS AND CHEMISTRY

At precisely the time Eddington was preparing his trip to Príncipe, in April of that year, Ernest Rutherford sent off a paper that, had he done nothing else, would have earned him a place in history. From then until 1932, when his student James Chadwick discovered the neutron, there occurred a second golden decade in physics—barely a year went by without some momentous experimental or theoretical breakthrough. All the seminal work of this golden decade was carried out in one of three places in Europe: the Cavendish Laboratory in Cambridge, England; Niels Bohr's Institute of Theoretical Physics in Copenhagen, Denmark; and the old university town of Göttingen, almost bang in the middle of Germany.

For Mark Oliphant, one of Rutherford's protégés in the 1920s, the main hallway of the Cavendish, where the director's office was, consisted of "uncarpeted floorboards, dingy varnished pine doors and stained, plastered walls, indifferently lit by a skylight with dirty glass." For C. P. Snow, who also trained there and described the lab in his first novel, *The Search*, the paint and the varnish and the dirty glass went unremarked. "I shall not easily forget those Wednesday meetings in the Cavendish. For me they were the essence of all the personal excitement in science . . . week after week I went away through the raw nights, with east winds howling from the fens down the old streets, full of a glow that I had seen and heard and been close to the leaders of the greatest movement in the world." Rutherford, who followed JJ as the director of the Cavendish in 1919, evidently agreed. At a meeting of the British Association in 1923 he startled colleagues by suddenly shouting out, "We are living in the heroic age of physics!"[1]

In some ways, Rutherford himself—now a rather florid man, with a mustache and a pipe that was always going out—embodied that heroic age. During World War I, particle physics had been on hold, more or less. Officially, Rutherford was working for the admiralty, looking into submarine detection. But he carried on his own research, when his duties allowed. And this eventually gave rise to his groundbreaking paper, with its bland title: "An Anomalous Effect in Nitrogen." As was usual in Rutherford experiments, the apparatus was simple to the point of being crude: a small glass tube inside a sealed brass box fitted at one end with a zinc sulphide scintillation screen. The brass box was filled with nitrogen and then through the glass tube was passed a source of alpha particles—helium nuclei—given off by radon, the radioactive gas of radium.[2]

The excitement came when Rutherford inspected the activity on the zinc sulphide screen. The scintillations were indistinguishable from those obtained from hydrogen. How could that be, since there was no hydrogen in the system? This led to the famously downbeat sentence in the fourth part of Rutherford's paper: "From the results so far obtained it is difficult to avoid the conclusion that the long-range atoms arising from the collision of [alpha] particles with nitrogen are not nitrogen atoms but probably atoms of hydrogen. . . . If this be the case, we must conclude that the nitrogen atom is disintegrated." The newspapers were not so cautious. Sir Ernest Rutherford, as he now was, had *"split the atom"* they shouted. He himself realized the importance of his work. His experiments had drawn him away, temporarily, from anti-submarine research. He defended himself to the overseers' committee: "If, as I have reason to believe, I have disintegrated the nucleus of the atom, this is of greater significance than the war."[3]

In a sense, Rutherford had finally achieved what the old alchemists had been aiming for, transmuting one element into another, nitrogen into hydrogen and oxygen. The mechanism whereby this artificial transmutation (the first ever) was achieved was clear. An alpha particle—a helium nucleus—has an atomic weight of 4. When it was bombarded onto a nitrogen atom, with an atomic weight of 14, it displaced a hydrogen nucleus (to which Rutherford soon gave the name proton). The arithmetic therefore became: $4 + 14 - 1 = 17$, the oxygen isotope, O^{17}.

The significance of the discovery, apart from the philosophical one of the transmutability of nature, lay in the new way it enabled the nucleus to be studied. Rutherford and Chadwick immediately began to probe other light atoms to see if they behaved in the same way. It turned out that they did—boron, fluorine, sodium, aluminum, phosphorus, each of these chemical elements had nuclei that could be probed. They were not just solid matter but had a structure.

All this work on light elements took five years, but then there was a problem. The heavier elements were, by definition, characterized by outer shells of many electrons that constituted a much stronger electrical barrier and would need a stronger source of alpha particles if they were to be penetrated. For James Chadwick and his young colleagues at the Cavendish, the way ahead was clear—they needed to explore means of accelerating particles to higher velocities. Rutherford wasn't convinced, preferring simple experimental tools. But elsewhere, especially in America, physicists realized that one way ahead lay with particle accelerators.[4]

The Music of the Spheres

Between 1924 and 1932, when Chadwick finally isolated the neutron, there were no breakthroughs in nuclear physics. Quantum physics, on the other hand, was a different matter. Niels Bohr's Institute of Theoretical Physics opened in Copenhagen on January 18, 1921. The large house, on four floors, shaped like an L, contained a lecture hall, library, and laboratories, as well as a table-tennis table, where Bohr also shone.

Bohr became a Danish hero twelve months later when he won the Nobel Prize, but in fact the year was dominated by something even more noteworthy—Bohr's final irrevocable linking of chemistry and physics. In 1922 Bohr showed how atomic structure was linked to the periodic table of elements drawn up by Mendeleyev. In his first breakthrough, just before World War I, the Dane had explained how electrons orbit the nucleus only in certain formations, and how this helped explain the characteristic spectra of light emitted by crystals of different substances (Preface and chapter 3). This idea of natural orbits also married atomic structure to Max Planck's notion of quanta. Bohr now went on to argue that successive orbital shells of electrons could contain only a precise number of electrons. And he introduced the idea that elements that behave in a similar way chemically do so because they have a similar arrangement of electrons in their outer shells, which are the ones most used in chemical reactions. For example, he compared barium and radium, which are both alkaline earths but have very different atomic weights and occupy, respectively, the fifty-sixth and eighty-eighth place in the periodic table. Bohr explained this by showing that barium has electron shells filled successively by 2, 8, 18, 18, 8, and 2 (= 56) electrons. By the same token, radium has electron shells filled successively by 2, 8, 18, 32, 18, 8, and 2 (= 88) electrons.

Besides explaining their position on the periodic table, the fact that the outer shell of each element has two electrons means that barium and radium are chemically similar despite their considerable other differences. As Einstein said, "This is the highest form of musicality in the sphere of thought."[5]

The Link Between Mathematics and Atomic Structure

During the 1920s the center of gravity of physics—certainly of quantum physics—shifted to Copenhagen, largely because of Bohr. A big man in every sense, he was generous, avuncular, and completely devoid of those instincts for rivalry that can so easily sour relations. But the success of Copenhagen also had to do with the fact that Denmark was a small, neutral country where national rivalries could be forgotten. Among the sixty-three physicists of renown who studied at Copenhagen in the 1920s were Paul Dirac (British), Wolfgang Pauli and Werner Heisenberg (German), and Lev Landau (Russian).

This international mixing was unusual—virtually unique— because, in many areas of science, the aftermath of World War I lasted for years. In 1919 the Allies established an International Research Council but Germany and Austria were excluded. Out of 275 international science conferences held between 1919 and 1925, 165 were without German participation.[6] Not until 1925 and the Locarno Pact was this rule relaxed, but even then German and Austrian scientists spurned the olive branch.

But, somehow, Bohr managed to keep his Copenhagen institute an international affair. The Swiss-Austrian Wolfgang Pauli was one of those who took advantage of a visit to Copenhagen. In 1924 Pauli was a small, sullen-looking, pudgy twenty-three-year-old with hooded eyes who was given to pranks and prone to depression when scientific problems defeated him.

He came from a well-connected Viennese academic background—his godfather was Ernst Mach—and Pauli grew up to know some of the members of the Vienna Circle (chapter 7). By night, however, he prowled red-light districts and frequented risqué cabarets that catered largely to a rough clientele. He confessed in a letter to a friend that he enjoyed "the night, sexual excitement in the underworld—without feeling, without love, indeed without humanity."[7]

His ambition and his abilities in physics and mathematics

were such, however, that Bohr invited him to Copenhagen. One problem in particular had set him prowling the streets of the Danish capital. It was something that vexed Bohr too, and it arose from the fact that no one at the time understood why all the electrons in orbit around the nucleus didn't just crowd in on the inner shell. This, as we have seen, is what should have happened, with the electrons emitting energy in the form of light. What was known by now, however, was that each shell of electrons was arranged so that the inner shell always contains just one orbit, whereas the next shell out contains four. Pauli's contribution was to show that no orbit could contain more than two electrons. Once it had two, an orbit was "full," and other electrons were excluded, forced to the next orbit out. This meant that the inner shell (one orbit) could not contain more than two electrons, and that the next shell out (four orbits) could not contain more than eight. This became known as Pauli's "exclusion principle," a term coined by Dirac.[8]

Part of its beauty lay in the way it expanded Bohr's explanation of chemical behavior. Hydrogen, for example, with one electron in the first orbit, is chemically active. Helium, however, with two electrons in the first orbit (i.e., an orbit that is "full" or "complete"), is virtually inert. Lithium, the third element, has two electrons in the inner shell and one in the next, and is chemically very active. Neon, however, which has ten electrons, two in the inner shell (filling it) and eight in the four outer orbits of the second shell (again, filling those orbits), is also inert. So together Bohr and Pauli had shown how the chemical properties of elements are determined not only by the number of electrons the atom possesses but also by the dispersal of those electrons through the orbital shells.[9]

The next year, 1925, was the high point of this second golden age, and the center of activity moved to Göttingen. Before World War I, British and American students regularly went to Germany to complete their studies, and Göttingen was a frequent stopping-off place. Bohr gave a lecture there in 1922 and was taken to task by a young student who corrected a point in his argument. Bohr, being Bohr, hadn't minded. "At the end of the discussion he came over to me and asked me to join him that afternoon on a walk over the Hain mountain," Werner Heisenberg wrote later. "My real scientific career only began that afternoon." [10]

Born in Würzburg in 1901, the son of a professor of Byzantine history, Werner Heisenberg was an even better musician than Einstein, and as accomplished on the piano as Planck, being able to play the classics at the age of thirteen. Mathematics was his strength as a young man (he taught himself calculus) but not his only interest. He was a fervent nationalist and in the turbulent wake of the war took part in more than one street fight against the Communists (this involvement would become relevant later). An organization called the Heisenberggruppe was formed at his school in Munich where, as the name indicates, Werner was the leader—with the group meeting in his home. Heisenberg was also a legendary chess player, who sometimes played his opponents without his queen, "to give them a chance."

At Munich University, he continued to shine and bewilder. Arnold Sommerfeld, a professor there, who well appreciated his brilliance, nonetheless ordered him to give up chess, since he was spending too much time on it.[11] It was Sommerfeld who was responsible for taking Heisenberg to Göttingen to hear Bohr, where Heisenberg made his crucial intervention, after which he and Bohr went for their stroll.

In fact, it turned into more than a stroll, for Bohr invited the young Bavarian to Copenhagen, where they immediately set about tackling yet another problem of quantum theory, what Bohr called "correspondence." This stemmed from the observation that, at low frequencies, quantum physics and classical physics came together. But how could that be? According to quantum theory, energy—like light—was emitted in tiny packets; yet according to classical physics, it was emitted continuously. Heisenberg returned to Göttingen enthused but also confused.

And so, when, toward the end of May 1925, he suffered one of his many attacks of hay fever, he took two weeks' holiday in Heligoland, a narrow island off the German coast in the North Sea, where there was next to no pollen. Here he cleared his head with long walks and bracing dips in the sea.[12]

The idea that came to Heisenberg in that cold, fresh environment was one of the best examples of what has come to be called quantum weirdness. Heisenberg took the view that we should stop trying to visualize what goes on inside an atom, as it is impossible to observe directly something so small. All we can do is measure its properties. And so, if something is measured as continuous at one point, and discrete at another, that is the way of reality. If the two measurements exist, it makes no sense to say that they disagree: they are just measurements.[13]

This was Heisenberg's central insight, but in a hectic three weeks he went further, developing a method of mathematics known as matrix math. It originated from an idea of the German mathematician David Hilbert, in which the measurements obtained are grouped in a two-dimensional table of numbers where two matrices can be multiplied together to give another matrix. In Heisenberg's scheme, each atom would be represented by one matrix, each "rule" by another. If one multiplied the "sodium" matrix by the "spectral line matrix," the result should give the matrix of wavelengths of sodium's spectral lines. To Heisenberg's and Bohr's great satisfaction, it did. "For the first time atomic structure had a genuine, though very surprising, mathematical base." Heisenberg called his creation/discovery quantum mechanics.[14]

The acceptance of Heisenberg's idea was made easier by a new theory of Louis de Broglie in Paris, also published in 1925. De Broglie was, in prerevolutionary terms, a French prince (his great-great-grandfather had been guillotined during the Terror). But he was interested now in light. Both Planck and Einstein had argued that light, hitherto regarded as a wave, could sometimes behave as a particle. De Broglie reversed this idea, arguing that

particles could sometimes behave like waves. He had introduced
the idea as part of his doctoral dissertation, and the examining
committee in Paris at first didn't know what to make of it. In
fact, they were on the verge of rejecting it when they heard from
Einstein—one of the examining committee had sent him the
paper, seeking his opinion. Einstein's reply was emphatic: the
notion was brilliant. De Broglie got his doctorate.[15] No sooner
had de Broglie broached this theory than experimentation proved
him right. The *wave-particle duality* of matter was the second weird
notion of physics, but it caught on quickly.

One reason was the work of yet another genius, the Austrian Erwin Schrödinger, who was disturbed by Heisenberg's
idea and fascinated by de Broglie's. Schrödinger was originally
a well-heeled Viennese, but suffered as a result of the notorious
runaway inflation of the Weimar years, which destroyed the
family fortune: money was always a problem for him. Despite all
this, he was an inveterate womanizer (who kept a diary, in code,
of his sexual conquests).

In 1926 he was thirty-nine, quite "old" for a physicist, when
he came up with a defining equation which explained how waves
change over time and incorporated the notion that the electron,
in its orbit around the nucleus, is not like a planet but moves in
a wave. Moreover, this wave pattern determines the size of the
orbit, because to form a complete circuit the wave must conform
to a whole number, not a fraction (otherwise the wave would
descend into chaos). In turn this determined the distance of the
orbit from the nucleus and shaped chemical action.

Against Generalizing

The final layer of weirdness came in 1927, again from Heisenberg. It was late February, and Bohr had gone off to Norway to
ski. Heisenberg paced the streets of Copenhagen on his own. Late
one evening, in his room high up in Bohr's institute, a remark
of Einstein's stirred something deep in Heisenberg's brain: "It

is theory which decides what we can observe." It was well after midnight, but Heisenberg decided he needed some air, so he went out and trudged across some muddy soccer fields nearby. Could it be, Heisenberg asked himself, as he walked, that at the level of the atom there was a limit to what could be known? To identify the position of a particle, it must impact on a zinc sulphide screen. But this alters its velocity, which means that it cannot be measured at the crucial moment. Conversely, when the velocity of a particle is measured—by scattering gamma rays from it, say—it is knocked into a different path, and its exact position at the point of measurement is changed. Set out in a twenty-seven-page document, Heisenberg's "uncertainty principle," as it came to be called, posited that the exact position and precise velocity of an electron could not be determined at the same time.

This was disturbing both practically and philosophically, because it implied that in the subatomic world, cause and effect could never be measured. The only way to understand electron behavior was statistical, using the rules of probability. "Even in principle," Heisenberg said, "we cannot know the present in all detail. For that reason everything observed is a selection from a plenitude of possibilities and a limitation on what is possible in the future. . . . Uncertainty is embedded throughout nature . . . it is always there, inescapable." [16]

It sounds ominous, but on the macroscopic scale in which we habitually live, it does not make too much difference. We can still fly around the world confident that we will actually arrive at where we are intending to go. There will be minor discrepancies but not enough to make a practical difference. It is more that there is, underneath it all, a philosophical discomfort that there is an "inherent inaccuracy" in everything we do. This doesn't stop us proceeding to make experiments, but it does tell us that there are limits to our understanding.

Nonetheless, as Heisenberg made clear, the classical concept of causality had to be abandoned, and this led to Bohr's idea of complementarity. This is one of the most contentious, vague, and ambiguous issues in physics. What Bohr meant by it was that there are mutually exclusive viewpoints about nature—such as the wave and the particle—and, it appears, it is the physicist's job to measure phenomena and make predictions based on those measurements without generalizing about the "reality" of nature behind those measurements. The fact that these ideas cannot be visualized or brought into common sense didn't and doesn't matter. Like relativity, particles/waves are difficult to visualize and their behavior goes *beyond* common sense. This, Bohr and Heisenberg were saying, is the way physics was going and had to be accepted. But this new understanding—of the way minuscule amounts of energy arrange themselves—would have knock-on effects for chemistry, astronomy, and other sciences later on.[17]

Many physicists found the wave-particle duality a difficult notion to swallow, and Einstein, for one, was never very happy with the basic notion of quantum theory, that the subatomic world could only be understood statistically (the statistical notion of radioactivity troubled many physicists for years). It remained a bone of contention between him and Bohr until the end of his life. In 1926 he wrote a famous letter to the physicist Max Born in Göttingen. "Quantum mechanics demands serious attention," he said. "But an inner voice tells me this is not the true Jacob. The theory accomplishes a lot, but it does not bring us closer to the secrets of the Old One. In any case, I am convinced that He does not play dice."[18]

Others agreed, Erwin Schrödinger's famous cat experiment being a case in point. Schrödinger proposed a thought experiment in which a cat, along with a weak radioactive source and a detector of radioactive particles, was sealed into a steel box. Also in the box was a small tube of poisonous gas and a hammer hooked up to a trigger. If the hammer is released it will break the glass tube and release the poison. The detector can be turned on only once and only for a minute. During that minute the radioactive source within the box has a 50 percent chance of emitting a particle and therefore a 50 percent chance of not emitting a particle. No one can see into the box.

According to strict quantum physics, until we open the box, after a minute, the cat is neither alive nor dead, but in an indeterminate state. Many think that Schrödinger devised this thought experiment to show that quantum physics does not apply to the macroworld, or that the inexactitudes it introduces are so small as, really, not to matter.

Nonetheless, in 1933, Heisenberg was awarded the Nobel Prize in physics, and when he and his mother arrived at the Stockholm railway station, they were greeted by Paul Dirac and Erwin Schrödinger, who were there to share the prize. It was the first time three physicists were chosen for their contribution to *theoretical* physics.[19]

Symmetry and the Neutron

For close to a decade, quantum mechanics had been making news. During that time experimental particle physics had been stalled. It is difficult at this distance to say why, for in 1920 Ernest Rutherford had made an extraordinary prediction. Delivering the Bakerian Lecture before the Royal Society of London, Rutherford gave an insider's account of his nitrogen experiment of the year before. But he also broached the possibility of a third major constituent of atoms in addition to electrons and protons. "Such a particle," he argued, "would have very novel properties. Its external [electrical] field would be practically zero, except very close to the nucleus, and in consequence it should be able to move freely through matter." Though difficult to discover, he said, it would be well worth finding. "It should readily enter the structure of atoms, and may either unite with the nucleus or be disintegrated by its intense field." If this constituent did indeed exist, he said, he proposed calling it the neutron.[20]

Just as James Chadwick had been present in 1911, in Manchester, at the "Lit. and Phil.," when Rutherford had revealed the structure of the atom, so he was in the audience for the Bakerian Lecture. By now he was Rutherford's right-hand man, but at the time of the lecture he did not really share his boss's enthusiasm for the neutron. The symmetry of the electron and the proton— negative and positive—seemed perfect and, moreover, complete. Throughout the late 1920s, however, anomalies built up. One of the more intriguing was the relationship between atomic weight and atomic number. The atomic number of a chemical element was derived from the nucleus's electrical charge and a count of the protons. Thus helium's atomic number was 2, but its atomic weight was 4. For silver the equivalent numbers were 47 and 107, for uranium 92 and 235 or 238. One popular theory was that there were additional protons in the nucleus, linked with electrons that neutralized them. But this only created another theoretical anomaly: particles as small and as light as electrons could only be kept within the nucleus by enormous quantities of energy. That energy should show itself when the nucleus was bombarded and had its structure changed—and that never happened. When these anomalies showed no sign of being satisfactorily resolved, Chadwick started to come round to Rutherford's view. Something like a neutron must exist.[21]

Chadwick was in physics by mistake. He had wanted to be a mathematician but turned to physics after he stood in the wrong queue at Manchester University and was impressed by the physicist who interviewed him. He had studied in Berlin under Hans Geiger but failed to leave early enough when World War I loomed and was interned in Germany for the duration. By the 1920s he was anxious to be on his way in his career.

To begin with, the experimental search for the neutron went nowhere. Believing it to be a close union of proton and electron, Rutherford and Chadwick devised ways of, as Richard Rhodes puts it, "torturing" hydrogen. The next development is complicated. First, between 1928 and 1930, a German physicist, Walther Bothe, studied the gamma radiation (an intense form of light) given off when light elements such as lithium and oxygen were bombarded by alpha particles. Curiously he found intense radiation given off not only by boron, magnesium, and aluminum—as he had expected, because alpha particles disintegrated those elements (as Rutherford and Chadwick had shown)—but also by beryllium, which was not disintegrated by alpha particles. Bothe's result was striking enough for Chadwick at Cambridge, and Irène Joliot-Curie and her husband Frédéric Joliot in Paris, to take up the German's approach.

Both labs soon found anomalies of their own. Hugh Webster, a student of Chadwick, discovered in spring 1931 that "the radiation [from beryllium] emitted in the same direction as the . . . alpha particles was harder [more penetrating] than the radiation emitted in a backward direction." This mattered because if the radiation was gamma rays—light—then it should spray equally in all directions, like the light that shines from a lightbulb. A particle, on the other hand, would behave differently. It might well be knocked forward in the direction of an incoming alpha. Chadwick thought, "Here's the neutron." [22]

Later that year, in December, Irène Joliot-Curie announced to the French Academy of Sciences that she had repeated Bothe's experiments with beryllium radiation but had standardized the measurements. Irène Joliot-Curie was Marie Curie's daughter, born in 1897. She had shown an early aptitude for mathematics, and part of her training was to help her mother, who, since the spring of 1919, had been paid $75 a week to teach US Army officers waiting to go home how to use X-ray equipment. Irène married Frédéric Joliot in October 1926. He was "the most brilliant and the most high-spirited of the workers at the Institute of Radium," who was also described by a colleague as "a handsome lady-killer." [23] They made an attractive and formidable scientific couple.

The standardization experiment carried out by the Joliot-Curies enabled them to calculate that the energy of the radiation given off was *three times* the energy of the bombarding alphas. This order of magnitude clearly meant that the radiation wasn't gamma; some other constituents must be involved. Unfortunately Irène

Joliot-Curie had not read Rutherford's Bakerian Lecture, and she took it for granted that the beryllium radiation was caused by protons. Barely two weeks later, in mid-January 1932, the Joliot-Curies published another paper. This time they announced that paraffin wax, when bombarded by beryllium radiation, emitted high-velocity protons.

When Chadwick read this account in the *Comptes rendus*, the French physics journal, in his morning mail in early February, he realized there was something very wrong with this description and interpretation. Any physicist worth his salt knew that a proton was 1,836 times heavier than an electron; it was all but impossible for the former to be dislodged by the latter. Later that morning, at their daily progress meeting, Chadwick discussed the paper with Rutherford. "As I told him about the Joliot-Curie observation and their views on it, I saw his growing amazement and he finally burst out 'I don't believe it.' Such an impatient remark was utterly out of character, and in all my long association with him I recall no similar occasion. . . . Of course, Rutherford agreed that one must believe the observations; the explanation was quite another matter."[24]

Chadwick lost no time in repeating the experiment. The first thing to excite him was that he found the beryllium radiation would pass unimpeded through a block of lead three-quarters of an inch thick. Next, he found that bombardment by the beryllium radiation knocked the protons out of some elements by up to forty centimeters, fully sixteen inches. Whatever the radiation was, it was huge—and in terms of electrical charge, it was neutral. Finally, Chadwick took away the paraffin sheet that the Joliot-Curies had used to see what happened when elements were bombarded directly by beryllium radiation. Using an oscilloscope to measure the radiation, he found first that beryllium radiation displaced protons whatever the element and, crucially, that the energies of the displaced protons were just too huge to have been produced by gamma rays. Chadwick had learned a thing or two from Rutherford by now, including the habit of understatement. In the paper entitled "Possible Existence of a Neutron," which he rushed to *Nature*, he wrote, "It is evident that we must either relinquish the application of the conservation of energy and momentum in these collisions or adopt another hypothesis about the nature of radiation." Adding that his experiment appeared to be the first evidence of a particle with "no net charge," he concluded, "We may suppose it to be the 'neutron' discussed by Rutherford in his Bakerian Lecture."[25]

Chadwick, who had worked day and night for ten days to make sure he was first, was awarded the Nobel Prize for his discovery. The neutral electrical charge of the new particle would allow the nucleus to be probed in a far more intimate way. Other physicists were in fact already looking beyond his discovery—and in some cases they didn't like what they saw.

"Birth Pangs" of the Elements

About this time, observations from outer space drew attention to other aspects of nuclear physics. Cosmic rays ("the poor man's laboratory") were observed and found to contain particles, and the idea began to form that such rays were "the birth pangs" of the elements—showing how, in the early universe, the heavier elements were built up from the lighter ones.[26] This would become a major factor in the evolving understanding of cosmology later in the century, and be another important dimension of the converging order.

All this went together with the realization, especially in America, that one way forward lay with the development of "particle accelerators." Linear accelerators were used at first, by Ernest Lawrence and his colleagues in Berkeley, California, then the "cyclotron" was devised, the word being first used as a "sort of laboratory slang," though it had been conceived, at least in theory, by Leo Szilard, the Hungarian-American physicist who also conceived the idea of a nuclear chain reaction. Szilard apart, this was very much an American idea and, together with the work of Linus Pauling (see below, immediately), marks the entry of American physics onto the world scene, as well as the advent of "big science." The cyclotron, Helge Kragh says, proved extremely useful in a broad range of areas, extending from pure research on nuclear reactions to industrial and medical applications.[27]

At a meeting of the BAAS in Bristol in September 1930, Paul Dirac gave a lecture in which he said it had always been the dream of philosophers to have matter built up out of one fundamental kind of particle, so that it was not "altogether satisfactory" to have two in the then current theory—the electron and the proton. It was ironic therefore that, not long after his Bristol address, Dirac and his physicist colleagues were speaking about *four* particles. Chadwick had identified the neutron, Pauli had posited the neutrino, and Dirac himself had given particle physics a heavy push with his idea of anti-particle (a positively charged electron, called the negatron at one point, later identified in cosmic rays as the highly penetrating positron, midway in mass between the proton and the electron). Finally (finally before World War II, that is) the meson—aka the mesotron, the baryton, the yukon or heavy electron, the m-meson, and now known as the muon—was discovered in 1937, and Dirac's dream of a one-particle—or even a two-particle—universe was well and truly over.[28]

The Zone Between Physics and Chemistry

After the insights of Bohr, Rutherford, and the other physicists of the quantum revolution, the greatest breakthrough in theoretical chemistry in modern times was achieved by one man: Linus Pauling. His idea about the nature of the chemical bond was as fundamental and as unifying as the gene and the quantum because it showed how physics governed molecular structure and how that structure was related to the properties—and even the appearance—of the chemical elements. Pauling's work explained why some substances are yellow liquids, others white powders, and still others red solids.

Born the son of a pharmacist, near Portland, Oregon, in 1901, Pauling was blessed with a dose of self-confidence that matched Wolfgang Pauli's. As a young graduate he spurned an offer from Harvard, preferring instead an institution that had started life as Throop Polytechnic but in 1922 was renamed the California Institute of Technology, or Caltech. Partly because of Pauling, Caltech developed into a major center of science, but when he arrived there were only three buildings, surrounded by thirty acres of weedy fields, scrub oak, and an old orange grove.

He initially wanted to work on a new technique that could show the relationship between the distinctively shaped crystals into which chemicals formed and the actual architecture of the molecules that made up the crystals. It had been found earlier in the century that if a finely focused beam of monochromatic X-rays (discovered, remember, in 1895) was sprayed at a crystal, the beam would disperse—diffract—in a particular way. Since many substances—salts, metals, minerals, drugs, proteins, vitamins—formed crystals, the X-rays revealed the arrangement and density of the atoms and molecules and even the length of the bonds between them. Suddenly a technique for examining three-dimensional chemical *structure* was possible. Added to what Bohr had discovered, this further linked physics and chemistry.

X-rays are composed of photons, a concept introduced by Einstein but not fully accepted until 1922, and so X-ray crystallography, as it was called, was barely out of its infancy when Pauling got his PhD. Even so, he quickly realized that neither his mathematics nor his physics were anywhere near good enough to make the most of the new techniques. He decided to go to Europe to meet the great scientists of the day: Niels Bohr, Erwin Schrödinger, Werner Heisenberg, among others, although he studied most under Arnold Sommerfeld at Munich. As he wrote later, "I had something of a shock when I went to Europe in 1926 and discovered that there were a good number of people around that I thought were smarter than me."[29]

So far as his own interest was concerned—the nature of the chemical bond—his visit to Zurich was the most profitable. There he came across two less famous Germans, Walter Heitler and Fritz London, who had developed an idea about how electrons and wave functions applied to chemical reactions. At its simplest, imagine the following: Two hydrogen atoms are approaching one another. Each is comprised of one nucleus (a proton) and one electron. As the two atoms get closer to one another, "the electron of one would find itself drawn to the nucleus of the other and vice versa. At a certain point the electron of one would jump to the new atom, and the same would happen with the electron of the other atom." They called this an "electron exchange," adding that it would take place billions of times a second. In a sense, the electrons would be homeless, the exchange forming the "cement" that held the two atoms together, "setting up a chemical bond with a definite length." It was a very neat piece of work, but from the intensely ambitious Pauling's point of view there was one drawback about this idea: it wasn't his. In order to make his name he needed to push the idea forward.

By the time Pauling returned to America from Europe, Caltech had made considerable progress. Negotiations were under way to build the world's biggest telescope at Mount Wilson, a jet propulsion laboratory was planned, and T. H. Morgan, the famous geneticist, was about to arrive, to initiate a biology laboratory. Pauling was determined to outshine them all.[30]

With his valuable European experience in his pocket, so to speak, Pauling grasped that the field in which there was scope to do new and significant work lay in the zone between the new physics and chemistry—quantum mechanics promised a whole new area of chemical applications. For Pauling the opportunity lay in the fact that the individuals he had met in Europe were, for the most part, physicists, not chemists, and therefore could not be expected to grasp fully how their concepts related to the rich and varied landscape of chemical phenomena. A whole new world of "quantum chemistry" was waiting to be born.[31]

The Order of the Chemical Bond

Throughout the early 1930s Pauling released report after report, all having to do with the chemical bond. His early experiments on carbon, the basic constituent of life, and then on silicates, showed that the elements could be systematically grouped according to their electronic relationships. He showed that some bonds were weaker than others and that this helped explain chemical properties. Mica, for example, is a silicate that, as all chemists know, splits into thin, transparent sheets. Pauling was able to show that mica's crystals have strong bonds in two directions and a weak one in a third direction, exactly corresponding to observation. In a second instance, another silicate we all know as talc is characterized by weak bonds all around, so that it crumbles instead of splitting, and forms a powder. A third group of silicates, known as zeolites, were characterized by their ability to absorb certain gases, such as water vapor, but not others. Pauling was able to show that zeolites were "honeycombed" with passages so minuscule that they operated like molecular sieves, letting through only very small molecules and keeping others out. He showed that here too their structure was as relevant as their chemical and quantum properties.[32]

And so, here at last was an atomic, electronic explanation for the observable properties of well-known molecules. The century had begun with the discovery of the fundamentals that applied to physics. Now the same was happening in chemistry, and once more knowledge was beginning to come together. Pauling published *The Nature of the Chemical Bond and the Structure of Molecules and Crystals* in 1939.

But Pauling did even more than this. He made his mark further by arguing that quantum mechanics could explain the tetrahedral binding of carbon. Carbon was and is the jewel in the crown of organic chemistry. Carbon atoms form the backbone of proteins, fats, and starches: carbon chemistry is essentially the chemistry of life. At the time, however, there was a problem, in that physicists and chemists could not agree on the element's electronic structure. It was accepted that each carbon atom contained six electrons and that the first two played no role in forming bonds, because they paired with each other, to form the two-electron inner-shell structure of helium, as originally outlined by Bohr. The four electrons left over should be at the next energy level out, in the second "shell," or orbit, of the atom (again as Bohr and Pauli had shown). Chemists accepted that carbon had four bonds to offer to other atoms and that in nature these formed the four corners of a three-sided pyramid, or tetrahedron.

The problem was that the physicists disagreed. Their research, they pointed out, showed that carbon's four outer electrons actually existed in two different energy levels, or sub-shells. Pauli's exclusion principle showed that the two lower-level electrons should pair with each other, leaving just two over to form bonds with other atoms. "Carbon, the physicists said, should have a valence of two, though there were rare cases, like carbon monoxide, where the carbon is double-bonded to a single oxygen atom."[33] Reconciling this discrepancy was a major challenge.

The "Joining Patterns" of Molecules

Pauling resolved it with a novel idea, based on Heitler and London's exchange energy theory. Carbon atoms in molecular form, he argued, were structurally different from a single carbon atom. When the carbon atoms came together, the exchange energy generated by their proximity was sufficient to break carbon's four binding electrons out of the physicists' sub-shells, when a tetrahedron was the most parsimonious structural arrangement for four electrons.

This might be an impressive idea, but it could only be substantiated with considerable mathematical backup. And here Pauling set his own pattern. He submitted his paper on carbon and exchange energy to the *Journal of the American Chemical Society*, promising that he would send in a "detailed account" of the

mathematics at a later date. This was something Pauling did
several times, announcing ideas in print—to establish priority—
though in fact, in this case as in others, it would be some time
before his "detailed account" appeared.[34]

Nonetheless, the idea proved influential with others, made
Pauling better known, and led to a further idea that was still more
transformative. This was an idea that he called complementarity—
not to be confused with Niels Bohr's idea of complementarity.
Pauling's complementarity referred to the fact that the *structure* of
atoms—revealed by X-ray crystallography—had a great deal to
do with the way the different elements combined, that chemical
reactivity was not simply a question of which electrons were
available in which shells, but that the three-dimensional *shape* of
an atom determined which other atoms it could form a relationship with.

This again was a situation in which most crystallographers were
physicists, whereas Pauling was a chemist. One effect of this was
that the physicists treated each element afresh, as a one-off. They
studied its crystalline structure and drew what conclusions they
could about its properties and behavior. But Pauling knew much
more about the elements than the physicists did, and so he had an
intimate knowledge of how they behaved, what similarities they
had with other elements (silicon, for instance, behaved much like
carbon), and what the differences were, and what combined with
what. And so Pauling was able to anticipate a good proportion of
the crystallography findings. This gave him a much deeper grasp
of the landscape, and he therefore moved ahead more quickly
and decisively than the pure physicists.[35] He was able to distill
what he knew about quantum mechanics, ionic sizes, and crystal
structures, and put that together with a traditional understanding
of the habits of the elements, all wrapped up into a set of rules for
indicating which "joining patterns" were most likely.

He first went public with the patterns in late 1928 as part of
a Festschrift written in honor of Arnold Sommerfeld's sixtieth
birthday. The following year, Pauling described them in more

detail in the *JACS*, as a result of which they became widely known among crystallographers as "Pauling's Rules." Their synthesis of the new physics and classical chemistry worked very well, the rules helping to extend our understanding of what was being revealed under X-ray crystallography, so that more and more complex substances could be deciphered.

Pauling wasn't done yet, however. In fact, he was on the cusp of making one of the seminal conceptual breakthroughs of the scientific century. He was to make no fewer than three other advances that affect the theme of this book.

The first occurred on a second visit he made to Europe, in 1929, in this case to Ludwigshafen, a few hours' train ride from Munich. There he visited Herman Mark, a Viennese chemist he had met on his earlier visit. Mark was, by the time of Pauling's second visit, a distinguished crystallographer, working especially on organic molecules, and he had been snapped up by the giant German chemical firm IG Farben to head its research into such commercially promising products as plastic and synthetic rubber. Mark had been provided with excellent facilities, but for Pauling, the high point of the visit came when the Viennese told him that one of his assistants had "developed a way to shoot a beam of electrons through a jet of gas in a vacuum tube."[36] The molecules of gas, this man had found, appeared to diffract the electrons in systematic ways, scattering them "into patterns of concentric rings, the intensities and relative positions of which could be related to the distances between atoms in the molecules." This "electron diffraction" was not of obvious immediate use to Mark, but for Pauling it was a different matter entirely. The Farben apparatus, focusing as it did on separate molecules in a gas, offered a way to simplify X-ray crystallographic calculations: the sheer speed with which the electron-diffraction photographs could be produced (a few tenths of a second rather than several hours for X-ray crystallography) meant that a range of volatile substances—especially organic compounds difficult to hold in crystalline form—now came within the realm of possible investigation.

When Pauling got back to Caltech in fall 1930, he immediately had a new graduate student start to build an electron-diffraction machine. It took two years, but the machine eventually became the "workhorse" of the laboratory, and over the next twenty years Pauling and his coworkers identified the structure of 225 molecules.[37]

Meanwhile, while Pauling had been away, a colleague, John C. Slater, at MIT, had produced an important mathematical breakthrough: an important simplification of the famous Schrödinger wave equation, which made it much easier to account theoretically for carbon's four binding electrons. Pauling focused on Slater's mathematics, and was eventually able to show how, based on quantum mechanics, four equal electron orbitals could be "oriented precisely" at the corners of a tetrahedron. Slater had had the same idea, and submitted a similar calculation to the *JACS* slightly ahead of Pauling, but the coincidence of them both submitting such similar work only strengthened their joint conclusion. The carbon situation was now much clearer.

It was another example of "simultaneous discovery," to put alongside the conservation of energy, and the conception of the periodic table, but it was even more than that. It was now that the physicist Victor Weisskopf noted that quantum mechanics had finally united the two great fields of physics and chemistry.[38] "By using the rules of the new physics to explain the bonding of atoms into molecules, Slater and Pauling consummated the marriage."

A New Explanation for the Relationship Between the Elements

But if that was the climax of one amalgamation, it was also the beginning of something else. One problem that had raised its ugly head amid all the beautiful unification that was going on was that there were two seemingly different types of bonds between atoms—ionic and covalent—and it was by no means clear how they were related. A covalent bond occurred between two atoms that shared a pair of electrons equally, as Heitler and London proposed. A bond was ionic when one atom "pulled" the entire electron pair to itself, which meant there was a net negative charge on one atom, with a positive charge on the other— the resulting bond then consisted of the electrostatic attraction between the two.[39] This clearly added to the complexity of the chemical bond, and the question now became whether ionic and covalent bonds were entirely separate entities or different points along a continuum.

Using quantum mechanics, Pauling was able to show that ionic and covalent bonds were not totally separate entities but were indeed different points on a continuum. "Partial ionic" bonds existed, and there were links between atoms that had both ionic and covalent characteristics. Pauling termed this relationship "resonance," and its importance lay in the fact that these "hybrid" bonds were very stable. As a case in point, hydrogen chloride could be understood either as a hydrogen atom linked to a chlorine atom through a Heitler-London type covalent bond, or else as a positively charged hydrogen ion and a negatively charged chloride ion. The actual molecule, Pauling argued, "resonates" between these two alternatives (analogous to, but not the same as, the wave-particle duality). In the example given, HCl was both ionic and covalent in the ratio 20:80.[40]

All this was important because, at the time Pauling developed his concept of resonance, he now adapted an idea from biology. Having based his chemistry for so long on quantum physics, Pauling had started attending biology seminars at Caltech in the department headed by T. H. Morgan. He was in particular interested in how geneticists identified the location of genes on chromosomes, which they did indirectly, by inference, by measuring how frequently two independent traits were inherited together. The principle established here was that the closer two genes were physically on the chromosome, the greater the probability they would stay together during genetic crossover in reproduction. Pauling now adapted this idea to resonance, to create his own scale of the relationship between pairs of elements, according to how ionic or covalent their bonds were. He found that the more ionic the bonds were between the atoms, the greater was the difference in their ability to attract electrons. It was a new way to understand the chemical behavior of the elements.[41]

The power of this idea was clarified when he used it to solve one of the most stubborn problems in organic chemistry: the benzene conundrum. It had been known since the nineteenth century that benzene is composed of six carbon atoms and six hydrogen atoms, that it is highly stable but extremely reactive in forming all manner of compounds, and that it largely accounts for the huge diversity of organic substances. How was it that benzene had both a highly stable backbone and yet was so "eager," and so able, to form so many different—and often complex—molecules?

During the winter of 1932–33, Pauling and his student George Wheland set themselves the task of solving the benzene puzzle according to the principles of resonance. And what they found was revolutionary. They showed that benzene resonates between *five* structures, some more ionic, others more covalent, and while the fact that all the bonds display some form of resonance explained the stability of benzene, the *range* of bonds—between ionic and covalent—explained its great reactivity.

This was an extremely clever and fruitful advance on traditional thinking, but in the long run, the real significance of the benzene paper was that it marked an important expansion by Pauling—into the realm of organic chemistry. We have seen already that he was interested in carbon, that he had taken some lectures in biology at Caltech itself, and that his new technique of electron diffraction, taken from Germany, was especially suitable for the investigation of organic substances. Now he had improved our understanding of benzene, with carbon the very basis of organic chemistry. A good deal of subsequent work would be carried out on other benzene-based aromatic molecules, like naphthalene, with George Wheland publishing an influential book in 1944, *The Theory of Resonance and Its Application to Organic Chemistry.*[42]

But, more than that, Pauling and Caltech comprised one of two teams in the world where an important scientific crossover was taking place (we shall come to the other one in chapter 9). In 1936 a colleague of Pauling gave this crossover a name. Until that point, Pauling and his colleagues had been operating in a world which they defined as "quantum chemistry." In 1936 a new phrase was coined: "molecular biology."

Three Great Unifiers

We can now see more clearly that the story of science, as well as being a history of discovery, is also the narrative of convergence. Einstein's unifications of mass and energy and space and time were the first great convergence event in science after the two original unifying theories of the 1850s. Bohr's linking of the nucleus to radioactivity and of electron orbits to chemical properties of atoms was the second. Pauling was the third great unifier of the early twentieth century. He discovered the order among the chemical bonds and showed that the architecture— the shape—of molecules was also relevant. This meant, crucially, that molecules were as important as atoms in the understanding of matter. Molecules were not just the sum of their parts. This view—not immediately appreciated—would grow in importance as the century drew on.

The change from quantum chemistry to molecular biology was more than just a change in terminology. It was a change in perspective. While physics had been developing and deepening, so had biology. And as physics and chemistry had drawn closer together, so, over the same period, had chemistry and biology. The discovery of the gene, although it had been made almost coincidentally with the theory of evolution, had not been immediately linked to it. It was, in effect, a failure of—or a delay in—convergence. But the developing science of genetics—which was also being pursued at Caltech—would be an important factor in leading to the conception of molecular biology. The link between physics and biology—via quantum chemistry—was a unification of profound significance and would resonate throughout the twentieth century, right up until the present time.

Nor did that exhaust the other aspects of unification that were occurring. It was not immediately obvious how Einstein's unifications married with Bohr's and Pauling's, but that too would emerge in time. Curiously enough, it would also turn out that the proliferation in the number of particles being discovered, though it seemed counterintuitive at first, would also help unify our understanding of certain aspects of the universe that had been beyond comprehension beforehand.

6

THE INTERPLAY OF CHEMISTRY AND BIOLOGY: "THE INTIMATE CONNECTION BETWEEN TWO KINGDOMS"

While the great story of evolution was the main biological unification of the nineteenth century (chapter 2), it was not, in fact, the only one. The most important, certainly in the early part of the century, was the discovery of cells and, no less important, the realization that both animals and plants are made up of cells. This idea, that all forms of life are composed of "independent, but cooperative" units, ranks as one of the seminal discoveries in biology. This provided a fundamental link between zoology and botany, over and above the fact that they both concern living things.

The first person to observe cells was Robert Hooke (1635–1703), curator of experiments at the Royal Society, whose *Micrographia*—an early examination of microscopic life—appeared in 1665. In later centuries many others, benefiting from ever-improving microscopes, observed "globules" or "vesicles" of different sizes and shapes, in both animal and vegetable tissue. We know from a letter that Antonie van Leeuwenhoek of Delft wrote to Robert Hooke, in March 1682, that he had already observed a darker body inside cells which would come to be called the nucleus.

By the end of the eighteenth century, most botanists accepted that *plants* were composed largely of cells, with Caspar Friedrich Wolff (1733–94) one of the first to advocate that the fundamental subunit of all tissues—animal and vegetable—was a vesicle or globule, which, like others before him, he sometimes called a cell. However, no one had ever suggested—in print anyway—that plant cells and animal cells were homologous, and no one knew how cells divided or how new cells were formed. In 1805, Lorenz Oken (1779–1851) put forward the view that all living forms, plants as well as animals, were composed of "infusoria"—these being simple organisms like bacteria or protozoa, in other words, the simplest and most primitive forms of life then known.

But the first man to advance thought to its modern understanding was Jan Evangelista Purkyně. Strictly speaking, Purkyně was Czech, not German. However, since their defeat in the Battle of White Mountain in 1620, the inhabitants of Bohemia had been "inundated" by waves of Germanization—which led to Czech speakers being gradually reduced to menial positions. The Charles University in Prague, founded by Charles IV in 1348 and originally open to Czechs, Germans, and Poles, was, by the time Mozart made his celebrated journey to that city in 1787, a German-speaking institution.[1]

Purkinje, as he is spelled in the German literature, was educated as a choirboy in Mikulov (Nikolsburg) in Moravia, but left his order and took a medical and philosophy degree at the Charles University. From very early on, he had entertained the notion that there were fundamental parallels between animal and plant cells. The 1830s saw more progress, with several experiments clarifying the structure of such animal tissues as skin and bone— these papers referred to "granules," "*Körnchen*," "*Körperchen*," and "*Zellen*." The idea that there was "homology" between *some* plant cells and *some* animal cells, says Henry Harris in his history of early biology, was gaining strength. Then there was the fact that Franz Bauer, an Austrian who was a superb botanical artist, highlighted the nucleus in his drawings. These had been made as early

as 1802 but were not released until the 1830s, when Bauer made
it plain he regarded the nucleus as a regular feature of cells. The
nucleus was actually so named by Robert Brown, custodian of the
botanical collections at the British Museum (and the man who
identified "Brownian motion"), but his suggestion was made the
most of in Germany, the word being used as an alternative to *Kern*
("kernel"). The nucleolus, within the nucleus, was first observed
by Rudolf Wagner in 1835, though to begin with he called it a
"fleck," and then "the germinative spot" (*macula germinative*).

Purkyně's advances were not due simply to improved
microscopy. He was able to make thinner sections than anyone
before him and he perfected new staining techniques. He and
his colleagues alluded several times in print to the similarity
between animal and plant cells, and in a lecture he gave to the
Society of German Naturalists and Doctors, meeting in Prague
in September 1837, Purkyně made a *tour d'horizon* of the animal
tissues in which "*Körnchen*"—with a central nucleus—had been
observed: salivary glands, pancreas, the wax glands of the ear,
kidneys, and testes. "The animal organism can be almost entirely
reduced to three principal elementary components: fluids, cells,
and fibres. . . . The basic cellular tissue is again clearly analogous
to that of plants which, as is well known, is almost entirely composed of granules or cells."[2]

"The Common Principle Governing Animals and Plants"

In November 1832, Karl Asmund Rudolphi, professor of anatomy
and physiology at the University of Berlin, died. The vacant chair
was occupied the following year by a man who was to become
one of the more famous nineteenth-century biologists: Johannes
Müller. In 1835 Müller published a monograph on the comparative anatomy of the *Myxinidae* (hagfish), in which he described
the similarity between cells in the notochord (the neural channel
in the spine) and plant cells. This was a crucial observation, all
the more so as Theodor Schwann became Müller's assistant.

Schwann would capitalize on Müller's insight but only after his momentous meeting with the botanist Matthias Jakob Schleiden.

Schleiden's career had followed a familiar pattern for a German scholar. He first took up legal studies, obtaining a doctorate at Heidelberg University in 1827. He didn't enjoy legal work, however, and changed professions, beginning a degree in natural science at Göttingen in 1833, subsequently transferring to Berlin. Schleiden was invited to work in Müller's laboratory and it was there that he met Theodor Schwann.

Though a late convert to botany, Schleiden was always very keen on the microscope, and in 1838 he released "*Beiträge zur Phytogenesis*" in *Müller's Archiv*, a journal that the Berlin professor had started and which had become one of the most respected periodicals of the time. This article, immediately translated into English and French, was the first airing of the cell theory which, according to tradition, was conceived in a conversation between Schleiden and Schwann on the subject of phytogenesis (the origin and development of plants).

Schleiden was impressed by Robert Brown's identification of the cell nucleus (1832) and used that as his starting point. The nucleus was then called the cytoblast, and according to Schleiden, "as soon as the cytoblast reaches its final size, a fine, transparent vesicle forms around it: this is the new cell." Schleiden described this cell as "the foundation of the vegetable world."

While this paper clearly announced "the advent of plant cytology," Schleiden did so by asserting that cells are "*crystallised* inside an amorphous primary substance," which was quite wrong (italics added). Nevertheless, his botany textbook, published in 1842, the *Grundzge der wissenschaftlichen Botanik*, gave over a large section to plant cytology, and in so doing transformed the teaching of botany, attracting many people to what they felt was a new science.

His friend and colleague Theodor Schwann was a biologist for fifty years yet devoted only five of those (1834–39) to the subject for which he is best known. (These were the years he was an assistant to Müller. In 1839 he was offered a chair in Louvain, accepted, and moved away from the intellectual center of gravity in Berlin.) Schwann's most famous monograph was published in 1838, the very same year in which Schleiden released his "*Beiträge*" article. It began by outlining the structure and growth of the cells of the notochord and of cartilage. He did so, Schwann said, because their architecture "most closely resembles" that of plants and because cell formation from the "Cytoblastem" is clearly demonstrated. The second section bore a title that reflected his argument and tone: "On cells as the foundation of all tissue in the animal body." Purkyně and others had of course described cells in many tissues and had *speculated* they might be fundamental entities; but Schwann was the first to assert categorically that cells were basic.[3]

After that, his book surveyed what was by then a lot of histological evidence to support the thesis. He discussed cells attached to each other (the epithelium, nails, feathers, and the crystalline lens); cells where the walls are amalgamated with the intercellular substance (cartilage, bone, and teeth); cells giving rise to fibers (connective and tendinous tissue). Not everything he had to say was accurate, but as he said in the foreword:

> The aim of the present treatise is to establish the intimate connection between the two kingdoms of the organic world by demonstrating the identity of the laws governing the development of the elementary subunits of animals and plants. The main outcome of the investigation is that a common principle underlies the development of all the individual elementary subunits of all organisms, much as the same laws govern the formation of crystals despite their differences in shape.

The reference to crystals was of course similar to what Schleiden had said and, again, quite wrong. This error—an important one—was compounded, in the eyes of many, because Schwann hardly referred in his own work to others who had made contributions in the same field.

Inorganic–Organic

Throughout the eighteenth and nineteenth centuries, organic chemistry and physiology were muddied by the concept of the "vital force"—the belief that living organisms could not be explained by physical laws alone, that there must be some "special influence" at work. This view was reinforced by the sheer extent and diversity of organic matter, which, it was thought, only a deity could have envisaged. Yet, as more substances were analyzed and found to be made of carbon, nitrogen, and water only, the mystery deepened.

It was in this intellectual/religious climate that, in 1828, Friedrich Wöhler performed the experiment for which he will always be remembered. By treating silver cyanate with ammonium chloride he was hoping to derive the ammonium salt of cyanic acid. However, after he had filtered off the (insoluble) silver chloride and evaporated the residual solution, he found he had "colourless, clear crystals in the form of slender four-sided dull-pointed prisms." To his astonishment, they resembled nothing so much as urea. "This similarity . . . induced me to carry out comparative experiments with completely pure urea isolated from urine, from which it was plainly apparent that [urea and] this crystalline substance, of cyanate of ammonia, if one can so call it, are completely identical compounds." In fact, the two compounds were not identical—they were isomers, but even so Wöhler's was an iconic experiment. He had manufactured an organic substance, urea, hitherto the product solely of animals, out of inorganic materials *and without any intervention of vital force.* "[Justus von] Liebig and his successors regarded [this] experiment as [the] beginning of a truly scientific organic chemistry."[4]

The Link Between Dyes and Drugs

In May 1862, Queen Victoria attended the London International Exhibition in South Kensington wearing a vivid mauve gown.

This choice was significant for two reasons. The 1862 exhibition, like the 1851 Great Exhibition before it, had been spearheaded by her husband, Albert, the prince consort. But he had died of typhoid in December the previous year, so the queen was officially still in mourning, which could last for up to two years. Her presence at the exhibition in mauve and not black signified that she had moved—early—into "half-mourning," then a recognized stage of the grieving process. But there was more to it than that. She was also making a nationalistic statement. One of the main exhibits at the show was a massive pillar of purple dye. "Sitting next to the pile was its inventor/discoverer, William Perkin."[5]

Perkin, always interested in both engineering and chemistry, had been a student at the new Royal College of Chemistry (founded in 1845, now Imperial College). This was established as a consequence of the growing awareness in Britain that its science was lagging behind that of its continental competitors, Germany in particular. Among the benefactors of the college was Prince Albert, who had persuaded his fellow German August Wilhelm von Hofmann (then only twenty-eight) to be the first professor at the Royal College. Hofmann, born at Giessen in 1818, was a cultured man who grew up to speak four European languages fluently, who studied philology as well as chemistry, and whose obituary notices, when he died in 1892, were collected into *three* volumes. He traveled to Britain having been given a two-year leave of absence by Bonn University in case the Royal College job didn't work out. But it did work out, at least for a while, thanks largely to Hofmann's enthusiasm. Perkin started as one of Hofmann's students but by 1856 had been appointed his personal laboratory assistant.

Hofmann began this collaboration by suggesting that Perkin try to synthesize quinine. Virtually every professional chemist had been trying to do this for years, to create a synthetic cure for malaria (vital in an age of colonial expansion). Like everyone else, Perkin failed but then he toyed with a substance, an aniline derivative called allyl toluidine, extracted from coal tar, a plentiful waste product of the mining industry. And, "by one of those flukes of science," the aniline Perkin used contained impurities. Entirely unexpectedly, he found that the black sludge left behind after the failed quinine experiment, when washed with water, turned a vivid purple.

Until that point in history, people had had little choice in the colors available for their clothes. Derived from animal, vegetable, and mineral substances, the "earth colors"—reds, browns, and yellows—were by far the most common, and the cheapest. As a result, the rarer colors were much sought after, blue and purple in particular. Perkin—who had a copious Victorian beard to rival Hofmann's—therefore took out a patent on his purple dye and set up a factory on the banks of the Grand Union Canal northwest of London, where, it was said, the waters of the canal changed color every week depending on what dyes were being made at the time. By the time he was thirty-five, Perkin was rich.[6]

The Germans, through Hofmann, had had a hand in Perkin's education. Now they saw their chance to take a dividend. With abundant coal in the Ruhr, and more chemists than anywhere else, coal-tar dye companies sprang up all over Germany. A plethora of new synthetic dyes was rapidly discovered, and in no time, German dye companies led the world. Coal-tar dyes expanded so quickly that, within a few decades, they had virtually eliminated natural colors from the market.

The new color industry also owed its life to the simultaneous development of two other industrial/scientific innovations. One was the large-scale manufacture of illuminating gas, a by-product of which was tar. Second was the rise of systematic organic chemistry in the laboratory. The starting point here occurred in 1843 when Justus von Liebig instructed one of his assistants to analyze some light coal oil sent to him by a former student, Ernest Sell. The assistant chosen by Liebig to analyze Sell's oil was none other than Hofmann, who had just secured his doctorate at Giessen. Hofmann's analysis revealed that coal-tar oil contained aniline and benzene, two substances that would themselves go on to become important industrially and commercially. Hofmann described aniline as his great love affair and was soon able to explain the structural relation between aniline yellow, aniline blue, and imperial purple, all of which had recently been dis-covered. These dyes—mauve, fuchsin (magenta), aniline blue, yellow, and imperial purple—"were the most important coal tar colors that the young aniline dye industry produced."[7]

Commercially, these dyes were very successful, but one other factor contributed to the German preeminence. This was the creation of factory research laboratories. The creation of the factory laboratory was an event "whose historical significance . . . lies in the changes it brought about in the techniques of scientific research—changes that accelerated man's control over nature to such an extent that every major institution has since been affected."[8] And the most significant achievement of the industrial or factory laboratory was the way it transformed the coal-tar dye industry into the pharmaceuticals industry.

Pharmaceuticals came into their own during the 1880s and 1890s, partly because it was now that anesthetics began to be generally used—chloroform and ether becoming profitable substances for the dye companies to manufacture. And partly because, with the conception of the germ theory of disease, there arose a need for antiseptics. These were almost all phenols (chemicals bonded to aromatic hydrocarbons), which the dye companies had for years been using as dyestuff components.

Antipyretics (which reduce fever) and analgesics (which relieve pain) were discovered much as mauve was—by accident in the search for something else. Dr. Ludwig Knorr at Erlangen was yet another of those looking for a quinine substitute when he found that the pyrazolone compound he had just manufac-tured had pain-killing and fever-lowering properties. Höchst, the chemical company founded in 1863, bought the rights to this drug in 1883, and it was quickly followed by similarly acting substances, of which the most notable were "Antifebrin," pure acetanilide (1886); phenacetin or p-Ethoxyacetanilide (1887);

dimethylaminoantipyrine, sold as "Pyramidon" by Höchst (1893); and aspirin (1897). Sedatives appeared in the 1890s: "Sulfonal" and "Trional" (manufactured by Bayer, also founded in 1863, this time in Barmen) and "Hypnal" and "Valyl" (Höchst). The work of Koch and Pasteur on immunology led Höchst into the large-scale production of serums and vaccines to treat such dreaded diseases as diphtheria, typhus, cholera, and tetanus.

Following Bayer's lead, interest in pharmaceuticals snowballed, various firms employing bacteriologists, veterinarians, and other specialists. The new field of insecticides saw the building of laboratory-greenhouses, where botanists and entomologists tested the killing power of pesticides. These various developments—in botany, zoology, and chemistry—were brought together toward the end of the century in the work of Rudolf Virchow, Robert Koch, and Robert Ehrlich.

Life as the Sum of Physical and Chemical Action

Rudolf Carl Virchow was one of the most successful physicians of the nineteenth century. His long career epitomized the rise of experimental medicine after 1840, an ascendancy that transformed a discipline that was until that point still largely clinical and prescientific. It is not often recognized that scientific medicine only took off after biology and chemistry started to come together.[9]

From a farming family, Virchow was born in 1821 in a small market town in Pomerania. A short, round-faced man with staring eyes, because of his abilities he received in 1839 a military fellowship to study medicine at the Friedrich-Wilhelm Institute in Berlin. This was an institution specifically designed to provide an education for those who would not normally be able to afford one, in return for which they joined the army medical service for a specified time (von Helmholtz had a similar scholarship).

Always outspoken, in 1845 Virchow delivered two speeches before an influential audience at the Friedrich-Wilhelm Institute in which he dispensed with all transcendental influences in medicine (and the vital force) and argued that progress would only come from three main directions: clinical observations, "including the examination of the patient with the aid of physicochemical methods"; animal experimentation "to test specific aetiologies and study certain drug effects"; and pathological anatomy, especially at the microscopic level. "Life," he insisted, "was merely the sum of physical and chemical actions and essentially the expression of cell activity." On the strength of this, while still in his twenties, in 1847 he was appointed an instructor at the University of Berlin under the ubiquitous Johannes Müller.

But Virchow was never just a medical man. His political beliefs led him to take part in the uprisings of 1848 in Berlin, where he fought on the barricades, afterward becoming a member of the Berlin Democratic Congress and editor of a weekly entitled *Die Medizinische Reform.* This was a heady time, but in 1849, as a result of his political involvements, he was suspended from his academic positions. He quit Berlin and took up the recently created chair in pathological anatomy at the University of Würzburg, the first of its kind in Germany. While there, he was for a time distanced from political activity, and it was now that he achieved his greatest scientific contributions, establishing in particular his concept of "cellular pathology." It was in Würzburg that he did his most productive microscopic investigations of vascular inflammation, and the problems of thrombosis and embolism, terms that he coined. He was also the first to recognize leukemia cells and founded the field of comparative pathology—the comparison of diseases common to humans and animals, especially mammals. In 1856 Virchow returned to Berlin, as professor of pathological anatomy and director of the newly created Pathological Institute.

Virchow's supreme skill lay in proselytizing the idea of cellular pathology, that the way forward in medicine lay in an ever-deeper understanding of what happens to the cell during disease, and that this meant observation and experimentation, not speculation. For Virchow the microscope now became the central tool for understanding.[10]

The same was true of Robert Koch (1843–1910), the man who devised so many of the basic principles and techniques of modern bacteriology. He was one of thirteen children, and by the time he was ready for the local primary school, he had taught himself to read and write. At university he first thought of studying philology at Göttingen but enrolled instead in natural sciences. No bacteriology was yet taught at Göttingen, but after graduation in 1866 Koch attended Rudolf Virchow's course on pathology at the Charité hospital in Berlin. But his successes surely owed a great deal—more even than in Virchow's case—to the scale of his microscopic investigations. He installed a laboratory in his own home, where he had an excellent microscope by Edmund Hartnack of Potsdam. He began by studying anthrax.

It had been known for some time that anthrax was caused by rod-like microorganisms observed in the blood of infected sheep. Koch's first contribution was to invent techniques for culturing them in samples of cattle blood, which enabled him to study the microorganisms under his microscope. He found that although the bacilli were relatively short-lived, the spores remained infective for years. He proved that anthrax developed in mice only when the inoculum contained viable rods or spores of *Bacillus anthracis*, publishing his results in 1877. He accompanied these with a technical paper that detailed his method of fixing thin films of bacterial culture on glass slides, enabling them to be stained with aniline dyes.

Koch's next move was to equip his microscope with Ernst Abbe's new condenser and oil-immersion system, which enabled him to detect organisms significantly smaller than *B. anthracis*. As a result (using mice and rabbits), he identified six transmissible infections that were pathologically and bacteriologically distinctive. He deduced that human diseases would derive from similarly pathogenic bacteria.[11]

On the strength of this, and despite some unpleasant

disagreements with Louis Pasteur over the priority in anthrax studies, which had nationalistic overtones, in 1880 Koch was made government adviser (*Regierungsrat*) in the Imperial Department of Health (*Kaiserlichen Reichsgesundheitsamt*) in Berlin. There he shared a small laboratory with his assistants, Friedrich Loeffler and Georg Gaffky, both army doctors. They were charged with developing methods to isolate and cultivate pathogenic bacteria and to establish scientific principles that would improve hygiene and public health.

In 1881 Koch turned his attention to tuberculosis. Inside six months, "working alone and without a hint to colleagues," he confirmed that the disease was transmissible (which not everyone accepted) and isolated from a number of tuberculous specimens of human and animal origin a bacillus, which had characteristic staining properties. He then induced TB by inoculating several species of animals with pure cultures of this bacterium. His lecture, to the Physiological Society in Berlin on March 24, 1882, was described by Paul Ehrlich as the "greatest scientific event." The demonstration of the tubercle bacillus in the sputum was soon accepted as of crucial diagnostic significance.[12]

In the same year there was an outbreak of cholera in the Nile Delta. The French government, alerted by Louis Pasteur to the possibility that the epidemic could reach Europe, and told that the cause of cholera "was probably microbial," sent a four-man scientific mission to Alexandria. Koch arrived just over a week later, leading an official German commission. Within days he had observed colonies of tiny rods in the walls of the small intestine in ten bodies of people who had died of cholera. He also found the same again in about twenty cholera patients. Though promising, this organism failed to induce cholera when fed to or injected into monkeys and other animals. However, Koch's observations in Egypt were confirmed in Bengal, where his commission traveled next, and where cholera was endemic. In the spring of 1884 he identified village ponds, used for drinking water and all other domestic purposes, as the reasons why cholera was endemic in Bengal.

Although Koch and his work caught the eye (and continue to do so), the bacilli of swine erysipelas, glanders (an infectious disease of horses), and diphtheria were actually isolated by Loeffler, and the typhoid bacillus by Gaffky. Advances were being made at such a rate that additional institutes of public health were established in Prussia, and in 1885, Koch was appointed to the new chair of hygiene at the University of Berlin and an Institute for Infectious Diseases was created in Berlin. The circle around Koch was by now more impressive than that around Virchow, and included Paul Ehrlich and August von Wassermann. As a result of Koch's work, a communicable diseases control law was passed in 1900, the year in which his institute moved to larger quarters, adjoining the Rudolf Virchow hospital, making it the most famous medical complex in the world.[13]

The coming together of dyes and drugs was the first remarkable instance of the interplay of chemistry and biology. (In some cases, the dye companies used the waste products of the dye industry as experimental material for pharmaceuticals research.) And the coincident identification of cells as homologous in both plants and animals, their role in pathology, and the very fact that dyes stained different pathogens differently, was a second great coming together in this field, something that could not have been conceived beforehand. Coal was, to begin with, regarded as a mineral, but coal tar turned out to be an organic substance: here was another link between supposedly inorganic and organic substances, as Wöhler had shown in his famous experiment.

The link between dyes and drugs, together with the identification of cells, was therefore an all-important ingredient in the convergence of the sciences. The extremely useful propensity of pathological entities to react differently to different dyes, so that cellular pathology could be recognized according to the range of colors produced by staining, was a unification that would lead to a new understanding of basic body chemistry.

The Discovery of Antibiotics and the Human Immune Response

Despite the stirring achievements of Virchow and Koch, which would take time to work through their effects, at the beginning of the twentieth century people's health was still dominated by a "savage trinity" of diseases: tuberculosis, alcoholism, and syphilis. TB lent itself to drama and fiction. It afflicted the young as well as the old, the well-off and the poor, and it was for the most part a slow, lingering death. As "consumption" it features in *La Bohème*, *La Traviata*, *Death in Venice*, and *The Magic Mountain*. Anton Chekhov, Katherine Mansfield, and Franz Kafka all died of the disease.

The fear and moral disapproval surrounding syphilis a century and more ago mingled so much that despite the extent of the problem it was scarcely talked about. Eventually, however, in Brussels in 1899, Dr. Jean Alfred Fournier established the medical speciality of syphilology, using epidemiological and statistical techniques to underline the fact that the disease affected not just the "demi-monde" but all levels of society, that women caught it earlier than men, and that it was "overwhelming" among girls whose poor background forced them into prostitution. Fournier divided syphilis into primary, secondary, and tertiary phases, and in doing so helped pave the way for clinical research.[14]

In March 1905, Fritz Schaudinn, a zoologist from Röseningken in East Prussia, noticed under the microscope "a very small spirochaete, mobile and very difficult to study" in a blood sample taken from a syphilitic. A week later Schaudinn and Eric Achille Hoffman, a bacteriologist and professor at Halle and Bonn, observed the same spirochaete in samples taken from different parts of the body of a patient who only later developed roseolae, the purple patches that disfigure the skin of syphilitics. Difficult as it was to study, because it was so small, the spirochaete was clearly the syphilis microbe, and it was labeled *Treponema* (it resembled a twisted thread) *pallidum* (a reference to its pale color). Before the year was out, a diagnostic test had been identified by August von Wassermann in Berlin. This meant that syphilis could now be identified early, which helped prevent its spread. A cure was still needed.

The man who found it was Paul Ehrlich (1854–1915). Born in Strehlen, Upper Silesia (on the Polish and Czech borders), his crucial observation was that, as one bacillus after another was discovered, associated with different diseases, the cells that had been infected also *varied* in their response to staining techniques. Clearly, the biochemistry of these cells was affected according to the bacillus that had been introduced. This result was in some ways the culmination of the convergence of cellular pathology research and the organic dye research that had been developed in the preceding decades. For this deduction gave Ehrlich the idea of the antitoxin—what he called the "magic bullet"—a special substance secreted by the body to *counteract* invasions.[15]

By 1907 Ehrlich had produced no fewer than 606 different substances or "magic bullets" designed to counteract a variety of diseases. Most of them worked no magic at all, but "Preparation 606," as it was known in Ehrlich's laboratory, was eventually found to be a treatment for syphilis. This was the hydrochloride of dioxydiaminoarsenobenzene, in other words an arsenic-based salt. Though it had severe toxic side effects, arsenic was a traditional remedy for syphilis, and doctors had for some time been experimenting with different compounds with an arsenic base.

During the year, Ehrlich's study, always untidy, became clogged with photographs of chickens, mice, and rabbits, all of which had been deliberately infected with syphilis to begin with and, after being given Preparation 606, showed progressive healing. To be on the safe side, Ehrlich sent Preparation 606 to several other labs to see if different researchers would get the same results. Boxes were sent to St. Petersburg, Sicily, and Magdeburg.

At the Congress for Internal Medicine held at Wiesbaden on April 19, 1910, Ehrlich told the delegates that in October 1909 twenty-four human syphilitics had been successfully treated with

Preparation 606. Ehrlich called this magic bullet Salvarsan, which had the chemical name of arsphenamine.[16]

These discoveries fit neatly into our theme. Joseph Fruton, professor of biochemistry at Yale, has drawn attention to "the interplay of chemistry and biology" (his phrase) in a much more extensive way than considered here. He explores fermentation, enzymes, proteins, muscle activity, the citric acid cycle, glucose and glycogen, steroids as well as gene activity. Indeed, medicine itself which, as we have noted, only became truly scientific in the 1840s, may be regarded as the beneficiary of the interplay of chemistry and biology for the good of mankind. In fact, it is the direct beneficiary of nineteenth-century developments in six separate sciences—histology, cellular pathology, aniline dyestuffs, pharmaceuticals, microscope technology, and photography—all of which enabled scientists to study pathological substances and organisms at the microscopic level.

It didn't stop there. For about fifty years, biologists had been observing under the microscope a certain characteristic behavior of cells undergoing reproduction. They saw a number of minute threads forming part of the nuclei of cells, which separated out during reproduction. As early as 1882, Walther Flemming, professor of biology at Kiel, recorded that, if stained with dye, these threads turned a deeper color than the rest of the cell. This led to speculation that the threads were composed of a special substance, labeled chromatin because it colored the threads. The threads were soon called chromosomes, but it was nine years before Hermann Henking, in Leipzig, in 1891, examining the testicles of the firebug, *Pyrrhocoris*, made the next crucial observation. This was that during meiosis (cell division), half the spermatozoa received eleven chromosomes while the other half received not only these eleven but an additional body that responded even more strongly to staining. Henking could not be sure that this extra body was a chromosome at all, so he simply called it "X." It never crossed his mind that, because half received it and half didn't, the "X body" might determine what sex an insect was. But others soon drew this conclusion.[17]

After Henking's observation, it was also soon confirmed that the same chromosomes appear in the same configuration in successive generations, and Walter Sutton, at Columbia, showed in 1902 that during reproduction similar chromosomes come together, then separate. This configuration suggested that chromosomes behaved according to a statistical law put forward by a Moravian monk called Gregor Mendel. Having so long interplayed with chemistry, biology was now lining up with mathematics.[18]

The discovery of the importance of cells, and the way dyestuffs led to pharmaceuticals, and the way staining techniques, within cells, helped to reveal the nature of disease, was a remarkable instance of the way discoveries in one science can quickly lead to advances elsewhere. In this field of activity there was no seminal moment, like Clausius's papers in the *Annalen*, or Darwin's book of 1859, but the transformations taking place, to create modern scientific medicine, did occur at more or less the same time. These developments may not have impacted quite so suddenly as those other developments did, but the changes considered here underline that the mid-nineteenth century did see a marked change in intellectual climate. Wherever you looked, one science overlapped with, converged upon—and supported—another.

7

THE UNITY OF SCIENCE
MOVEMENT: "INTEGRATION
IS THE NEW AIM"

I n 1924 a group of scientists and philosophers in Vienna began to meet every Thursday. Originally organized as the Ernst Mach Society, in 1929 they changed their name to the *Wiener Kreis*, or Vienna Circle. Under this title they became what was arguably the most important philosophical movement of the twentieth century. One of their main projects was an exploration of the very topic we are considering: the unity of science.

The guiding spirit of the circle was Moritz Schlick (1882–1936), professor of philosophy at the University of Vienna from 1922. Like many members of the *Kreis*, he had trained as a scientist, in his case as a physicist under Max Planck, from 1900 to 1904, exactly the time Planck was conceiving his notion of the quantum. The twenty-odd members of the circle that Schlick put together included: Otto Neurath, a native of Vienna, and a remarkable Jewish polymath; Rudolf Carnap, a mathematician who had been a pupil of Gottlob Frege at Jena; Philipp Frank, another physicist; Heinz Hartmann, a psychoanalyst; Kurt Gödel, a mathematician; and at times Karl Popper, who would become an influential philosopher after World War II. Schlick's original label for the kind of philosophy that evolved in Vienna in the

1920s was *Konsequenter Empirismus,* or consistent empiricism. However, after he visited America in 1929, and again in 1931–32, the term logical positivism emerged—and stuck.[1]

The logical positivists made a spirited attack on metaphysics. They were against any suggestion that "there might be a world beyond the ordinary world of science and common sense, the world revealed to us by our senses." For the logical positivists, any statement that wasn't empirically testable—verifiable—or a statement derived from logic or mathematics was nonsensical. And so vast areas of theology, aesthetics, and politics were dismissed. Under them, philosophy became the handmaiden of science and a "second-order subject." First-order subjects talk about the world (like physics and biology); second-order subjects talk about their talk about the world. All philosophy can do, therefore, on this account, is analyze and criticize the concepts and theories of science so as to refine them and make them more accurate and useful.[2]

One aspect of this approach, as noted, was an exploration of the unity of science, "to relate and harmonise the achievements of individual researchers in the various branches of science."[3] The "Unity of Science Movement," as it came to be called, began formally with a congress at the Charles University in Prague in 1934, at which it was decided to hold a Unity of Science congress every year, beginning in Paris the following summer. At the third of these congresses, the *International Encyclopaedia of Unified Science* was launched. Otto Neurath led the way here (Schlick had been assassinated in 1936, shot on the steps of Vienna University by a jealous student). But the first volume of the encyclopedia, published in 1938, had articles by an extremely impressive raft of scientists, philosophers, and mathematicians—Niels Bohr, John Dewey, Bertrand Russell, and Rudolf Carnap. How Mary Somerville would have enjoyed being part of this group.

Neurath set out clearly the movement's aims. He began by arguing that a "universal scientific attitude" was an important prerequisite for a stable and progressive future (this was the eve of World War II, remember). He went on to argue that a scientific synthesis was a great "rational" aim, "in the hope that science will help to ameliorate personal and social life." He looked forward to the increased cooperation of physicists with biologists, biologists with social scientists, and with logicians and mathematicians.[4] He saw the new encyclopedia as updating the aims of the French encyclopedists of the eighteenth century.

An "All-Embracing Vision and Thought Is an Old Desire of Humanity"

For the time being, Neurath thought, there would be a "mosaic" of sciences, as they combined new observations and new logical constructions of diverse character and origin. But he insisted that an "all-embracing vision and thought is an old desire of humanity," going all the way back to the *Summa* of Aquinas. In the nineteenth century, Comte, Spencer, and Mill had had much the same idea, of the connections between physics, biology, and the social sciences. But lately, he said, one science after another had become separated from the "mother-philosophy," something he regretted because "the mosaic pattern of science progressively shows more marked interconnections than in the times in which empirical studies were relatively isolated. Scientific analysis of the sciences [had] led to the observation that an increase of logical intercorrelation between statements of the same science and between statements of different sciences is a historical fact. . . . The evolving of all such logical connections and the integration of science is a new aim of science."[5]

He went on to describe the movement for the unity of science, which was not just limited to the Vienna Circle—he referred also to groups with similar aims in Berlin and Scandinavia, as well as *Scientia* in Italy and the *Centre international de synthèse* in France. Scientists, he said, were now intent on building what he called "systematical bridges" from science to science, "analysing concepts which are used in different sciences, considering all questions dealing with classification, order, etc." Axiomatization of science, as he put it, "seems to give an opportunity to make the use of fundamental terms more precise and to prepare the combination of different sciences." But he accepted that "we cannot anticipate a 'final axiomatisation.'"

Nonetheless, the "thesis of physicalism," he insisted, emphasizes that "it is possible to reduce all terms to well-known terms of our language of daily life." He thought that science was supplying its own "integrating glue," but he rejected the idea of a "super-science" emerging someday and he wondered openly, "What is the maximum of scientific co-ordination which remains possible?" In fact, he said, "the most one can achieve in integration of scientific work seems to be an encyclopedia, constructed by scientists in cooperation. . . . Encyclopedism may be regarded as a special attitude."[6] He hoped that the encyclopedia would find new ways to classify the sciences, "a new way to assemble systematically all the special sciences," one idea being an "Isotype Thesaurus," which would show the relationship between facts, observations, and theories across different disciplines.

Niels Bohr, in a very short essay, suggested that the aim of human understanding should be unity and that "the history of science teaches us again and again how the extension of our knowledge may lead to the recognition of relations between formerly unconnected groups of phenomena, the harmonious synthesis of which demands a renewed revision of the presuppositions for the unambiguous application of even our most elementary concepts."[7] John Dewey asked, pertinently, "What sort of unity is feasible or desirable?" Dewey thought that the mere *effort* to unite human experience was in itself morally desirable, that "convergence to a common centre will be effected most readily and most vitally through the reciprocal exchange which attends genuine cooperative effort." It seemed to him, he said, that the greatest need was "the linkage of the physico-chemical sciences with psychological and social fields . . . through the intermediary of biology."[8]

Bertrand Russell made two points. One, that mathematical logic had at last begun to be appreciated on a wider scale as having a greater role in understanding the various sciences, reinforcing Mary Somerville's arguments. And two, that the unity of *method* was one of the most valuable lessons science had for wider society, being a step on the way to greater reciprocity in society as a whole.

Finally, Rudolf Carnap gave a fairly orthodox account of reductionism, that "biology presupposes physics but not vice versa."[9] Physics and logico-mathematical terms described the fundamental world, and the rest (in a wider sense, as he put it) "was biology." He tried to inject some clarity into reductionism. "We do not ask 'Is the world one?' 'Are all events fundamentally of one kind?' 'Are the so-called mental processes really physical processes or not?' 'Are the so-called physical processes spiritual or not?' It seems doubtful whether we can find any theoretical content in such philosophical questions. . . . When we ask whether there is a unity in science, we mean this as a question of logic, concerning the logical relationships between the terms and the laws of the various branches of science."[10] He thought there was no need for a detailed look at the social sciences "because it is easy to see that every term of this field is reducible to terms of the other fields."[11]

Examining the various laws of science, he said he didn't believe, at that stage, that it was possible to derive biological laws from the physical ones and he acknowledged that some philosophers "believe that such a derivation is forever impossible." Nonetheless, he insisted that "in the meantime the efforts towards derivation of more and more biological laws from physical laws—in the customary formulation: explanation of more and more processes in organisms with the help of physics and chemistry—will be, as it has been, a very fruitful tendency in biological research." At the same time he thought that the laws of psychology and social science could not be derived from those of biology or physics but that there was no scientific reason which would mean that such a derivation "should be in principle and forever impossible." Therefore, he concluded, that though there was at that time *no unity of laws*, the construction of one homogenous system of laws for all of science was an aim for the future, which "cannot be shown to be unattainable."[12]

In a book on the subject, *The Unity of Science*, which he published in 1934, Carnap pressed his arguments further. His central point in the book was totally reductionist—"all statements in science can be translated into physical language."[13] He thought that, for the moment anyway, science was no unity but could be separated into the formal sciences (logic and mathematics) and the empirical sciences. He acknowledged that there was a division of opinion regarding the sciences; that, on the one hand, they are fundamentally distinct as regards subject matter, and, on the other, that all empirical statements "can be expressed in a single language, all states of affairs are of one kind and are known by the same method." There was no cleavage between philosophy and the sciences, he said, because "analysis of language has ultimately shown that Philosophy cannot be a distinct system of statements, equal or superior in rank to the empirical sciences." "Observation," he said, is only a method, designed to help bring about understanding.[14]

All qualitative statements, he went on, if they mean anything, can be translated into physical statements of a particularly simple form and can be established experimentally in other sensory fields. This he called *"physicalising."* Physicists, he said, believe that agreement can be reached "to any degree of exactitude attainable by single investigators; and that when such agreement is not found in practice, technical difficulties (imperfection of instruments, lack of time, etc.) are the cause. *Physical determinations are valid inter-subjectively.*"[15]

All this was strong enough but then, "It will be demonstrated that all other languages used in science (e.g. Biology, Psychology, or the social sciences) can be reduced to the physical language. *Apart from the physical language* (and its sub-languages) *no inter-subjective language is known."* Carnap thought that chemistry, geology, and astronomy presented no threat to this argument. In his view, in biology, once it had shed its disputes over vitalism ("still violently controversial at this time"), the question would clarify itself as to whether the natural laws which explain "all inorganic phenomena can also be a sufficient explanation in the region of the organic. . . . The Viennese Circle is of the opinion that biological research in its present form is not adequate to answer the question." But he went on to say that he expected an "affirmative" answer in the future. Once you divided biology into its various parts—fertilization, metabolism, egg, cell division, growth, regeneration—he thought the case for reduction became clearer, that every statement in biology could indeed be translated into physical language. He could see why psychology provoked more controversy, he said, but thought that the substance of psychology being translated into physical science was "well grounded" and that sociology—to include all historical, cultural, and economic phenomena—"easily follows," provided "pseudo-concepts" were kept out ("objective spirit," "the meaning of history").[16]

At the end he advocated that we should try, as far as possible, to use a language in everyday life that was as close to the findings and practice of science as could be. That language should be precise, limited to what could be empirically demonstrated. "Every statement in the new language could be interpreted as having the same sense as a statement of the physical language, i.e., every statement of the new language would refer to physical facts, to spatiotemporal events." And finally: "Because the physical language is thus the basic language of Science *the whole of Science becomes Physics.*"[17]

The Dictator Principle

In subsequent issues, the *Encyclopaedia of Unified Science* tackled the foundations of various aspects of science—of mathematics, physics, biology, and psychology. In all, nineteen monographs were published between 1938 and 1969, the best known probably being Thomas Kuhn's *The Structure of Scientific Revolutions*. But overall they cannot be said to have lived up to the aims of the first volume and advanced in any fundamental way the unity of the sciences, though Carl Hempel's *Concept Formation in Empirical Science* came closest. One reason for the failure was that, although the journal lasted until 1969, the sixth international congress, held in Chicago in 1941, not long before the Japanese attacked Pearl Harbor, proved to be the last.

But where the movement *was* successful was in its emphasis on the potential link between mathematics and the sciences. For around that time there were two other—more specific— attempts to explore this relationship. One was abortive in an interesting way, the other extremely successful, with long-term consequences. They carried Mary Somerville's reasoning way beyond what she could ever have imagined, though she would no doubt have wholly approved of their aims.

The late 1920s and early 1930s, as we have seen, were exciting years in physics—the development of quantum mechanics, the discovery of the neutron, new particles in cosmic rays. It was also a time when new ideas began to emerge in biology, and they were not entirely unrelated.

The vitalist controversy had taken a long time to die in the nineteenth century, but was eventually replaced by a more mechanist view of what life is. Not completely, however, for many biologists well into the twentieth century clung to the idea that biology could not be reduced to chemistry and physics but required its own set of laws.[18] Under these circumstances, a new breed of biologist, known as "organicists," emerged, who were neither vitalists nor mechanists, but who insisted there was something not yet grasped about life and who thought it would eventually be explained by "yet to be discovered" laws of physics and chemistry. Probably the best-known organicist was the Austrian Ludwig von Bertalanffy, who was an expert on biological development and argued that some new biological principles were needed to capture fully what life is. He did not think, for example, that the classical laws of thermodynamics applied to "open systems"—i.e., living organisms as opposed to mechanical or electrodynamical systems—and believed that only a more interdisciplinary approach would come close to understanding life.

Among those influenced by von Bertalanffy was a German quantum physicist, Pascual Jordan, from Hanover. Jordan studied under Max Born in Göttingen and in 1925 he and Born published a classic paper, "On Quantum Mechanics." A year later, a second paper was published by Jordan, Born, and Heisenberg, which built on Heisenberg's original discovery to explain mathematically the behavior of the atomic world.[19]

In 1927 Jordan went to Copenhagen to work with Niels Bohr, and around 1929, the two men began to discuss whether quantum mechanics might have some application in the field of biology. For the next two years, by which time Jordan had returned to Germany, to a post at the University of Rostock, he kept up a correspondence with Bohr, exploring further how the convergence between quantum mechanics and biology might work. Their ideas culminated in what, say Jim Al-Khalili and Johnjoe McFadden, "is arguably the first scientific paper on quantum biology." Published in 1932, it was entitled "Quantum Mechanics and the Fundamental Problems of Biology and Psychology."

At this point, however, Jordan seems to have been politicized, as so many German scientists were, by Nazi ideology, and his scientific theorizing became increasingly speculative, even claiming that the concept of a dictatorial leader was a central principle in life. (He argued that in bacteria, for example, a number of "special molecules endowed with dictatorial authority over the total organism" form a "steering centre" of the living cell.) "Absorption of a light quantum in the steering centre of the cell can bring the entire organism to death and dissolution—similar to the way a successfully executed assault against a leading statesman can set an entire nation into a profound process of dissolution."

As Al-Khalili and McFadden say, this mixing of Nazi ideology with biology is fascinating, chilling, and embarrassing all at the same time. In fact, there was a germ of an idea that was to crop up much later. At the time, however, because Germany was defeated in World War II, Jordan's Nazi affiliation saw him and his theories widely discredited among colleagues and his ideas were "scattered to the four winds."[20]

Spontaneous Order in the Kingdom of Mathematics

While the Vienna Circle was launching its "Unity in Science" movement, and while the Nazis were busy doing their best to divert the course of science for their own ends, a Scotsman in far away Fife was preparing the second edition of a book that was to have a profound influence on the integration of different sciences as the decades passed. D'Arcy Wentworth Thompson was a colorful professor of zoology at St. Andrews University, who had collected animals and plants as far afield as the Bering Strait, and who walked the cobbled streets of the city in his tennis pumps, with a parrot on his shoulder. (The cobbles marked where Protestant martyrs were burned at the stake.) He had published the first volume of *On Growth and Form* in 1917. It was a substantial book even then, consisting of 793 pages, but the definitive second edition, published in 1942, had 1,116 pages (and this at a time when, because of wartime paper rationing, most books were on the short side).

Because he was a classics scholar as well as a biologist, Thompson's book was full of classical allusions (it is said to have influenced Claude Lévi-Strauss, Salvador Dalí, Henry Moore, and Jackson Pollock among others). But its main theme was that Thompson thought the biologists of his day were too much in thrall to Darwin's theory of evolution and paid insufficient attention to the role that physics and mechanics (and therefore mathematics) had played in determining the forms that living organisms take. He underscored his argument at great length, via hundreds of examples, and in doing so produced ideas that for many people—biologists, other scientists, and general readers alike—were totally original.

At its most simple, the book showed how the forms that jellyfish take match the forms that droplets of liquid *spontaneously* adopt when falling into a viscous fluid. He showed how the spiral structures of plants follow the Fibonacci sequence, an ancient progression of numbers in which any one integer in the sequence is the sum of the previous two: 0, 1, 1, 2, 3, 5, 8, 13, 21, 34, and so on. He showed that the branching of blood vessels in the body is designed so as to minimize the amount of energy needed to propel the blood around the body, and offered a mathematical equation to describe the pressure of the liquid, the tension of the vessel walls, and the radius of curvature. He showed that the path of an insect as it approaches a source of light is mathematical. "Owing to the structure of their compound eyes, these insects do not look straight ahead but make for a light which they see abeam, at a certain angle. As they continually adjust their path to this constant angle, a spiral pathway brings them to their destination at last."[21] He showed how in animals size and speed, size and jumping ability, size and vision are related, and in an extraordinary series of precise drawings he showed how, for example, two different species of fish, *Argyropelecus olfersi* and *Sternoptyx diaphana*, represent "precisely the same outline" except that the latter's oblique coordinates have axes that are inclined at an angle of seventy degrees. He added that this alteration "is precisely analogous to the simplest and commonest kind of deformation to which fossils are subject as the result of shearing-stress in the solid rock."[22] No less interesting, he pointed out that the skulls of chimpanzees, baboons, and humans are also mathematically related and that their coordinates can be calculated.

Thompson himself was less interested in the unity of science than in the beauty of mathematics, and he thought he had found an area that would repay a great deal of mathematical research.

Quoting a fellow scientist, he concluded, "Conterminous with space and coeval with time is the kingdom of mathematics; within this range her dominion is supreme; otherwise than according to her order nothing can exist, and nothing takes place in contradiction to her laws."[23]

In other words, Thompson was identifying a fundamental link between biological forms and mechanics, physics and mathematics. In doing so he was—to an extent—aligning himself with Rudolf Carnap's argument, that physics and mathematics provide a universal language, which underlines the unity of science. For many years Thompson was regarded as a maverick. But, given time, that would change.

8

HUBBLE, HITLER, HIROSHIMA: EINSTEIN'S UNIFICATIONS VINDICATED

When Einstein had originally produced his theory of relativity, most scientists took it for granted that the universe was static. But relativity had a surprise for astronomers: Einstein's equations predicted that the universe must either be expanding or contracting. So weird did it appear—even to Einstein himself—that he tinkered with his calculations to make his theoretical universe stand still.

Alexander Friedmann, a young Russian scientist from St. Petersburg (and a part-time balloonist), only learned of Einstein's theories after World War I ended (he was a bomber pilot during the conflict). Even so he was the first man to cause Einstein to think again. Friedmann taught himself relativity, during which time he realized Einstein had made a mistake and that the universe must be either expanding or contracting. He found this such an exciting idea that he dared to improve on Einstein's work, developing a mathematical model to underline his conviction, and sent it to the German. By the early 1920s, however, the great man had become famous and was snowed under with letters. Friedmann's ideas were lost in the avalanche.

It was only when he was given an introduction by a mutual

colleague that Einstein finally got to grips with the Russian's ideas. As a result he began to have second thoughts about his "cosmological constant"—and its implications. But it wasn't Einstein who pushed Friedmann's ideas forward. A Belgian cosmologist, Georges Lemaître, and a number of others, built on his ideas so that, as the 1920s advanced, a fully realized geometric description of a homogenous and *expanding* universe was fleshed out.[1] Lemaître even suggested that the universe may have begun at a point in the remote past with a "primal atom" or "cosmic egg," what would later be called the Big Bang theory.

A theory was one thing. But planets and stars and galaxies are not exactly small entities. Surely, if the universe really was expanding, it could be observed? One way to do this was by observation of what were then called "spiral nebulae." Nowadays we know that nebulae are distant galaxies, but then, with the telescopes of the time, they were simply indistinct smudges in the sky.

It was then discovered that the light emanating from spiral nebulae is shifted toward the red end of the spectrum. One way of illustrating the significance of this red shift is by analogy to the Doppler effect—named after Christian Doppler, the Austrian physicist who first explained the observation in 1842. When a train or a motorcycle comes toward us, say, its noise changes, and then, as it goes past and away, the noise changes a second time. The explanation is simple: As the train or bike approaches, the sound waves reach the observer closer and closer together—the intervals get shorter. As the train or bike recedes, the source of the noise is further away all the time, and the interval gets longer and longer. Much the same happens with light: where the source of light is approaching, the light is shifted to the blue end of the spectrum, while light where the source is receding is shifted to the red end.

The first crucial tests were made in 1922, by Vesto Slipher at the Lowell Observatory in Flagstaff, Arizona. (The Lowell had originally been built in 1893 to investigate the "canals" on

Mars.) In this case, Slipher anticipated finding red shifts on one side of the nebulae spirals (the part swirling away from the observer) and blue shifts on the other side (because the spiral was swirling toward earth). Instead, he found that all but four of the forty nebulae he examined produced red shifts only.[2] Why? The confusion probably arose because Slipher could not be certain of exactly how far away the nebulae were. This made his correlation of red shift and distance problematic. But the results were highly suggestive.

Three years elapsed before the situation was finally clarified. In 1929, Edwin Hubble, using the largest telescope of the day, the one-hundred-inch reflector scope at Mount Wilson, near Los Angeles, managed to identify individual stars in the spiral arms of a number of nebulae, thereby confirming the suspicions of many astronomers that "nebulae" were in fact entire galaxies.

Hubble, six feet two inches tall, with "patrician contours" lining his face, was an athletic if rather awkward and inward character ("Study has been my middle name," as he once wrote to his grandfather).[3] A graduate of Chicago University, he was a Rhodes Scholar at Oxford, where he acquired a number of British affectations—wearing a cape, carrying a cane, and smoking British pipes—which irked his colleagues, despite his obvious brilliance.

It was Hubble who also located a number of "Cepheid variable" stars. Cepheid variables—stars that vary in brightness in a regular way (periods that range from one to fifty days)—had been known since the late eighteenth century. But it was only in 1908 that Henrietta Leavitt at Harvard (hired as a female "computer," to catalogue stars) showed that there is a mathematical relationship between the average brightness of a star, its size, and its distance from earth.[4] Using the Cepheid variables that he could now see, Hubble calculated how far away a score of nebulae were. His next step was to correlate those distances with their corresponding red shifts. Altogether, Hubble collected information on twenty-four different galaxies, and his results were simple and sensational: the farther away a galaxy was, the more its light was red-shifted.

This became known as Hubble's law, and although his original observations were made on twenty-four galaxies, since 1929 the law has proved to apply to thousands more.[5]

Once more then, one of Einstein's predictions had proved correct: the universe was indeed expanding. For many people this took some getting used to. It involved implications about the beginning of the universe, its character, the very meaning of time.

De-convergence: Aryan Physics and Jewish Science

Einstein's other big idea, $E=mc^2$, took longer to be explored, partly because it became mired in the politics of the Third Reich.

The dismissal of scientists began almost immediately after Hitler became chancellor, in the spring of 1933. For the most part, one would think that science—especially the "hard" sciences of physics, chemistry, mathematics, and geology—would be un-affected by political regimes. It is, after all, generally agreed that research into the fundamental building blocks of nature is as free from political overtones as intellectual work can be. But in Nazi Germany nothing could be taken for granted.

At first, some Jewish academics were exempt from dismissal, if they had been employed before World War I, or had fought in the war, or had fathers or sons who had done so. But such exemption had to be applied for, and James Franck, a veteran of the earlier war (he had worked with the chemist Fritz Haber on poison gas and, like Haber, was Jewish), felt that to apply would amount to collusion with the regime. He therefore resigned his position as head of the Second Physics Institute in Göttingen, whereupon he was roundly criticized "for stirring up anti-German propa-ganda."[6]

In contrast, at Heidelberg, the physics institute had been renamed after Philipp Lenard, a prominent advocate of "Aryan physics." What Lenard and others meant by this phrase was that "German natural sciences" differed from "Jewish science," the former consisting of "observation and experimentation and not excessive theorising and reliance on abstract mathematical constructions," unlike, say, relativity theory.

Einstein's persecution had begun early. He had come under attack largely because of the international acclaim he received after Arthur Eddington's announcement, in November 1919, that he had obtained experimental confirmation for the predictions of general relativity theory. Nazi students disrupted a Berlin lecture as early as 1920. The award of the Nobel Prize in 1921 added both to his fame and the bitterness with which he was regarded in some quarters in Germany, as did his involvement with Zionism. Whenever Einstein traveled abroad, an official from the local embassy or consulate would send secret reports to Berlin.[7]

When the Nazis achieved power, ten years later, action was not long delayed. In January 1933 Einstein was away from Berlin on a visit to the United States. Despite facing a number of personal problems, he made a point of announcing he would not return to his positions at the university in Berlin and the Kaiser-Wilhelm Gesellschaft as long as the Nazis were in charge.* The Nazis repaid the compliment by freezing his bank account, searching his house for weapons allegedly hidden there by Communists, and publicly burning copies of a popular book of his on relativity. Later in the spring, the regime issued a catalogue of "state enemies": Einstein's picture headed the list, and below the photograph was the text "Not yet hanged." He eventually found a berth at the newly established Institute for Advanced Study at Princeton. When the news was released, one newspaper in Germany ran the headline: GOOD NEWS FROM EINSTEIN—HE IS NOT COMING BACK. On March 28, 1933 Einstein resigned his membership of the Prussian Academy of Sciences, preventing

---

* The Kaiser Wilhelm Society for the Advancement of Science funded scientific research institutes outside universities in Germany, so that leading figures were free of the burden of teaching. It was renamed the Max Planck Society after World War II.

any Nazi from firing him, and was distressed that none of his former colleagues—not even Max Planck—made any attempt to protest his treatment.[8]

Diaspora, with Diamonds

But, as the 1930s passed, physics—and especially the implications of $E=mc^2$—began to take on an almost apocalyptic significance. In 1933, as Hitler came to power, Einstein was not the only German physicist abroad in the United States: Otto Hahn was lecturing in Cornell. That left Lise Meitner in charge at the Kaiser Wilhelm Institute for Chemistry back in Berlin. She was Jewish but Austrian and so, for the time being, was not affected by the racial laws. She watched as former colleagues were dismissed or left on their own initiative, including Otto Frisch, her nephew, with whom she had often played piano. He was dismissed from his post in Hamburg, and so too was Leo Szilard, Hungarian-Jewish, who left for England "with his life's savings hidden in his shoes."

In 1936 Hahn and Meitner were nominated for the Nobel Prize by Max Planck, Heisenberg, and Max von Laue, apparently in an attempt to protect their Jewish colleagues.* This was already going against the grain. Meitner had been dismissed from her position at the University of Berlin in 1933, barred from speaking at scientific meetings, and all her joint discoveries with Hahn attributed to him alone. Yet she was able to engage in active research until 1938. But when the Anschluss occurred, in March 1938, her protection was removed at a stroke. Nazi sympathizers at the KWIC no longer bothered to moderate their language. The fanatical Nazi chemist Kurt Hess, who had worked next to Meitner for years, proclaimed that "the Jewess endangers this institute."[9]

As director, Hahn had to tell her she must leave, though it

---

* After Carl von Ossietzky was awarded the Nobel Peace Prize in 1936, no German was allowed to accept any of the Nobel awards.

wasn't as simple as that. Meitner was arguably the best nuclear scientist in Germany at the time, and neither Ernst Telschow, director general of the KWG, nor Carl Bosch, Nobel Prize–winning chemist, who founded IG Farben, the world's largest chemical company, wanted her to leave. Others saw the danger before she did. She hung on in Germany, despite offers to lecture in Switzerland and Copenhagen, which were disguises to enable her to escape.

When, eventually, she decided to go to Copenhagen, where Frisch was working with Bohr, it was too late: she was refused a visa for Denmark. The regime had decided it didn't want Jews going abroad and spreading bad propaganda about the Reich, and Meitner's case, we now know, had reached as high as Himmler. Peter Debye, director of the KWI in Berlin, wrote to Bohr in Copenhagen and to colleagues in Holland, looking for an opening that would seem genuine, before travel restrictions came into force.

Eventually an unsalaried post was found for her at Leiden, courtesy of the Dutch government. Even then, her problems weren't over. Only four people knew of the plan to spirit her away: Debye, Hahn, Laue, and Paul Rosbaud, a scientific advisor to the Springer publishing firm, and also a British spy. On July 12 she worked normally at the institute until 8 p.m. She then left for home and packed two small suitcases, transferring to Hahn's house, to spend the night (in case anyone should "come looking" for her) and the daytime hours of the thirteenth. After dark on the thirteenth, Hahn made up for his having to dismiss her by giving her a diamond ring so she had something to sell. Then Rosbaud drove her to the station where she met Dirk Coster, an old Dutch friend, a physicist who had been instrumental in securing the Leiden position for her. Together they traveled to the border. Here they were confronted by a Nazi military patrol of five men, who took away her Austrian passport for ten minutes. She said later that it felt like ten hours, because the passport was out of date. But the men returned and handed back the passport without a word.

It was a close-run thing, though. Kurt Hess, the fanatic who had thought that a Jewess endangered the institute, had spotted her sudden absence and sent a note to alert the authorities.[10] He was too late, and Lise Meitner's escape was to prove decisive.

In the Gothenburg Woods: Fission

The climax of post-Einstein and post-Rutherford physics was achieved by these personalities in the run-up to and in the immediate wake of the outbreak of World War II.

The first crucial result was obtained in Berlin, where Otto Hahn found that if he bombarded uranium with neutrons he repeatedly got barium. In a letter he shared these bewildering results with Meitner, who had now moved on to exile in Gothenburg, in Sweden. As luck would have it, Meitner was visited in the Christmas of 1938 by her nephew, Otto Frisch, another nuclear physicist in exile, in his case with Niels Bohr in Copenhagen. The pair went cross-country skiing in the woods. Meitner told her nephew about Hahn's letter, and they turned the barium problem over in their minds as they moved between the trees. Until then, physicists had considered that when the nucleus was bombarded, it was so stable that at most the odd particle could be chipped off. Now, huddled on a fallen tree in the Gothenburg woods, Meitner and Frisch wondered whether, instead of being chipped away by neutrons, a nucleus could in certain circumstances be cleaved in two.[11]

They had been in the cold woods for three hours. Nonetheless, they did the calculations before turning for home. What the arithmetic showed was that if the uranium atom *did* split, as they thought it might, it could produce barium (56 protons) and krypton (36): 56 + 36 = 92. As this news sank in around the world, people realized that, as the nucleus split apart, it released energy, as heat. If that energy was in the form of neutrons, and in sufficient quantity, then a chain reaction, and an atomic bomb, might be possible. But how much ^{235}U was needed?[12]

The pitiful irony of this predicament was that it was still early 1939. Hitler's aggression was growing, but the world was, technically, still at peace. The Hahn-Meitner-Frisch results were published openly in *Nature*, and thus read by physicists in Nazi Germany, in Soviet Russia, and in Japan, as well as in Britain, France, Italy, and the United States. The problem that now faced the physicists was: how likely was a chain reaction?[13]

On March 18, 1939, the French scientists, the Joliot-Curies, who had missed out on the discovery of the neutron (chapter 5), insisted on publication, in *Nature*, of their vital observation that nuclear fission emitted on average 2.42 neutrons for every neutron absorbed, meaning that energy was released in sufficient quantities to maintain a chain reaction. In Germany, the article was read by Paul Harteck, a thirty-seven-year-old chemist at Hamburg University and an expert on neutrons. Harteck, who had spent a year in Cambridge in 1932, the very year the neutron was identified, immediately recognized the implications of the paper and approached the weapons research office of the German Army Ordnance to say that a weapon of mass destruction, derived from uranium fission, was a distinct possibility.[14]

By then, Werner Heisenberg had already discussed the possibilities of an atomic bomb, and Abraham Esau, a physicist in Bernhard Rust's Education Department, had called a meeting to set up a "Uranium Club," prompted by physicists at Göttingen who also saw the potential for nuclear power in uranium. A second, more important, meeting was held at the Army Ordnance office in Berlin in September 1939, the month war broke out, at which Werner Heisenberg, Otto Hahn, Hans Geiger, Carl Friedrich von Weizsäcker, and Paul Harteck were all in attendance. Heisenberg came to dominate this group and wrote a report for Army Ordnance on the feasibility of liberating energy by controlled fission in a uranium machine. This, he said, could be used to provide heat to power tanks and submarines. A further memo also pointed out that if the uranium was sufficiently enriched in uranium-235 then the chain reaction could become a runaway process, releasing all the energy at once, and the fissile material would become an explosive "more than ten times as powerful as existing explosives." As a result the Kaiser Wilhelm Institute for Physics, in Berlin, was requisitioned for war work.[15]

This sounds decisive, but the uranium team in Germany was never to exceed a hundred members, compared with tens of thousands in the Manhattan Project at Los Alamos in the United States. Whereas Germany had the largest supply of uranium reserves, at the Joachimsthal mines in recently occupied Czechoslovakia, it had no cyclotron, with which the properties of nuclear reactions could be studied.

The development of the German bomb—or rather the pace at which its research developed—has been the subject of much controversy (not least in Michael Frayn's play *Copenhagen*). Just who knew what, and when, among German physicists has proved to be a long-running saga/mystery that has occasioned numerous books, fueled not least by the return to Germany in 2002 of certain documents seized by the Russians in 1945. These documents suggest that the Germans were further along than previous accounts suggest and show that the efforts to discredit Einstein did not affect the course of high-energy physics, even within Nazi Germany.[16]

E=mc²: Beyond Theory

But if there was a single moment when an atomic bomb moved out of the realm of theory and became a practical option, then it occurred one night in 1940, in Birmingham, England. The Blitz was in full spate, there were blackouts every night, when no lights were allowed, and at times Otto Frisch and Rudolf Peierls must have wondered whether they had made the right decision in emigrating to Britain.

Frisch, as we have seen, was Lise Meitner's nephew. While she had gone into exile in Sweden in 1938, after the Anschluss, he had remained in Copenhagen with Niels Bohr. As war approached,

Frisch grew more and more apprehensive. Should the Nazis invade Denmark, he might well be sent to the camps, however valuable he was as a scientist. Frisch was also an accomplished pianist, and his chief consolation was in being able to play. But then, in the summer of 1939, Marcus Oliphant, joint inventor of the cavity magnetometer (from which radar was developed), who by now had become professor of physics at Birmingham, invited Frisch to Britain, ostensibly for scientific discussions. Frisch packed a couple of bags, as one would do for a weekend away. Once he was in England, however, Oliphant made it clear to Frisch he could stay if he wished. The professor had made no elaborate plans, but he could read the situation as well as anyone, and he realized that physical safety was what counted above all else. While Frisch was in Birmingham, war was declared, so he just stayed. All his possessions, including his beloved piano, were lost.[17]

Peierls was already in Birmingham, and had been for some time. A Berliner, the son of a Jewish director of AEG, he was a small, bespectacled, intense-looking man, and had certainly been an intense child, breaking off from rollicking with his playmates "to think." Peierls had the classic education of prewar—even nineteenth-century—Germany, studying at several universities: Munich (under Sommerfeld), Leipzig (Heisenberg), Zurich (Pauli), later (on a Rockefeller scholarship) at Rome (Fermi) and Manchester (Bethe). Peierls was in Manchester when the purge of German universities began—he could afford to stay away, so he did. He would become a naturalized British citizen (after some difficulty) in February 1940, but for five months, from September 3, 1939, onward, he and Frisch were technically enemy aliens. They got round this "inconvenience" in their conversations with Oliphant by pretending that they were discussing only theoretical problems.

Until Frisch joined Peierls in Birmingham, the chief argument against an atomic bomb had been the amount of uranium needed to "go critical"—to start a sustainable chain reaction and cause an explosion. Estimates had varied hugely, from thirteen to forty-four tons and even to one hundred tons. It was Frisch and Peierls, walking through the blacked-out streets of Edgbaston, the leafy university quarter in Birmingham, who first grasped that the previous calculations had been wildly inaccurate. Frisch worked out that, in fact, not much more than a kilogram of material was needed. Peierls's reckoning confirmed how explosive the bomb would be. This meant calculating the available time before the expanding material separated enough to stop the chain reaction proceeding. The figure Peierls came up with was about four-millionths of a second, during which there would be eighty neutron generations (i.e., 1 would produce 2 would produce 4, 8, 16, 32 . . . and so on). Peierls worked out that eighty generations would give temperatures as hot as the interior of the sun and "pressures greater than the centre of the Earth where iron flows as a liquid."[18]

A kilogram of uranium, which is a heavy metal, is about the size of a golf ball—surprisingly small. Frisch and Peierls rechecked their calculations, and did them again, with the same results. And so, as rare as ^{235}U is in nature (in the proportions 1:139 of ^{238}U), they dared to hope that enough material might be separated out—for a bomb and a trial bomb—in a matter of months rather than years.

They took their calculations to Oliphant. He, like them, recognized immediately that a threshold had been crossed. He had them prepare a report—just three pages—and took it personally to Henry Tizard in London, who headed a committee that advised the government on scientific applications useful for warfare.

A Crucial Difference Between Heavy and Light Elements

Since 1932, when James Chadwick identified the neutron, atomic physics had been primarily devoted to obtaining two things: a deeper understanding of radioactivity, and a clearer picture of the structure of the atomic nucleus. In 1933 the Joliot-Curies, in France, had produced important work that won them the Nobel Prize. By bombarding medium-weight elements with alpha particles from polonium, they had, as we have seen, found a way of making matter artificially radioactive. In other words, they could now transmute elements into other elements almost at will. As Rutherford had foreseen, the crucial particle here was the neutron, which interacted with the nucleus, forcing it to give up some of its energy in radioactive decay.

Also in 1933 the Italian physicist Enrico Fermi had burst onto the scene with his theory of beta decay (despite *Nature* turning down one of his papers). This too related to the way the nucleus gave up energy in the form of electrons, and it was in this theory that Fermi introduced the idea of the "weak interaction." This was a new type of force, bringing the number of basic forces known in nature to four: gravity and electromagnetism, operating at great distances, and the strong and weak forces, operating at the subatomic level (this would prove important for ideas about the way the elements formed in the early universe—chapter 11). Although theoretical, Fermi's paper was based on extensive research, which led him to show that although lighter elements, when bombarded, were transmuted to still lighter elements by the emission of either a proton or an alpha particle, heavier elements acted in the opposite way. That is to say, their stronger electrical barriers *captured* the incoming neutron, making them heavier. However, being now unstable, they decayed to an element with one more unit of atomic number.[19]

This raised a fascinating possibility. Uranium was the heaviest element known in nature, the top of the periodic table, with an atomic number of 92. If *it* was bombarded with neutrons, and captured one, it should produce a heavier isotope: ^{238}U should become ^{239}U. This should then decay to an element that was entirely new, never before seen on earth, with the atomic number 93. It would take a while to produce what would be called "transuranic" elements, but when they did arrive, Fermi was awarded the 1938 Nobel Prize.[20] Since his wife was Jewish, he used the opportunity of the awards ceremony in Stockholm to go on to safety in the United States.

All these various discoveries—by Fermi, Meitner-Hahn, Frisch-Peierls—were gathering pace. So much so that in the summer of 1939 a handful of British physicists recommended that the government acquire the uranium in the Belgian Congo, if only to stop others. In America, three Hungarian refugees, Leo Szilard, Eugene Wigner, and Edward Teller, had the same idea and thought to go to Einstein, who knew the queen of Belgium, to ask her to set the ball rolling. In the end they approached President Roosevelt instead, judging that Einstein was so famous that his inquiry would not remain secret. However, an intermediary was used, who took six weeks to get in to see the president. Even then, nothing happened. It was only after Frisch and Peierls's calculations, and the three-page paper they wrote as a result, that movement began. By that stage the Joliot-Curies had produced another vital paper—showing that each bombardment of a ^{235}U atom released, on average, 3.5 neutrons. That was nearly twice what Peierls had originally thought.

Going Critical

The Frisch-Peierls memorandum was considered by a small subcommittee brought into being by Henry Tizard, which met for the first time in the offices of the Royal Society, in the east wing of Burlington House, off Piccadilly, in April 1940. This committee came to the conclusion that the chances of making a bomb in time to have an impact on the war were good, and from then on the development of an atomic bomb became British policy.

In America, a "Uranium Committee" had been established whose chairman was Vannevar Bush, a dual-doctorate engineer from MIT. Oliphant and another physicist, John Cockroft (who would share the Nobel Prize in 1951, for splitting the atomic nucleus), traveled to America and persuaded Bush to convey some of the urgency they felt to Roosevelt. Strapped by war,

Britain did not have the funds for such a project, and any location, however secret, might be bombed. Roosevelt would not commit the United States to build a bomb, but he did agree to explore whether a bomb could be built.[21]

The National Academy of Sciences report, produced as a result of Vannevar Bush's October conversation with the president, was ready in a matter of weeks and was considered at a meeting chaired by Bush in Washington on Saturday, December 6, 1941. The report concluded that a bomb was possible and should be pursued. By that stage, American scientists had managed to produce two "transuranic" elements—called neptunium and plutonium (because they were the next heavenly bodies beyond Uranus in the night sky)—and which were by definition unstable. Plutonium in particular looked promising as an alternative source of chain-reaction neutrons to ^{235}U. Bush's committee also decided which outfits in America would pursue the different methods of isotope separation—electromagnetic or by centrifuge. Once that was settled, the meeting broke up around lunchtime, the various participants agreeing to meet again in two weeks. The very next morning—December 7, 1941, the day "that will live in infamy"—the Japanese attacked Pearl Harbor, and America, like Britain, was at war. As Richard Rhodes put it, the lack of urgency in the United States was no longer a problem.[22]

The early months of 1942 were spent trying to calculate which method of ^{235}U separation would work best, and in the summer a special study session of theoretical physicists, now known as the Manhattan Project, was called at Berkeley. The results of the deliberations showed that much more uranium would be needed than previous calculations had suggested, but also that the bomb—the energy released—would be far more powerful. Bush realized that university physics departments in big cities were no longer enough. A secret, isolated location, dedicated to the manufacture of an actual bomb, was needed.

When Colonel Leslie Groves, commander of the US Army Corps of Engineers, was offered the job of finding the site, he was standing in a corridor of the House of Representatives building in Washington, DC. He exploded. The job offer meant staying in Washington, there was a war on, he'd only ever had "desk" commands, and he wanted some foreign travel. When he found that as part of the package he was to be promoted to brigadier, his attitude started to change. He quickly saw that if a bomb *were* to be produced, and it did decide the war, here was a chance for him to play a far more important role than in any overseas assignment. Accepting the challenge, he immediately went off on a tour of the project's laboratories. When he returned to Washington he singled out Major John Dudley as the man to find what was at first called Site Y. Dudley's instructions were very specific: the site had to accommodate 265 people; it should be west of the Mississippi, and at least two hundred miles from the Mexican or Canadian border; it should have some buildings already, and be in a natural bowl.

Dudley came up with, first, Oak City, Utah, but too many people needed evicting. Then he produced Jemez Springs, New Mexico, but its canyon was too confining. Further up the canyon, however, on the top of the mesa, was a boys' school on a piece of land that looked ideal. It was called Los Alamos.[23]

As the first moves to convert Los Alamos were being made, Enrico Fermi was taking the initial step toward the nuclear age in a disused squash court in Chicago. By now, no one had any doubt that a bomb could be made, but it was still necessary to confirm Leo Szilard's original idea of a nuclear chain reaction. Throughout November 1942, therefore, Fermi assembled what he called a "pile" in the squash court. This consisted of six tons of uranium, fifty tons of uranium oxide, and four hundred tons of graphite blocks. The material was built up in an approximate sphere shape in fifty-seven layers and in all was about twenty-four feet wide and nearly as high. This virtually filled the squash court, and Fermi and his colleagues had to use the viewing gallery as their office.

The day of the experiment, December 2, was bitterly cold,

below zero. That morning the first news had been received about the two million Jews who had perished in Europe, with millions more in danger. Fermi gathered with his colleagues in the gallery of the squash court, wearing their gray lab coats, "now black with graphite." The gallery was filled with machines to measure the neutron emission and devices to drop safety rods into the pile in case of emergency (these rods would rapidly absorb neutrons and kill the reactions). The crucial part of the experiment began around 10 a.m. as, one by one, the cadmium absorption rods were pulled out, six inches at a time. With each movement, the clicking of the neutron records increased and then leveled off, in sync and exactly on cue.

This went on all through the morning and early afternoon, with a short break for lunch. Just after a quarter to four, Fermi ordered the rods pulled out enough for the pile to go critical. This time the clicks on the neutron counter did not level off but rose in pitch to a roar, at which point Fermi switched to a chart recorder. Even then they had to keep changing the scale of the recorder, to accommodate the increasing intensity of the neutrons. At 3:52 p.m., Fermi ordered the rods put back in: the pile had been self-sustaining for more than four minutes. He raised his hand and said, "The pile has gone critical."[24]

The Largest Research Project in History

Intellectually, the central job of Los Alamos was to work on three processes designed to produce enough fissile material for a bomb. Two of these concerned uranium, one plutonium. The first uranium method was known as gaseous diffusion. Metal uranium reacts with fluorine to produce a gas, uranium hexafluoride. This is composed of two kinds of molecule, one with ^{238}U and another with ^{235}U. The heavier molecule, ^{238}U, is slightly slower than its half sister, so when they are passed through a filter, ^{235}U tends to go first, and gas on the far side of the filter is richer in that isotope. When the process is repeated, the mixture is even richer: repeat it often enough (several thousand times), and the 90 percent level the Los Alamos people needed can be obtained. It was an arduous process, but it worked.

The other method involved stripping uranium atoms of their electrons in a vacuum and then giving them an electrical charge that made them susceptible to outside fields. These were then passed in a beam that curved within an electrical field so that the heavy isotope would take a wider course than the lighter form, and become separated. In plutonium production, ^{235}U was bombarded with neutrons, to create a new, transuranic element, plutonium-239, which did indeed prove fissile, as the theoreticians had predicted.[25]

At its height, fifty thousand people were employed at Los Alamos on the Manhattan Project, and it was costing $2 billion a year, the largest research project in history. The aim was to produce one uranium and one plutonium bomb by late summer 1945.

The Journey of Death

On April 12, 1945, President Roosevelt died of a massive cerebral hemorrhage. Within twenty-four hours his successor, Harry Truman, had been told about the atomic bomb. Inside a month, on May 8, the war in Europe was at an end. But the Japanese hung on, and Truman, a newcomer to office, was faced with the prospect of being the man to issue the instruction to use the awesome weapon. By VE Day, the target researchers for the atomic bombs had already selected Hiroshima and Nagasaki, the delivery system had been perfected, the crews chosen, and the aeronautical procedure for actually dropping the mechanism tried out and improved. Critical amounts of uranium and plutonium became available after May 31, and a test explosion was set for 0550 hours on July 16 in the deserts of Alamogordo (near Rio Grande, the border with Mexico), in an area known locally as *Jornada del Muerto* ("the Journey of Death").

The test explosion went exactly according to plan. J. Robert Oppenheimer, the scientific director of Los Alamos, watched with his brother Frank as the clouds turned "brilliant purple" and the echo of the explosion went on and on and on. The scientists were split among themselves as to whether the Russians should be told, whether the Japanese should be warned, and whether the first bomb should be dropped in the sea nearby. In the end, total secrecy was maintained, one important reason for doing so being the fear that the Japanese might move thousands of captured American servicemen into any potential target area as a deterrent.[26]

The ^{235}U bomb was dropped on Hiroshima shortly before nine a.m. local time, on August 6. It was three meters long, twenty-eight inches in diameter, and carried sixty-four kilograms of uranium, of which only 0.6 grams were changed into energy. That energy reached 7,200 degrees Fahrenheit, killing more than 66,000 people immediately, within a radius of two miles, with 200,000 eventually succumbing from lingering ailments. In the time it took for the bomb to fall, the *Enola Gay*, the aircraft it had been carried in, had traveled eleven and a half miles away. Even so, the light of the explosion filled the cockpit, and the aircraft's frame "crackled and crinkled" with the blast.[27] The plutonium version fell on Nagasaki three days later. Six days after that, Emperor Hirohito announced Japan's surrender. In that sense, the bombs worked. And there could now be no doubt at all that E did indeed $=mc^2$.

The vindication of the unifications proposed by Einstein— between space and time and between mass and energy—had been achieved in spectacular fashion, by any measure. Moreover, by the time of Hiroshima and Nagasaki it was realized that both aspects of his ideas, an expanding universe and thermonuclear power, were themselves related, albeit distantly for the moment. Many of the physicists who worked at Los Alamos realized that the type of explosion they had helped to create must be not dissimilar to those taking place all the time in stars like the sun. In the interwar years many cosmic rays, made up of particles which did not exist naturally on earth, had been observed arriving from outer space. All this brought with it the astounding idea that the universe itself had evolved, supporting Lemaître's ideas of a Big Bang; and that the formation of, first, radiation, then particles, had generated atoms of the different elements in a certain order; and that gravity had gradually brought these early forms of matter together, to form nebulae, stars, and planets. The design and manufacture of the atom bomb, and then the creation of the hydrogen bomb, helped confirm the behavior of particles (involving complex mathematics, carried out by early computers, specially designed for the role) that would lead, in the second half of the twentieth century, to a very finely worked out cosmology, with an elaborate but consistent chronology. This was itself nothing less than the unification of physics, astronomy, and mathematics.

The path from the conservation of energy to $E=mc^2$ to the expanding universe to thermonuclear weapons to our modern understanding of the universe we inhabit was not the main purpose as to why these experiments and observations were carried out. That is why the story is so extraordinary. However breathtaking and difficult, it is intellectually consistent, part of an ordered story that would eventually be shown to have endured for nearly 14 billion years.

PART THREE

"The Friendly Invasion of the Biological Sciences by the Physical Sciences"

While Einstein's ideas were being triumphantly (and tragically) vindicated, quantum physics was—quietly at first—transforming chemistry, and opening up a link to biology that would help explain in unexpected and fresh detail Darwin's and Mendel's theories. After Bohr's linking of physics to chemistry, and Einstein's unifications, this was the next great scientific convergence of modern times. It clarified our understanding of the order underlying the sciences, but at the same time introduced fresh levels of complexity, which would resonate in important ways later in the twentieth century, when new ideas about order in the living world became a major preoccupation.

   This convergence had its origins in the 1920s, a decade of enormous growth in science research. There were two reasons for this. World War I had shown what benefits could accrue from the systematic application of research-based ideas to practical problems. This resulted in the establishment—on the part of several of the belligerents—of government science organizations, with their associated budgets. And in the United States, in particular, the Rockefeller and Carnegie organizations—devoted as they were to the betterment of mankind—realized, because of the war, that science offered the surest path to successful results. Harvard, MIT, Columbia, Johns Hopkins, and Caltech all benefited from these new arrangements.

Two men in particular who were part of this trend are relevant for us. The first is Arthur Amos Noyes. A chemist who had trained in Germany and was for two years president of MIT, before moving to Caltech, Noyes had strong views not just about chemistry and chemists but also about how science should be run. He was more than adept at championing chemistry at Caltech, and at obtaining sufficient funds for the highest-quality research, but there was one weak spot—organic chemistry. And this is where the second man came in. Warren Weaver was "a second-rate scientist with a first-rate knack for knowing the right people." Because of those contacts, he was in 1932 plucked from the relative obscurity of a teaching job in Wisconsin to run the Natural Sciences Division of the Rockefeller Foundation, "the single most important scientific funding agency in the world. He would have the power to open new areas of research, make or break careers, dispense millions of dollars, change the course of scientific history."

When he took up his appointment, Weaver was especially enthusiastic about a new form of biology. Like Noyes, Weaver believed it was necessary for the methods of the more "successful" natural sciences—mathematics, physics, and chemistry—to be applied to biology. He called this "the friendly invasion of the biological sciences by the physical sciences." Impressed by the way quantum physics was helping to explain chemistry—as shown, for example, in the work of Linus Pauling—Weaver was convinced that the new biology was going to change the way mankind thought about the living world. Whereas the old biology focused on whole organisms, "molecular biology"—the new phrase he dreamed up to describe this new discipline—would concentrate on the "unknown world" inside individual cells.

Weaver had some ambitious ideas. He told the Rockefeller trustees that molecular biology would eventually help explain "violence, unhappiness, irrationality, and sexual problems." The Rockefeller trustees found his arguments—bloated though some of them were—irresistible, and from that point on, "the Rockefeller Foundation stopped awarding grants for mathematics, physics and chemistry that did not relate directly to the life sciences."

As part of this, there was also an ambitious plan for the Rockefeller Foundation to cooperate with Niels Bohr's institute in Copenhagen. Like Weaver, Bohr and his colleagues believed that physics and chemistry were more advanced than biology and that the new techniques developed by physicists could help biologists investigate such processes as metabolism and organic growth at the molecular and atomic level. In particular, the Copenhagen scientists were interested in using radioactive isotopes to follow the detailed behavior of chemicals in living organisms. Radioactive phosphorus, carbon, and heavy water were thought to be especially promising in this regard.

But, in fact, it didn't really work out. The progress of nuclear physics, Bohr's more long-term interest in Copenhagen, was too rapid, too exciting—and too threatening—in the second half of the 1930s. Bohr played a central role in the development of atomic and nuclear physics in the run-up to World War II, but the centers of excellence in molecular biology did not include Copenhagen. The major advances were made elsewhere, in two other places—Caltech and the Cavendish Laboratory. The work there epitomized the transition—intellectual and geographical—that took place in biology in the 1930s and 1940s.

Despite what did or didn't happen in Copenhagen, this transition—the effects of the physical sciences' invasion of biology at Caltech and the Cavendish—cannot be exaggerated.

CALTECH AND THE CAVENDISH: FROM ATOMIC PHYSICS TO MOLECULAR BIOLOGY VIA QUANTUM CHEMISTRY

What was Warren Weaver so excited about? In order to understand, we need now to back up. In science the twentieth century had opened with a bang. In physics, the electron, the quantum, radioactivity, radio waves, and X-rays had all just been identified. In chemistry the dyestuffs and pharmaceuticals revolutions (themselves linked) were maturing; and in psychology Sigmund Freud had revealed the unconscious, a revolutionary notion that changed the way people thought about mental illness and about themselves. There were equivalent developments in biology.[1]

Even after all this time, the coincidence in the rediscovery of the work of the botanist-monk Gregor Mendel makes for moving reading. Between October 1899 and March 1900, three other botanists—two Germans (Carl Correns and Erich von Tschermak) and the Dutchman Hugo de Vries—published papers about plant biology, each of which (in a footnote) referred to Mendel's priority in discovering the principles of what we now call genetics.[2] Thanks to this coincidence, and their scrupulousness in acknowledging his achievement, Mendel—once forgotten—is now a household name.[3]

Johann Gregor Mendel was born in 1822 in Heinzendorf in what was then Austria and is now Hynice in the Czech Republic. In 1843, Johann entered the Augustinian monastery in Brno, where he adopted the name Gregor. Mendel had no real Christian vocation, but the environment freed him economically and gave him the peace of mind he needed to pursue his studies. As this makes him sound, he was in some ways a paradoxical figure. He was not a playful genius like Picasso, say. Instead, he was very timid, suffering badly from exam fright, for example, but he was passionate about learning, and dogged. At one point he kept wild mice in his monastery rooms, breeding them to see how their coat color varied. But the abbot objected to the smell and the fact that a monk who had taken a vow of celibacy was conducting experiments involving rodent sex. Whereupon, as Mendel put it, "I turned from animal breeding to plant breeding. You see, the bishop did not understand that plants also have sex."[4]

He began experimenting with peas. The results for which we remember him were the fruit of ten years of "tedious experiments" in plant growing and crossing, seed gathering, careful labeling, sorting, and counting—almost thirty thousand plants were involved. As the *Dictionary of Scientific Biography* notes: "It is hardly conceivable that it could have been accomplished without a precise plan and a preconceived idea of the results to be expected."[5] In other words, his experiments were designed to test a specific hypothesis.

From 1856 to 1863, Mendel cultivated seven pairs of characteristics, suspecting that heredity "is particulate," contrary to the ideas of "blending inheritance" to which many others subscribed. This was an important insight, which would in time link up with advances in chemistry and quantum physics. Mendel observed that, with seven pairs of characteristics, in the first generation all hybrids are alike—and the parental characteristics (e.g., round seed shape) are unchanged. This characteristic he called "dominant." The other characteristics (e.g., angular shape), which only appear in the next generation, he called "recessive." What he called "elements" determine each paired character and pass in the germ cells of the hybrids, *without influencing each other.* In hybrid progeny both parental forms appear again and this, he realized, could be represented mathematically/statistically, *A* denoting dominant round seed shape, and *a* denoting the recessive angular shape. Were they to meet at random, he said, the resulting combination would be:

$$\tfrac{1}{4}AA + \tfrac{1}{4}Aa + \tfrac{1}{4}aA + \tfrac{1}{4}aa$$

After 1900, this was known as Mendel's law (or principle) of segregation and can be simplified mathematically as:

$$A + 2Aa + a$$

He also observed that with seven alternative characteristics 128 associations were found—in other words, 2^7. He therefore concluded that the "behaviour of each of the different traits in a hybrid association is independent of all other differences in the two parental plants." This principle was later called Mendel's law of independent assortment.[6] It was another illustration of the particulate nature of matter, which Planck and Bohr were explaining in their respective fields, though it was not realized as such at the time.

The Link Between Cytology and Mendel's Mathematics

Mendel's employment of large populations of plants was new, and it was this which enabled him to extract "laws" from otherwise apparently random behavior—statistics had come of age in biology. He attempted to sum up the significance of his work in *Versuche ber Pflanzenhybriden* (1866). This memoir, his magnum opus and one of the most important papers in the history of biology, was the foundation of genetic studies. It was never truly appreciated at the time because he had difficulty following up his pea work—his experiments with bees failing because of the complex problems involved in the controlled mating of queen bees. He did show that hybrids of *Matthiola, Zea,* and *Mirabilis* "behave exactly like those of *Pisum*" but colleagues such as Carl Nägeli, to whom he wrote a series of letters, remained doubtful.[7]

Mendel *had* read *On the Origin of Species.* A copy of the German translation, with Mendel's marginalia, is preserved in the Mendelianum in Brno. These marginalia show his readiness to accept the theory of natural selection. Darwin, however, never seems to have grasped that hybridization provided an explanation as to the causes of variation.* As a result, Mendel died a lonely, unrecognized genius.

The rediscovery of Mendel was not without its controversial side. None of the three men who discovered that Mendel had beaten them to it—de Vries, Correns, and von Tschermak—can have been wholly happy to find themselves runners-up. Moreover, in his first paper on the subject de Vries did not mention Mendel, but as his paper had appeared before Correns's—just— Correns, miffed, lost no time in pointing out that de Vries had used the very same terms, "dominant" and "recessive," that Mendel had used, and so de Vries was forced to own up to the fact that he *had* read Mendel's work.

In fact, the dispute between Correns and de Vries sparked rather more than heat over their rival claims to priority: there has been a wholesale reevaluation of Mendel himself. Augustine Brannigan, in an account published in 1981, argued that the general view—that Mendel was a genius who was overlooked for more than three decades—is quite wrong. He showed that many of Mendel's findings had been reported previously—in some cases a long time previously—by such figures as Thomas

---

* Although one of the rare books which refers to Mendel, S. O. Focke's *Die Pflanzen Mischlinge* (1881), did pass through Darwin's hands.

Andrew Knight, John Goss, Alexander Seton, Augustin Sageret, and Johann Dzierzon. All of these individuals, however, worked in the fields of hybridization and their findings were published in such journals as the *Transactions of the Horticultural Society of London*. Mendel, Brannigan shows, was aware of their work and his own contribution, far from being revolutionary, continued in the line they had established, which sought to show that new species could—though with difficulty—be created by crossbreeding.

None of which changes the fact that, since de Vries's rediscovery of Mendel's laws, the basic mechanism of heredity has been confirmed many times. However, the approach taken by Mendel and de Vries was statistical, centering on that 3:1 ratio in the variability of offspring. The more the ratio was confirmed, the more people realized there had to be a physical, biological, and cytological grounding for the mechanism identified. And there was one structure that immediately suggested itself. For about fifty years, biologists had been observing under the microscope a certain characteristic behavior of cells undergoing reproduction. They saw a number of minute threads forming part of the nuclei of cells, which separated out during reproduction. Walther Flemming had already identified chromatin (in 1882), and Hermann Henking, in 1891, had identified the "X" body, without suspecting it might determine sex, though others soon drew this conclusion (see chapter 6).[8]

After Henking's observation, it was also soon confirmed that the same chromosomes appear in the same configuration in successive generations. And Walter Sutton, at Columbia, showed in 1902 that during reproduction, similar chromosomes come together, then separate. In other words, chromosomes behaved in exactly the way Mendel's law suggested.[9]

The Vinegar Fly Aligns Mendel with Darwin

But the man who coined the term "genetics" was William Bateson. Bateson was a large, "stoop-shouldered" man, a don at St

John's College, Cambridge, a zoologist with a handlebar mustache, and one of the chief combatants in the end-of-century controversy over evolution. Not at all a Little Englander, Bateson considered anyone who couldn't converse in French a philistine and subscribed to German newspapers to maintain his colloquial skills in that language. He invented the word "genetics" in 1905 when he was asked to devise a plan for a new Institute for the Study of Heredity and Variation at Cambridge. That plan came to nothing but he was more successful in getting the Royal Horticultural Society—sponsor of the International Conference on Plant Breeding and Hybridization—to change its name, for its third conference, in 1906, to the International Conference on Genetics. The term "gene" itself came four years later, coined by Wilhelm Johannsen, a professor of plant physiology at Copenhagen Agricultural College, who also came up with "phenotype" and "genotype."[10]

A year after his success with getting the conference name changed, Bateson was in America to deliver a lecture series, when he met Thomas Hunt Morgan at Columbia University. Morgan was already making his name as a leading genetics researcher in the United States, but the two men didn't hit it off. "T. H. Morgan is a thickhead," Bateson once wrote to his wife, but it wasn't true, far from it. Morgan, as much by luck as judgment, had hit on a particular organism for study, much as Mendel had hit on peas. In Morgan's case it was the so-called fruit fly, *Drosophila melanogaster*, more properly called the vinegar fly because of its preference for overripe fruit "with some tang to it."[11] Morgan studied this fly because his quarters in Columbia were cramped and already filled with live pigeons, chickens, starfish, yellow mice, and rats. The fruit fly was cheap to feed and house—all that was needed were some ripe bananas and some milk bottles for breeding. Morgan bought the bananas, but he and his assistants "swiped" empty milk bottles from the stoops of upper Manhattan on their morning walk to work.

The advantage of vinegar flies is that they have nice big

chromosomes, which could be easily seen under the primitive microscopes available to Morgan, and they reach sexual maturity within a week, with the females bearing several hundred baby flies at a time. They are an experimenter's dream.

Even so, it took a while for the fruit flies to bear fruit, so to speak. For two years Morgan subjected his flies to toxins, chemicals, and X-rays in the hope of inducing some interesting genetic change. Only after that time did he notice, one day, that, among a crowd of normal red-eyed vinegar flies, there appeared a single fly whose eyes were white.

It would take a week for the fly to reach sexual maturity, and that week was a difficult time. Morgan brought it home each evening, in a jar which he kept by his bed while, as it happened, his wife was in hospital having a baby. According to the family legend, when Morgan went to visit his wife, she would ask, "How is the white-eyed fly?" well before he would ask, "How is the baby?"[12]

The following week the white-eyed fly was ready to breed. Over the next few months, Morgan and his team mated thousands and thousands of flies in their laboratory at Columbia University (this is how the "fly room" got its name). The sheer bulk of Morgan's results enabled him to conclude that mutations formed in vinegar flies at a steady pace. By 1912, more than twenty recessive mutants had been discovered, including one they called "rudimentary wings" and another which produced "yellow body colour." But that wasn't all. The mutations only ever occurred in one sex, males or females, never in both. This observation, that mutations are always sex-linked, was significant because it supported the idea of *particulate* inheritance. The only *physical* difference between the cells of the male fly and the female lay in the "X body." It followed, therefore, that the "X body" *was* a chromosome, that it determined the sex of the adult fly, and that the various mutations observed in the fly room were also carried on this body.

Morgan published a paper on *Drosophila* as early as July 1910 in

Science, but the full force of his argument was made in 1915 in *The Mechanism of Mendelian Heredity*, the first book to properly air the concept of the "gene." For Morgan and his colleagues the gene was to be understood as "a particular segment of the chromosome, which influences growth in a definite way and therefore governs a specific character in the adult organism."[13] Morgan argued that the gene was self-replicating, transmitted unchanged from parent to offspring, mutation being the only way new genes could arise, producing new characteristics. Most importantly, mutation was a random, accidental process that could not be affected in any way by the needs of the organism. As Darwin had argued.

There were of course complications. Morgan conceded, for example, that a single adult characteristic can be controlled by more than one gene, while at the same time a single gene can affect several traits. Also important was the position of the gene on the chromosome, since its effects could occasionally be modified by neighboring genes.

But genetics had come a long way in fifteen years, and not just empirically but philosophically too. In some ways the gene was a more potent fundamental particle than either the electron or the atom, since it was far more directly linked to man's humanity. The accidental and uncontrollable nature of mutation as the sole mechanism for evolutionary change, under the "indifferent control of natural selection," was considered by critics—philosophers and religious authorities—as a bleak imposition of banal forces without meaning, yet another low point in man's descent from the high ground he had occupied when religious views had ruled the world.

But for people like Morgan and Warren Weaver, the gene was a world all to itself. Its structure might be fiendishly complicated, and other substances would need to be deciphered along the way—but this was where the secret of life lay, and no doubt the answer to many diseases as well. Just as atomic, nuclear, and quantum physics underlay chemistry, so quantum chemistry underlay molecular biology. And where else should Weaver look for world-class researchers other than Caltech, which had Linus Pauling, "the leading theoretical chemist of the world," and T. H. Morgan, the ranking authority on the gene, who was awarded the Nobel Prize in 1933?

The Evolutionary Synthesis

There was one other development that made genetics especially exciting in the 1930s. This related to four theoretical books, all published between 1937 and 1944, and thanks to which several nineteenth-century notions were finally laid to rest.

Between them, these studies created what is now known as the "evolutionary synthesis," which brought together our modern understanding of how evolution actually works. In chronological order these books were: *Genetics and the Origin of Species*, by Theodosius Dobzhansky (1937); *Evolution: The Modern Synthesis*, by Julian Huxley (1942); *Systematics and the Origin of Species*, by Ernst Mayr (also 1942); and *Tempo and Mode in Evolution*, by George Gaylord Simpson (1944). The essential problem they all sought to resolve was this: following the publication of Charles Darwin's *On the Origin of Species* in 1859, two of his theories were accepted relatively quickly, but two others were not. The idea of evolution itself—that species change—was readily grasped, as was the idea of "branching evolution," that all species are descended from a common ancestor. What was not accepted so easily was the idea of gradual change, or of natural selection as an engine of that change. In addition, Darwin, in spite of the title of his book, had failed to provide an account of speciation, how new species arise. All this made for three major areas of disagreement.[14]

The main arguments may be described as follows. First, many biologists believed then in "saltation"—that evolution proceeded not gradually but in large jumps. Only in this way, it was thought, could the great differences between species be accounted for. If evolution proceeded gradually, why wasn't this reflected in the fossil record? Why weren't "halfway" species ever found? Second,

there was the notion of "orthogenesis," that the direction of evolution was somehow preordained, that organisms somehow had
a final destiny toward which they were evolving. And third, there
was a widespread belief in "soft" inheritance, better known as
the inheritance of acquired characteristics, or Lamarckism. Julian
Huxley, grandson of T. H. Huxley, "Darwin's bulldog," and the
brother of Aldous, author of *Brave New World*, was the first to use
the word "synthesis," but he was the least original of the four.
What the others did between them was to bring together the latest
developments in genetics, cytology, embryology, paleontology,
systematics, and population studies, to show how the new discoveries fitted together under the umbrella of Darwinism.

Ernst Mayr, a German émigré who had been at the American
Museum of Natural History in New York since 1931, pointed
out how, contrary to what you might expect, even distinguished
early geneticists, like Hugo de Vries and T. H. Morgan himself, did not fully grasp evolution. In particular they failed to
appreciate the nature of species as *populations*.[15] Mayr argued
that the traditional view, that species consist of large numbers
of individuals and that each conforms to a basic archetype, was
wrong. Instead, species consist of populations, clusters of unique
individuals where there is *no* ideal type. For example, the human
races around the world are different, but also alike in certain
respects; above all, they can interbreed. Mayr advanced the view
that, in mammals at least, major geographical boundaries—like
mountains or seas—are needed for speciation to occur, for then
different populations become separated and begin developing
along separate lines. Again, as an example, this could be happening with different races, and may have been happening for several
thousand years—but it is a gradual process, and the races are still
nowhere near being "isolated genetic packages," which is the
definition of a species.

Dobzhansky, a Russian who had escaped to New York just
before Stalin's Great Break in 1928, to work with T. H. Morgan, covered broadly the same area but looked more closely at

genetics and paleontology. He was able to show that the spread of different fossilized species around the world was directly related to ancient geological and geographical events. Simpson, Mayr's colleague at the American Museum of Natural History, who studied in London and later became Alexander Agassiz professor at Harvard, looked at the pace of evolutionary change and the rates of mutation. He was able to confirm that the known rates of mutation in genes produced sufficient variation sufficiently often to account for the diversity we see on earth.[16]

An Attempted Marriage of Physics and Biology: "The Commonality of Life at the Molecular Level"

To begin with, Linus Pauling was not as keen on Warren Weaver's plans as he might have been, and he was slow to take up molecular biological research. He only did so when he became convinced that that was where the new research money was (which Weaver had been insisting on). By that time, moreover, Morgan, who had moved to Caltech from Columbia in 1928, was in his seventies and ready to retire. Pauling therefore teamed up with one of Morgan's brighter students, George Beadle.

We have seen already (chapter 5) that Pauling and George Wheland had extended their resonance research to organic substances, beginning with benzene and other aromatic compounds. The investigation of hemoglobin was one of the next moves the two men made. It was attractive for a number of reasons. For a start it was a protein, the most important class of molecules in the body (hair, skin, muscle, and tendons are proteins, as are the most important parts of the nerves and blood). Enzymes are proteins and so are chromosomes. "If there was a secret of life, it was thought, it would be found among the proteins."[17]

Practically, however, they were a nightmare. They were huge molecules, sometimes comprising tens of thousands of atoms. At least hemoglobin offered some advantages. Its supply was plentiful and it could be crystallized, meaning it must have some sort of regular structure. And, from Pauling's point of view, it was related to porphyrin, which he had studied for some years. This molecule had an unusual shape—it appeared to be a ring made up of smaller rings—and it was found right across nature, binding oxygen in the chlorophyll of plants and the hemoglobin of animals. "Porphyrin seemed to epitomize the molecular biology idea of the commonality of life at the molecular level: It showed up almost everywhere there was life."[18]

Pauling was inching forward in his study of several proteins, not just porphyrin and hemoglobin (he remained convinced that proteins were central to understanding biology), but in 1944 an event across the Atlantic had a marked effect on the whole discipline of molecular biology.

What Is Life?, published in Britain by Erwin Schrödinger, was not part of the evolutionary synthesis, but it played an equally important role in pushing biology—and the convergence—forward.

Schrödinger, as we have seen (chapter 5), was born in Vienna, in 1887, and worked as a physicist at the university there after graduating, then in Zurich, Jena, and Breslau, before succeeding Max Planck as professor of theoretical physics in Berlin. He had been awarded the 1933 Nobel Prize for his part (along with Werner Heisenberg and Paul Dirac) in the quantum mechanics revolution also considered in chapter 5. In the same year that he won the Nobel, however, Schrödinger left Germany in disgust at the Nazi regime. He had been elected a fellow of Magdalen College, Oxford, and taught in Belgium, but in October 1939 he moved on to Dublin, since in Britain on the outbreak of war he would have been forced to contend with his "enemy alien" status. Ireland remained neutral throughout World War II.

An added attraction of Dublin was its brand-new Institute for Advanced Studies, modeled on the IAS in Princeton and the brainchild of Éamon de Valera, the Irish *Taoiseach*, or prime minister. Schrödinger agreed to give the statutory public lectures for 1943 and took as his theme an attempted marriage between

physics and biology, especially as it related to the most fundamental aspects of life itself and heredity. He made it clear that
although science had become diversified, and "it has become next
to impossible for a single mind fully to command more than a
small specialized portion of it," he could see no other escape from
this dilemma "than that some of us should embark on a synthesis
of facts and theories . . . even at the risk of making fools of [our]
selves."

In the lectures, Schrödinger attempted two things. He first
considered how a physicist might define life. The answer he gave
was that a life system was one that took order from order, "drinking orderliness from a suitable environment." Such a procedure,
he said, could not be accommodated by the second law of thermodynamics, with its implications of entropy, and so he forecast
that although life processes would eventually be explicable by
physics, the explanation would be based on new laws of physics,
unknown at that time. Perhaps more interesting, and certainly
more influential, was his other argument. This was to look at the
hereditary structure, the chromosome, from the point of view of
the physicist.[19]

In the mid-1940s most biologists were unaware of both quantum physics and the latest developments in the chemical bond.
(Schrödinger had been in Zurich when Fritz London and Walter
Heitler discovered the bond; no reference is made in *What Is Life?*
to Linus Pauling.) Schrödinger showed that, from the physics
already known, the gene must be "an aperiodic crystal," that is, "a
regular array of repeating units in which the individual units are
not all the same." In other words, it was a structure half-familiar
already to science. He explained that the behavior of individual
atoms could only be known statistically. Therefore, for genes to
act with the very great precision and stability that they did, they
must be of a minimum size, with a minimum number of atoms.
Again using the latest physics, he showed that the dimensions of
individual genes along the chromosome could therefore be calculated (the figure he gave was 300 Å, or angstrom units). From

that, both the number of atoms in each gene and the amount of energy needed to create mutations could be worked out. The rate of mutation, he said, corresponded well with these calculations, as did the discrete character of the mutations themselves, which recalled the nature of quantum physics, where intermediate energy levels do not exist.[20]

All this was new for most biologists in 1943–44, but Schrödinger went further, inferring that the gene must consist of a long, highly stable molecule that contains a code. He compared this code to the Morse code, in the sense that even a small number of basic units would provide great diversity. Schrödinger was thus the first person to use the term "code," and it was this, and the fact that physics had something to say about biology, that attracted the attention of biologists and made his lectures and subsequent book so influential. On the basis of his reasoning, Schrödinger concluded that the gene must be "a large protein molecule, in which every atom, every radical, every heterocyclic ring, plays an individual role." The chromosome is a message written in code.

The book proved very influential: it has been described as "probably the most important work of biology written by a physicist." Timing also had something to do with the book's influence. Not a few physicists were turned off their own subject by the development of the atomic bomb. At any rate, among those who read *What Is Life?*, and were excited by its arguments, were Francis Crick, James Watson, and Maurice Wilkins, the first two at the Cavendish Laboratory in Cambridge, and the other in London.

Pauling was not so impressed. In particular he did not accept Schrödinger's view that life would be explained by *new* laws of physics, unknown at the time. He took this view because his research had shown that some organic substances formed crystals, which must mean that they obeyed the same "complementarity" rules that inorganic substances did and that, in enzymes' catalytic actions and in reproduction, his own idea of complementarity would also play a role, something not considered by Schrödinger.

But, even so, Pauling now at last did throw himself

wholeheartedly into molecular biology, and as the mid-1940s gave way to the late 1940s, he thought more and more about reproduction. In one lecture at that time, he said: "In general the use of a gene or virus as a template would lead to the formation of a molecule not with an identical structure but with a complementary structure. . . . If the structure that serves as a template (the gene or virus molecule) consists of, say, two parts, which are themselves complementary in structure, then each of these parts can serve as the mold for the production of a replica of the other part, and the complex of two complementary parts thus can serve as the mold for the production of duplicates of itself."[21] He was getting close, very close. But, for the moment, Pauling thought no more along these lines, and turned his interest to medicine.

The Boundaries and Borderlines of Chemistry and Biology

Although Warren Weaver had conceived the phrase "molecular biology" in 1938, in Britain it took time to catch on. This was not because the British were against the conjoining of either sciences or of the terms that described them. In 1938, J. D. Bernal had tried (and failed) to establish an Institute of Mathematico-Physico-Chemical Morphology in Cambridge, and a "central bureau of protein research" a year later, believing that the structure of proteins was "the major unsolved problem on the boundary of chemistry and biological substances."[22] In fact, a number of terms were used to describe broadly similar fields. Biomolecular research was one, "vital processes" another, and biophysics a third, more popular alternative. The "physics of life," as opposed to the "physics of death" (the atomic bomb), was especially popular in the immediate postwar world. An Institute of Biophysics had been proposed as early as 1944.

In Britain, biophysics comprised three different groups: the radiation group, which investigated the effects of radiation on the body and ways to protect it; the "nerve-muscle" group, which exploited new recording devices developed in the context of radar research; and the "structural group," which used a series of physical techniques, especially X-ray diffraction, decisively aided by the advent of electronic computers, to study complex biological structures.[23] Blood and its products had been extensively studied in wartime because of their relevance to battlefield casualties, and in Cambridge Max Perutz, an Austrian-born émigré, focused on hemoglobin.

Because of the war, and the development of operational research, a host of new instruments had helped to transform the life sciences: ultracentrifuges, spectroscopes, electron microscopes, and heavy and radioactive isotopes.[24] These included the first computers, which Pauling had early access to, though Britain soon had its own Hollerith machines, marketed by the British Tabulating Machine Company.

At Cambridge, the genetics department was led by R. A. Fisher, who was much involved in the statistical and mathematical approach to biology, and so it was not all that unusual when a Mathematical Laboratory was established in the university to offer computer services to X-ray crystallographers. It made use of an EDSAC (electronic delay storage automatic calculator), which itself followed the design of the stored-program computer, the EDVAC (electronic discrete variable automatic computer), described by John von Neumann at the IAS in Princeton (see chapter 17).[25]

In April 1946, Lawrence Bragg at the Cavendish had applied to the Royal Society for funds to be made available for "borderline subjects" (i.e., on the edge of at least two different traditional disciplines) to provide assistance for Max Perutz in his investigations of crystalline proteins (which may have seventy thousand atoms) by means of X-ray analysis.[26] Perutz, a graduate of Vienna University, had come to Cambridge to learn crystallography from Bernal who, a few years before, had obtained the first X-ray pictures of a crystalline protein.[27] Perutz later settled on hemoglobin as a research project for his PhD.

Bragg was criticized by many colleagues for backing "weird"

subjects like biophysics instead of nuclear physics which had provided a "grand tradition" in the past. But there were so many new instruments available now, after the war, that Bragg realized a greater flexibility had to be the way forward.[28]

Helixes: The Spontaneous Order of Proteins

In Britain, where there was a long tradition of X-ray crystallography, a view was beginning to form that long-chain molecules, whether in proteins or amino acids, or peptides, naturally (spontaneously) formed—for efficiency's sake—into spirals or, as they came to be called, helixes. Pauling was more than familiar with the helix concept (in fact, this was another Pauling coinage, after a doctoral fellow, Jack D. Dunitz, used it one day). His idea applied especially to proteins, and in 1948–49 Pauling had charged a student with reviewing all possible spiral models.[29] Pauling soon concluded that only two forms of spiral met the X-ray and chemical evidence. The tighter of the two numbered roughly 3.7 amino acids in each turn of the spiral, the looser one 5.1 amino acids. (This was when model building was in full flood; spheres and small rods were carved out of wood, and painted different colors.) Pauling labeled the tighter spiral the "alpha helix" and the looser one the "gamma helix," and showed that each individual amino acid along the chain should account for about 1.5 angstroms of the spiral's length. This level of accuracy was quite an achievement in itself, and in a similar vein, Pauling and his coworkers came up with models for various kinds of protein—collagen, gelatin, muscle. Pauling's protein work, says Thomas Hager, his biographer, had changed the landscape.

In the summer of 1951, Pauling turned his attention to DNA. This was the most common form of nucleic acid in chromosomes, and there was historical work showing it to be a long-chain molecule with a repeating pattern of just four subunits called nucleotides. This is, of course, exactly what Schrödinger had said was needed for reproduction, but Pauling had little time for the

German. (Or he said he did; it might have been pique because Schrödinger had not mentioned Pauling in *What Is Life?*) But DNA was not then the priority it appears in retrospect. DNA was an important component of chromosomes, but so too was protein, and it seemed to many researchers that protein was most likely to carry the genetic instructions.[30]

The only evidence to the contrary was a paper published in 1944 by Oswald Avery at the Rockefeller Institute in New York, who had found that DNA, "apparently by itself," could transfer new genetic traits between *Pneumococcus* bacteria. Pauling knew about this research, just as he knew Avery personally, but didn't attach much importance to his result, and so didn't follow it up.

One move Pauling did make was to write to Maurice Wilkins at King's College London, who had probably the best crystallographic photographs of DNA anywhere. Pauling was told about them by a visiting fellow at Caltech, who had been studying the effects of water on DNA.

When Wilkins received Pauling's letter in the late summer of 1951, he was unsure what to do. Wilkins knew that Pauling was a better scientist than he was, but he also knew that the best thing he had ever done in science until that point was to produce the best DNA crystal X-rays. So he was loath to part with them. He held onto Pauling's letter for a week, and then wrote saying that what he had was not good enough to release for the time being. Pauling wouldn't be put off just yet and wrote to Wilkins's superior. J. T. Randall wrote back more honestly, confirming that Wilkins and others were working on DNA, but he said it wouldn't be fair to them to let Pauling have sight of their material.[31]

Still, Pauling wasn't too fazed. He knew he worked faster than most people, and he knew too that he was due to attend a special meeting of the Royal Society in London in May 1952. It was not far off. He could afford to wait.

Biology Becomes as "Exact" as Physics and Chemistry

Pauling had won every race he had entered. DNA would be different. The first the public knew about the discovery, and that it wasn't Pauling who had made it, came on April 25, 1953, in *Nature*, in a nine-hundred-word paper entitled "Molecular Structure of Nucleic Acids." The paper followed the familiar, ordered layout of *Nature* articles. But although it was the paper that put molecular biology on everybody's lips, not just those of molecular biologists, it was the culmination of an intense two-year drama in which, if science really were the carefully ordered world that it is supposed to be, the wrong side won. And the focus of the drama was not Caltech but Cambridge and London.

Among the personalities, Francis Crick stands out: a "loud-mouth with a braying laugh," and an "irritating habit of doing other people's crosswords." To his fellow Cambridge colleague, the critic George Steiner, he had a voice "one goes grouse-shooting with." In *Who's Who*, after he was famous, he listed his recreation as "conversation, especially with pretty women."[32] So he was no shrinking violet. Born in Northampton in 1916, the son of a shoemaker (Northampton used to be famous for its shoe making), Crick graduated from London University (where he wore "exotic" suede shoes and was known as "Crackers"). During World War II, he worked at the admiralty, where he was interviewed by C. P. Snow and designed mines.

It was only in 1946, when Crick attended a lecture by none other than Linus Pauling, that his interest in biochemical research was kindled. He was also influenced by Erwin Schrödinger's *What Is Life?*, and its suggestion that quantum mechanics might be applied to biology. In 1949 he was taken on by the Cambridge Medical Research Council Unit at the Cavendish Laboratory, where he soon became known for his loud laugh and his habit of firing off theories on this or that at the drop of a hat (he had several jousts with Ludwig Wittgenstein).[33]

In 1951 an American joined the lab. James Dewey Watson,

the son of a bill collector, was a tall, stick-thin, socially ill-at-ease Chicagoan, twelve years younger than Crick but extremely self-confident intellectually. A child prodigy, he had also read Schrödinger's *What Is Life?*—while he was a zoology student at the University of Chicago—which influenced him toward microbiology. While working at Copenhagen, where he had been sent to learn nucleic acid chemistry, and unsatisfied in the Danish capital, he had traveled to Cambridge. Max Perutz recalled: "A strange head with a crew cut and bulging eyes popped through my door and asked me without so much as saying 'Hallo': 'Can I come and work here?'"[34]

As Paul Strathern tells the story, on a visit to Europe Watson had met Maurice Wilkins, at a scientific conference in Naples. Wilkins, described by one of his students as "a peevish, slightly old-maidy young man," then based at King's College in London, had worked on the Manhattan Project in World War II but became disillusioned and turned to biology (he too had read *What Is Life?*).[35] The British Medical Research Council had a biophysics unit at King's, which Wilkins then ran. One of his specialities, as we have seen, was X-ray diffraction pictures of DNA, and in Naples he generously showed Watson some of the results. It was this coincidence that shaped Watson's life. Then and there he seems to have decided that he would devote himself to discovering the structure of DNA. He knew there was a Nobel Prize in it, that molecular biology could not move ahead without such an advance, but that once the advance was made, the way would be open for genetic engineering, a whole new era of human experience. He arranged a transfer to the Cavendish. A few days after his twenty-third birthday, Watson arrived in Cambridge.

The history of DNA was not unlike that of the gene. Just as Mendel's insight was overlooked for thirty-five years, so too was "nuclein" neglected by geneticists after its discovery in 1869 in Tübingen by Friedrich Miescher. He purified a phosphorus-rich, acidic substance from the "pus-soaked bandages of wounded soldiers," calling it nuclein because it seemed to be ever-present in cell nuclei. Later, after moving to Basel, he got even purer samples from salmon eggs. Nuclein was renamed deoxyribose nucleic acid and later deoxyribonucleic acid, or DNA, and for most of the early twentieth century it was regarded as a form of scaffolding on which genes rested. It obviously had a sort of monotonous structure and was a large molecule, though Oswald Avery, at the Rockefeller Institute in New York, showed it had some of the properties of a gene in that it could change the nature of an organism and was heritable.[36]

When Watson arrived at the Cavendish, what he didn't know, however, was that the lab had a "gentleman's agreement" with King's. The Cambridge laboratory was studying the structure of protein, in particular hemoglobin, while London was studying DNA. That was only one of the problems. Although Watson hit it off immediately with Crick, and both shared an amazing self-confidence, that was virtually all they had in common. Crick was weak in biology, Watson in chemistry. Neither had any experience at all of X-ray diffraction, the technique developed by the leader of the lab, Lawrence Bragg, to determine atomic structure. None of this deterred them. The structure of DNA fascinated both men so much that virtually all their waking hours were spent discussing it—gossiping about it—mainly in the so-called RAF bar of the Eagle pub in Bene't Street, just off King's Parade, and about one hundred yards from the Cavendish.

As well as being self-confident, Watson and Crick were both competitive. Their main rivals came from King's, where Maurice Wilkins had recently hired the twenty-nine-year-old Rosalind Franklin ("Rosy," though never to her face). She was described as the "wilful daughter" of a cultured Jewish banking family, and her great-uncle Lord Samuels was a former home secretary who had drafted the memorandum that led to the Balfour Declaration and hence the creation of Israel.[37] She had just completed four years' X-ray diffraction work in Paris and was one of the world's top experts. When Franklin was hired by Wilkins, she thought she was to be his equal and that she would be in charge of the

X-ray diffraction work. Wilkins, on the other hand, thought that she was coming as his assistant. The misunderstanding did not make for a happy ship.

Despite this, Franklin made good progress and in the autumn of 1951 decided to give a seminar at King's to make known her findings. Remembering Watson's interest in the subject at their meeting in Naples, Wilkins invited the Cambridge man. At this seminar, Watson learned from Franklin that DNA almost certainly had a helical structure, each helix having a phosphate-sugar backbone, with attached bases: adenine, guanine, thymine, or cytosine. After the seminar, Watson (who described Franklin as a bluestocking with very red lipstick) took her for a Chinese dinner in Soho. There the conversation turned away from DNA to how miserable she was at King's. Wilkins, she said, was reserved, polite, but cold. In turn, this made Franklin on edge herself, a form of behavior she couldn't avoid but detested. At dinner Watson was outwardly sympathetic, but he returned to Cambridge convinced that the Wilkins-Franklin relationship "would never deliver the goods."

The Watson-Crick relationship meanwhile flourished, and this too was not unrelated to what happened subsequently. Because they were so different—in age, cultural and scientific background, and although they were in competition with King's—there was precious little rivalry between them. And because they were so conscious of their great ignorance on so many subjects relevant to their inquiry (they kept Pauling's *The Nature of the Chemical Bond* by their side, as a bible), they could slap down each other's ideas without feelings being hurt. It was light-years away from the Wilkins-Franklin ménage, and in the long run that may have been crucial.

In the short run there was disaster. In December 1952, Watson and Crick thought they had an answer to the puzzle, and invited Wilkins and Franklin to spend the day in Cambridge, to show them the Pauling-type wooden model they had built: a triple helix structure with the bases on the outside. Franklin savaged them, curtly grumbling that their model didn't fit any of her crystallography evidence, either for the helical structure or the position of the bases, which she said were on the *inside*. Nor did their model take any account of the fact that in nature DNA existed in association with water, which had a marked effect on its structure. She was genuinely appalled at their neglect of her research and complained that her day in Cambridge was a complete waste of time. For once, Watson and Crick's ebullient self-confidence let them down, even more so when word of the debacle reached the ears of their boss. Bragg called Crick into his office and put him firmly in his place. Crick, and by implication Watson, was accused of breaking the gentleman's agreement, and of endangering the lab's funding by doing so. They were expressly forbidden from continuing to work on the DNA problem.[38]

So far as Bragg was concerned, that was the end of the matter. But he had misjudged his men. Crick did stop working on DNA, but, as he told colleagues, no one could stop him *thinking* about it. Watson, for his part, continued work in secret, under cover of another project on the structure of the tobacco mosaic virus, which showed certain similarities with genes.

A new factor entered the situation when Peter Pauling, Linus's son, arrived at the Cavendish to do postgraduate research. He attracted a lot of beautiful women, much to Watson's satisfaction, but, more to the point, he was constantly in touch with his father and told his new colleagues that Linus had reached the point where he was putting together a model for DNA. Watson and Crick were devastated.

In fact, fate had intervened to their advantage. As we have seen, Pauling had been expecting to come to London earlier that year for a meeting on proteins at the Royal Society. But, owing to his reputation as an outspoken anti-nuclear pacifist, the US State Department had been prevailed upon by Senator Joe McCarthy to revoke his passport. Had Linus traveled to London as originally planned, he would surely have been able to refine his understanding of DNA.[39] As it was, when an advance copy of his paper arrived in Cambridge, Watson and Crick immediately saw that it had a fatal flaw. It described a triple-helix structure, with the bases on the outside—much like their own model that had been savaged by Franklin. In addition, Pauling had left out the ionization, meaning his structure would not hold together. Watson and Crick knew that it would be only a matter of time before Pauling himself realized his error, and they estimated they had six weeks to get in first. They took a risk, broke cover, and told Bragg what they were doing. This time he didn't object. There was no gentleman's agreement so far as Linus Pauling was concerned.

So began the most intense six weeks Watson and Crick had ever lived through. They now had permission to build more models (models were especially necessary in a three-dimensional world) and had developed their thinking about the way the four bases—adenine, guanine, thymine, and cytosine—were related to each other. They knew by now that adenine and guanine were attracted, as were thymine and cytosine. They knew this thanks to the work of the other prominent biochemist working on DNA in America, Erwin Chargaff at Columbia. Chargaff had found out one extremely interesting fact about DNA's nitrogenous bases— that, although the proportion of each base varied from species to species, there was usually a neat symmetry. The amount of adenine was the same as the amount of thymine, and the amount of cytosine was the same as the amount of guanine. Chargoff didn't know what this implied, but Crick did.

Also, by now, from Franklin's latest crystallography, which gave much more accurate measurements of DNA's dimensions, Crick and Watson had a far better picture of their subject. In the photographs, they could see that, at the top and the bottom, there were very dense black smudges, which lay "precisely on layer line 10." This meant that each helix must have ten nucleotides per turn—ten phosphates, ten sugars, and ten bases. Since they now knew that the distance between two nucleotides was 3.4 angstroms, the pitch of the helix was not 27 angstroms but 34. This made for better model building. (Watson and Crick admitted that Pauling had been their inspiration for model building.)[40]

There were two "eureka" moments. The first was Crick's and the second Watson's. Crick's came when he realized that the repeating patterns occurred after 360 degrees, not 180 degrees as they had thought, making the whole structure more comfortable. The final eureka moment came when Watson realized they could have been making a simple error by using the wrong isomeric form of the bases. Each base came in two forms—*enol* and *keto*—and all the evidence so far had pointed to the *enol* form as being the correct one to use. But what if, as his colleague Jerry Donohue suggested, the *keto* form was tried? As soon as he followed this line, Watson immediately saw that the bases fitted together *on the inside*, to form the perfect double-helix structure. Even more important, when the two strands separated in reproduction, the mutual attraction of adenine to guanine, and of thymine to cytosine—their complementarity, as Pauling would have put it—meant that the new double helix was identical to the old one; the biological information contained in the genes was passed on unchanged, as it had to if the structure was to explain heredity.

They announced the new structure to their colleagues on March 7, 1953, and six weeks later their paper appeared in *Nature* (the order of their names was decided by a coin toss). Wilkins, says Strathern, was charitable toward Watson and Crick, calling them a couple of "old rogues." Franklin instantly accepted their model. Not everyone was as emollient. They were called "unscrupulous" and told they did not deserve the sole credit for what they had discovered.[41]

The drama was not yet over. In 1962, the Nobel Prize in physiology or medicine was awarded jointly to Crick, Watson, and Wilkins; and in the same year the prize for chemistry went to the head of the Cambridge X-ray diffraction unit, Max Perutz, and his assistant, John Kendrew, who had by then discovered the structure of myoglobin and hemoglobin. This meant that molecular biology had really come of age.

Rosalind Franklin got nothing. She had befriended the Cricks after the initial *froideur*, to the point where she convalesced with them after her surgery for ovarian cancer. But in 1958 that cancer killed her, at the age of thirty-seven.

In the summer of 1953, the physicist George Gamow wrote to Watson and Crick. "Your article," he said, "brings biology over into the group of 'exact' sciences."[42] As we shall now see, molecular biology (the study of proteins and nucleic acids) and evolution from here on rivaled particle and quantum physics as the dominant ideas in late twentieth-century science. And there is surely a case to be made for Linus Pauling, despite his hiccup over DNA, to be regarded as the most intellectually influential scientific figure of the modern world, the individual who did more to promote the convergence than anyone else.

10

BIOLOGY, THE "MOST UNIFYING" SCIENCE: THE SWITCH FROM REDUCTION TO COMPOSITION

The great switch to molecular biology involved more than the emergence of a brand-new and extremely powerful science. The structure of DNA was in itself as important as the discovery of the gene, the atom, the quantum, relativity, $E=mc^2$, and the Big Bang, because it was as *fundamental*. On top of which it was related much more closely to our humanity than most of the other entities. At the same time, although it was fundamental, it was hardly a simple structure. Everyone could see that molecular biology was as complex as it was important.

The way that the science had itself evolved, however, the way that atomic and nuclear physics had given rise to quantum chemistry, and then to molecular biology, naturally reinforced—and very powerfully—the reductionist idea that all the sciences were, at bottom, interconnected.

This theme was taken up at precisely this time—the late 1950s—in a seminal paper by two renowned philosophers of science, Paul Oppenheim and Hilary Putnam, in a paper entitled "Unity of Science as a Working Hypothesis." It was in some ways a reprise of the "Unity in Science" movement associated with the Vienna Circle. As Neurath had said, "An all-embracing vision and thought is an old desire of humanity." But of course science had moved on.

Paul Oppenheim (1885–1977) was a heroic figure—a chemist and a philosopher who grew up in Germany and witnessed the growing anti-Semitism. In 1933, when Hitler came to power, his elderly parents committed suicide together and he emigrated, first to Belgium then America. In Princeton he became friendly with Einstein and, being of independent means, used his funds to help less fortunate scientists escape the Nazis. Hilary Putnam (1926–2016) was born in Chicago but grew up in France, with a Communist father and Jewish mother. He went on to become professor of philosophy at Harvard and president of the American Philosophical Society.

In their paper the two men were anxious to show that, although the unity of science had not at that stage been achieved, "it does not follow, as some philosophers appear to think, that a tentative acceptance of the hypothesis that unitary science can be attained is therefore a mere 'act of faith'. We believe that this hypothesis is *credible*."

The assumption of unitary science, they said, recommended itself as a working hypothesis, because "we believe that it is in accord with the standards of reasonable scientific judgment to tentatively accept this hypothesis and to work on the assumption that further progress can be made in this direction, without claiming that its truth has been established, or denying that success might finally elude us."[1]

They began by arguing that there are six "reductive levels":

6. social groups
5. (multicellular) living things
4. cells
3. molecules
2. atoms
1. elementary particles

At the higher levels, they said, the reduction from human societies had not gone very far, partly because well-established theoretical knowledge at level six was still "rather rudimentary." However, in regard to certain very primitive groups of organisms, "astonishing successes" had been achieved.[2] "For instance, the differentiation into social castes among certain kinds of insects has been tentatively explained in terms of the so-called social hormones." And, they added, many scientists believed that there are some laws common to all forms of animal association, including that of humans (anticipating ethology, which science was in the process of establishing itself). Further, the most developed body of social theory concerning humans was economics "and this is at present entirely micro-reductionist in character."[3] Marx, Veblen, and Weber had all attempted some sort of theory reduction for economic determinism, none of which, Oppenheim and Putnam conceded, was universally accepted.

At the level of cell activity, they thought that the work of neurologists was especially promising—neuroanatomy, neurochemistry, and neurophysiology, including electroencephalography. As a result it was now possible, at the least, to advance theories in relation to memory, motivation, emotional disturbance, and such phenomena as learning, intelligence, and perception. Nerve networks, as originally proposed by Alan Turing, were potentially useful ideas, they said, with propositional logic perhaps finding parallels in such networks, though the models then in use, consisting of 10^4 elements, were a long way short of the 10^{10} neurons believed to comprise the brain. Purposeful behavior might be controlled by feedback mechanisms.

The explanation and reduction of genetics was the main subject tackled in the paper, certainly in terms of the amount of space devoted to it, and, given the recent rise of molecular biology, it is not hard to see why. The central elements, Oppenheim and Putnam said, were decoding, duplication, and mutation, and a reductive theory was needed for each. The problem of decoding reduces to an explanation of how the specific molecules that comprise the nucleus serve to specify the construction of specific protein catalysts. The problem of duplication "reduces to how the molecules of genetic material can be copied—like so many 'blueprints.' And the problem of mutation . . . reduces to how 'new' forms of genetic molecules can arise."[4] In this regard they looked with favor on the theory advanced by Max Delbrück as long ago as 1940—that different quantum levels within the atoms of the molecule correspond to different hereditary characteristics and that a mutation is simply a quantum jump of a rare type (with a high activation energy). (See chapter 19.) They felt that catalytic activity, so important in molecular biology, would be solved "in terms of quantum theory as it now exists."

Most important, they felt that evolution itself provides *indirect* factual support for the working hypothesis that unitary science "is attainable." Evolution, they said, is an "over-all phenomenon involving all levels from 1 through 6," and they went on to say that timescales have been worked out "by various scientists" showing the epochs when the first examples of each level appeared.[5]

Evolution as applied to the cosmos was also a fruitful undertaking, they argued, outlining how the first elementary particles may have formed (the subject of chapter 11 of this book). Reducibility of molecular and atomic phenomena "is today not open to doubt."

"To this day," they added, controversies existed concerning the dividing line between living and non-living things. "In particular, viruses are classified by some biologists as living, because they exhibit self-duplication and mutability." However, most biologists at that stage refused to apply the term "living" to viruses because they "exhibit these characteristic phenomena of life only due to activities of a living cell with which they are in contact."[6]

Wherever one draws the line, however, they insisted that "non-living molecules preceded primordial living substance and the latter evolved gradually into highly organized living units, the unicellular ancestors of all living things." And they quoted

Richard Benedict Goldschmidt (1878–1958), one of the first geneticists to theorize about evolution: "The first complex molecules endowed with the faculty of reproducing their own kind must have been synthesized—and with them the beginning of evolution in the Darwinian sense—a few billion years ago: all the facts of biology, geology, paleontology, biochemistry and radiology not only agree with this statement but actually prove it."

Cells come only from cells, as Virchow said, but the cellular "slime moulds" studied by John Tyler Bonner at Princeton showed that, at a certain stage, isolated amoebae clump together chemotactically (movement to or away from a chemical stimulus) and form simple multicellular organisms, a "sausage-like slug" that crawls with comparative rapidity and good coordination, and may be attracted by light. Young children begin as unsocialized, egocentric individuals and are only capable of social behavior—displaying concern for the welfare of others, for instance—later. This argument was chopping and changing, yet of a piece.

Further support for their hypothesis, they said, came from the *syntheses* that can take place in physics, chemistry, biology, and sociology. One can obtain an atom by bringing together the appropriate elementary particles: deuterium can be obtained by bombarding protons with neutrons (in, for example, hydrogen gas). In chemistry, under the influence of an electric spark the union of oxygen and hydrogen produces molecules of H_2O. On the borderline of life, while success at synthesizing a virus out of atoms was not yet in sight, synthesis out of nonliving highly complex macromolecules had been accomplished. Protein obtained from viruses was mixed with nucleic acid to obtain an active virus. Although non-infectious (so it was not *quite* true to life), the reconstituted virus had the same structure as a "natural" virus, and could produce the tobacco mosaic virus disease when applied to plants.

New human groups are forming all the time, they said—Boy Scouts, trades unions, professional associations, Israel. Synthesis is creative.

The Natural Hierarchy of the Sciences

In conclusion: "The possibility that all science may one day be reduced to microphysics (in the sense in which chemistry seems today to be reduced to it), and the presence of a unifying trend toward micro-reduction running through much of scientific activity, have often been noticed by specialists. . . . But these opinions have, in general, been expressed in more or less vague manner, and without very deep-going justification." Repeating what they said at the beginning, they offered at the end of their essay that such a hope is not a mere act of faith but that, "on the contrary, a tentative acceptance of this belief, an acceptance of it as a working hypothesis, is justified." It is credible, they said, "because there is really a large mass of direct and indirect evidence in its favour."[7] They even said there may be a "natural order" of sciences, that physics conventionally deals with their levels 1, 2, and 3, and biology with, at least, levels 4 and 5.

Of course, they said, the six levels, and the disciplines linked to them, are ideal creations: there is no discipline or activity concerned with molecules alone, say. The order from 1 to 6 is continuous, and the individual levels may not be settled for all time. But the continuous order, 1–6, is in the end Darwinian, "which is how it should be understood." And, to end, they called in aid a statement from Ludwig von Bertalanffy: "Reality, in the modern conception, appears as a tremendous hierarchical order of organized entities, leading, in a superposition of many levels, from physical and chemical to biological and sociological systems. Unity of Science is granted, not by any utopian reduction of all sciences to physics and chemistry, but by the structural uniformities of the different levels of reality."[8]

None of which really nails the overall aesthetic/emotional quality of their exercise. Surely Sir Arthur Eddington put it best. He is quoted in a footnote: "We shall not rest satisfied until we are able to represent all physical phenomena as an interplay of a vast number of structural units intrinsically alike."

A New Principle: Adaptive Organized Complexity

The importance of the advent of molecular biology was marked by a series of essays by George Gaylord Simpson, one of the creators of the evolutionary synthesis. The essays were gathered together in a book published in 1964, *This View of Life: The World of an Evolutionist.*[9] Simpson recognized the identification of the structure of DNA as a critical moment, not just in the history of science but in the history of the world.

The fact that biology had now entered the realm of the "exact" sciences, as George Gamow had appreciated, was the culmination of the process that Pauling and Schrödinger had begun, and Simpson clearly saw biology as taking over from physics as—in his words—the "focal point" of all science.[10] This was so, he said, because "if life is the most important thing about our world, the most important thing about life is its evolution."[11] Organic evolution, he insisted, is one of the "basic facts" about the world, because it is the process "by which the universe's greatest complexities arise and systematic organization culminates." He felt that this realization had been impeded by "the mistaken opinion that realism or objectivity requires reduction of biological phenomena to the physical level."[12]

At the same time, he argued that biology itself is the "most unifying" of sciences: that all living things are truly "physically related" in just the same way as parents and children and brothers and sisters although in very different degrees; that the doctrine of geological uniformitarianism widens the recognized reign of natural law; that evolution is "inherent" in the physical properties of the universe; that the natural mechanisms that gave rise to man "are now largely known and are probably entirely knowable in terms of the immanent physical laws of the universe"; and that our special human abilities are made possible "by the evolutionary intensification of awareness." The universe *is* orderly, he insisted, and its immanent processes have been always unchanging.[13]

He thought that the primacy of the physical sciences was

historical. The first sciences—"as we now strictly define science"—were the physical sciences, dating from a time when scientists considered themselves to be also, "or even primarily," philosophers; when "physics" was long synonymous with "natural philosophy." He thought it was a traditional "reductionist half-truth" that all phenomena were "ultimately explicable" in strictly physical terms, though other factors had been the prestige of the technology based on physics and that better minds had gone into the physical sciences than elsewhere. He didn't go the whole hog, though, dismissing such people as Sir James Jeans who, he thought, accepted the whole idea of orderlessness and acausality in the universe with what amounted to "mystical glee."[14] In fact, he thought that Bacon's idea of the unity of nature and Einstein's attempts to seek the unification of scientific concepts "in the form of principles of increasing generality" to have been by and large "worthy and fruitful."[15]

His central point was that biology offered the opportunity to add a new principle to our understanding. In addition to reduction, explanation in terms of physical, chemical, or mechanical principles, biology—which is more complicated than physics—invites us to introduce a second kind of explanation, what he called "compositionist." We need to understand structures not just in reductionist terms but also in terms of the adaptive usefulness of the structures and processes in the whole organism and the species of which it is part. And, still further, in terms of the ecological function in the communities in which the species occurs.[16]

The Crucial Level of Molecular Biology

His further, allied, point was that in the scale running from subatomic particles to multi-specific communities, a sharp "dichotomy somewhere along that scale" occurs at the level of molecular biology. This, and the *organization* of the organism, is the crucial—*central*—fact in science, from now on. Speaking of reduction in this context, he concluded, "I suggest that both the characterization of science as a whole and the unification of the various sciences can be most meaningfully sought in quite the opposite direction, not through principles that apply to all phenomena but through phenomena to which all principles apply." [17]

Expanding, he conceded that, for example, food chains in a community could be explained successively by the adaptations of specific populations, by the functioning of individual members, by the enzymes initiating and mediating the underlying chemical and physical properties, by DNA specification of the enzymes, by the structure of the DNA molecules, and, ultimately, by the atoms comprising such molecules. But he insisted that no predictions could be made on the basis of the atomic actions. Instead, he said, it is just as explanatory and just as essential for understanding to say that lion enzymes digest zebra meat *because* that enables the lion to survive and perpetuate the population of which he is a part. "One direction of explanation goes down the scale . . . the other goes up . . . reduction and composition." [18]

He had no doubt that the discovery of genetics and DNA was deepening the reductionist understanding, but so too were paleontological discoveries deepening the compositionist understanding, regularities constantly and repeatedly occurring in the fossil record. [19]

Simpson's book was as welcome as his contribution to the evolutionary synthesis. He identified clearly the shift from physics to biology and showed why it was important. Physics, as we shall see, didn't go away, but from now on biology no longer suffered from "physics envy." If anything, evolutionary theory invaded all manner of phenomena, in a way it had never done before.

Simpson was on the ball here too. As he put it, "I recently had to point out to some ethnologists that culture in general is biological adaptation and that they could resolve some of their squabbles and find the common theoretical basis that eludes them if they would just study culture from this point of view. The suggestion was not well received, but it is true just the same." One might say that physics imperialism was now being accompanied by biology imperialism. The new sciences of ethology and sociobiology were just around the corner.

More than that, though, in his essays Simpson had essentially put the arguments over reductionism to bed. His understanding of evolution was second to none and his idea of "composition" was so well informed and so clearly correct in detail that it ought to have caught on and been accepted much more quickly and widely than it was. It did catch on, eventually. But not then.

PART FOUR

The Continuum from Minerals to Man

While the first half of the twentieth century had seen the main sequence of the "hard" sciences—physics, chemistry, and biology—coming together, the second half would see many other aspects of the world around us being incorporated into one interlaced pattern, which together formed the greatest, most wide-ranging, coherent narrative there could ever be.

After World War II, when science had proved even more decisive for victory than in World War I, there was an explosion in the funding of science, helped by the advent of the Cold War, which maintained budgets at unprecedented levels. There were many knock-on effects of this. To give just one example, the United States needed to know about the ocean between it and its great rival, the Soviet Union. As a result, the Pacific and its rim were studied as never before. Oceanography, marine biology, hydrology, geology, climatology, even anthropology and archaeology benefited. Military (applied) research produced some stunning advances in fundamental science.

On top of that, the boost given to evolution by the advances in molecular biology meant that more and more areas of human activity were now viewed from an evolutionary perspective. One of the most important developments was that scientific advance enabled a more accurate understanding of past events—events in the universe, on earth, the emergence of life, of civilization, a new form of history: "Big History"—which showed that there was a coherent continuum through the ages. And as it became ever clearer that one story was unfolding, more and more it became the case that, as Patricia Churchland, the philosopher of science, put it, "It is now evident that where one discipline ends and the other begins no longer matters."

11

PHYSICS + ASTRONOMY = CHEMISTRY + COSMOLOGY: THE SECOND EVOLUTIONARY SYNTHESIS

After World War II and the Manhattan Project, most physicists were anxious to get back to "normal" work. Quite what normal work was now was settled at two big physics conferences: one at Shelter Island, off the coast of Long Island, near New York, in June 1947; and the other at Rochester, upstate New York, in 1956.

The high point of the Shelter Island Conference, on "The Foundations of Quantum Mechanics," was a report by Willis Lamb that presented evidence of small variations in the energy of hydrogen atoms that should not exist if Paul Dirac's equations linking relativity and quantum mechanics were absolutely correct. Lamb, a Californian, was a graduate of Berkeley, where he did his PhD under Robert Oppenheimer, director of the Manhattan Project. This "Lamb shift," outlined at the conference to Hans Bethe, Richard Feynman, and others, produced a revised mathematical account—quantum electrodynamics (QED)—that scientists have hailed as "the most accurate theory in physics." QED explains the way electrically charged particles interact with

one another and with magnetic fields, via the swapping of photons. John Gribbin describes this as the "jewel in the crown" of
quantum physics, "a theory that has been tested experimentally to
a large number of decimal places, and has passed every test." The
mathematics involved, as well as the ambition, would, no doubt,
have impressed even Mary Somerville.

Basically, QED explains all of chemistry: how explosions
occur, how a spring stretches, why the sky is blue. "Outside the
nucleus, on the scale of atoms and above, all that matters is QED
and gravity." Again quoting Gribbin, the classic illustration of
QED is its prediction of the magnetic momentum of the electron. Dirac's original equations produced a value of exactly 1, but
experiments produced a value of 1.00115965221, with an uncertainty of 4 in the final digit. QED, on the other hand, predicts
a value of 1.00115965246, with an uncertainty of 20 in the last
two digits. This is an accuracy of one part in ten decimal places
(0.00000001 percent). Richard Feynman brought its significance
home when he said it was equivalent to measuring the distance
from New York to Los Angeles with an accuracy to the width
of a human hair and that it was "the most precisely determined
agreement between theory and experiment for any theory and
any experiment ever carried out on Earth."[1]

In the same year as the conference, mathematically and physically trained cosmologists and astronomers began studying more
systematically than before cosmic rays arriving on earth from
the universe. They discovered new atomic particles that did not
behave exactly as predicted—for example, they did not decay into
other particles as fast as they "should" have done. This anomaly
gave rise to the next phase of particle physics, which dominated
the last half of the twentieth century; an amalgam of physics,
math, chemistry, astronomy, and—strange as it may seem—
history. Its two achievements were an understanding of how the
universe was formed, how and in which order the elements came
into being, and a systematic classification of particles even more
basic than electrons, protons, and neutrons.

Nucleosynthesis: The New Order of Elements and Particles

The study of elementary particles quickly leads back in time, to the very beginning of the universe. As we have seen, the "Big Bang" theory of the origin of the universe began in the 1920s with the work of Georges Lemaître and Edwin Hubble (chapter 8). Following the Shelter Island Conference, two Austrian émigrés in Britain, Hermann Bondi and Thomas Gold—together with Fred Hoyle, a professor at Cambridge—advanced a rival "steady state" theory, which envisaged matter being quietly formed throughout the universe, in localized "energetic events."

This was never taken seriously by more than a few scientists, especially as in the same year George Gamow, a Russian who had defected to the United States in the 1930s, presented new calculations showing how nuclear interactions taking place in the early moments of the fireball could have converted hydrogen into helium, explaining the proportions of these elements in very old stars. Gamow also said that there should be evidence of the initial explosion in the form of background radiation, at a low level of intensity, to be picked up wherever one looked for it in the universe.

Gamow's theories, as he developed them in his book *The Creation of the Universe* (1952), especially his chapter on "The Private Life of Stars," helped initiate a massive interest among physicists in "nucleosynthesis"—the ways in which the heavier elements are built up from hydrogen, the lightest element, and the role played by the various forms of elementary particles. This—in effect the evolution of the universe—is where the proliferating study of cosmic rays came in. Almost none of the new particles discovered since World War II exist naturally on earth, and they could only be studied by accelerating naturally occurring particles in particle accelerators and cyclotrons to make them collide with others. These were very large, very expensive pieces of equipment, and this is another reason why "Big Science" flourished most in America. Not only was it ahead intellectually, but America more than elsewhere had the appetite and the wherewithal to fund such ambition.[2] (And, in the Manhattan Project, a lot of relevant experience.)

Scores of particles were discovered in the decade following the Shelter Island Conference, but three stand out. The particles that did not behave as they should have done according to the earlier theories were christened "strange" by Murray Gell-Mann at Caltech in 1953 (the first example of a fashion for whimsical names for entities in physics). Various aspects of strangeness came under scrutiny at the second physics conference, in Rochester, in 1956, and were brought together by Gell-Mann in 1961 into a classification scheme for particles. This was a new order reminiscent of the periodic table (chapter 3), and which he called, maintaining the whimsy, the "Eightfold Way."

Gell-Mann was a New Yorker and yet another child prodigy. The son of Jewish-Austrian immigrants, he entered Yale when he was fifteen and had his PhD by the time he was twenty-two. After a year at the IAS at Princeton, he moved to Chicago, to work under Fermi, and then to Caltech, where he became friendly with—but also a competitor to—Richard Feynman. John Gribbin quotes an (unnamed) colleague of both who said: "Murray was clever, but you always felt that if you weren't so lazy and worked really hard, you could be as good as him; nobody ever felt that way about Dick."[3] In some ways Gell-Mann had a strange career ("strange" being the crucial word), because he made three major discoveries but always at much the same time as other physicists. Yet more simultaneous discovery.

The first occurred in 1953 when both Gell-Mann and the Japanese physicist Kazuhiko Nishijima (Osaka-based, but he had worked under Werner Heisenberg at Göttingen) independently hit upon the idea of explaining certain of the properties of the many fundamental particles that were being discovered at the time by assigning to them a quality called "strangeness." This was because such particles had, as mentioned earlier, strangely long lifetimes compared with other particles (though these lifetimes were still measured in minute fractions of a second).

Pondering this, in the early 1960s Gell-Mann and, this time, Yuval Ne'eman, an Israeli physicist, soldier, and politician—again both working independently—devised a classification scheme for particles which, as noted, came to be called the Eightfold Way. The Eightfold Way was based on mathematics rather than observation, and in 1964 mathematics led Gell-Mann (and, almost simultaneously, George Zweig) to introduce his third discovery, the concept of the "quark"—a particle more elementary still than electrons and from which all known matter is made. (Zweig called them "aces," but he was a student at the time. Gell-Mann was a professor, and the professor's "quark"—taken from James Joyce's *Finnegans Wake*—stuck.) The existence of quarks was not confirmed experimentally until 1977.[4]

Quarks come in six varieties, and were given entirely arbitrary names such as "up," "down," or "charmed." They have electrical charges that are fractions—plus or minus one-third or two-thirds of the charge on an electron—and it was this fragmentary charge that was so significant, further reducing the building blocks of nature. We now know that all matter is made up of two kinds of particle. "Baryons"—protons and neutrons, fairly heavy particles—are divisible into quarks. "Leptons," the other basic family, are much lighter, consisting of electrons, muons, the tau particle, and neutrinos, which are *not* broken down into quarks. A proton, for example, is comprised of two "up" quarks and one "down" quark, whereas a neutron is made up of two "down" quarks and one "up."

What follows in the next few paragraphs may be confusing to non-physicists, but keep in mind that the elementary particles that exist naturally on earth are exactly as identified in 1932: the electron, the proton, and the neutron. All the rest are found only in cosmic rays, arriving from space, or in the artificial circumstances of particle accelerators, which seek to approximate the conditions of the early universe.

Baryons are particles affected by the strong nuclear interaction and "baryonic material" is now often used to refer to "everyday atomic matter," made up of protons, neutrons, and electrons.[5]

Quarks and the particles they combine to form are known as hadrons, and they feel the strong force, though they can also be involved in weak interactions. The leptons feel the weak force but are never involved in strong interactions.

Then there are four bosons which carry the forces of nature: gluons, which carry the strong force; intermediate vector bosons, which carry the weak force; photons, bearers of electromagnetism; and gravitons, carriers of gravity.

Hadrons are particles that interact through the strong force and are therefore made up of quarks. Protons and neutrons are hadrons but so are a number of unstable particles, including the massless gluons that keep quarks together.[6]

It was the main aim of physicists to amalgamate all these discoveries into a grand synthesis that would have two elements. The first would explain the evolution of the universe, describe the creation of the chemical elements and their distribution among the planets and stars, and explain the creation of carbon, which had made life possible. The second would explain the fundamental forces that enable matter to form in the way that it forms. God apart, it would in effect explain everything.

Patterns in the Early Universe

In the first half of the twentieth century, apart from the development of the atomic bomb and relativity, the main achievement of physics was its unification with chemistry (as epitomized in the work of Niels Bohr and Linus Pauling). After the war, the discovery of yet more fundamental particles, especially quarks, brought about an equivalent unification—between physics and astronomy. The result of this consilience, as it would be called, was a much more complete explanation of how the heavens—the universe—began and evolved. It was, for those who do not find the reference blasphemous, an alternative Genesis.

Quarks, as we have just seen, were originally proposed by Murray Gell-Mann and George Zweig, almost simultaneously, in

1964. It is important to grasp that quarks do not exist in isolation in nature (at least on earth), but the significance of the quark (and certain other particles, isolated later) is that it helps explain conditions in the early moments of the universe, just after the Big Bang. The idea that the universe began at a finite moment in the past was accepted by most physicists, and many others, after Hubble's discovery of the red shift in 1929, but the 1960s saw renewed interest in the topic. This was partly as a result of Gell-Mann's theories about the quark but also because of an accidental discovery made in 1965 at the Bell Telephone Laboratories in New Jersey.

Since the year before, the Bell Labs had been in possession of a new kind of telescope. An antenna located on Crawford Hill at Holmdel, New Jersey, communicated with the skies via the *Echo* satellite. This meant the telescope was able to "see" into space without the distorting interference of the atmosphere, and that far more of the skies were accessible. As their first experiment, the scientists in charge of the new telescope, Arno Allan Penzias (Munich-born and yet another Jewish refugee from Nazi Germany) and the Texan Robert Woodrow Wilson, decided to study the radio waves being emitted by our own galaxy. This was essentially baseline research, the idea being that once they knew what pattern of radio waves *we* were emitting, it would be easier to study similar waves coming from elsewhere.[7]

Except that it wasn't that simple. Wherever they looked in the sky, Penzias and Wilson found a persistent source of interference—like static. At first they thought there was something wrong with their instruments. A pair of pigeons was nesting in the antenna, with the predictable result that there were droppings everywhere. The birds were captured and sent to another part of the Bell complex. They came back. This time, according to Steven Weinberg's account published later, they were dealt with "by more decisive means."[8] With the antenna cleaned up, the "static" was reduced, but only minimally, and it still appeared from all directions. Penzias discussed this mystery with another radio astronomer at MIT, Bernard Burke. Burke recalled that a colleague of his, Ken Turner of the Carnegie Institute of Technology, had mentioned a talk he had heard at Johns Hopkins University in Baltimore given by a young Canadian theorist from Princeton, P. J. E. Peebles, which might bear on the "static" mystery. Peebles's speciality was the early universe.

This was a relatively new field and still very speculative. As we saw a moment ago, George Gamow had begun to think about applying the new particle physics to the conditions that must have existed at the time of the Big Bang. He started with "primordial hydrogen," which, he said, would have been partly converted into helium, though the amount produced would have depended on the temperature of the Big Bang. He also said that the hot radiation corresponding to the enormous fireball would have thinned out and cooled as the universe expanded. He went on to argue that this radiation "should still exist in a highly 'red-shifted' form, as radio waves." This idea of "relict radiation" was taken up by others, some of whom calculated that such radiation should now have a temperature of 5K (i.e., five degrees above absolute zero). Curiously, with physics and astronomy only just beginning to come together, no physicist appeared to be aware that even then radio astronomy was far enough ahead to answer that question. So the experiment was never done. And when radio astronomers at Princeton, under Robert Dicke, began examining the skies for radiation, they never looked at the coolest kinds, not being aware of their significance. It was a classic case of the right hand not knowing what the left was doing. (Not something Mary Somerville would have allowed to happen.)[9]

When Peebles, from Winnipeg, started his PhD at Princeton in the late 1950s, he worked under Robert Dicke. Gamow's theories had been forgotten, or even overlooked, but, more to the point, Dicke himself seems to have forgotten his own earlier work. The result was that Peebles unknowingly repeated all the experiments and theorizing of those who had gone before. He arrived at the same conclusion—that the universe should now be filled with "a sea of background radiation" with a temperature of only a few K. Dicke, who either still failed to remember his earlier experiments or didn't realize their significance, liked Peebles's reasoning enough to suggest that they build a small radio telescope to look for the background radiation.

At this point, with the Princeton experiments ready to start, Penzias called Peebles and Dicke, an exchange that became famous in physics. Comparing what Peebles and Dicke knew about the evolution of background noise, and the observations of Penzias and Wilson, the two teams decided to publish in tandem a pair of papers in which Penzias and Wilson would describe their observations while Dicke and Peebles gave the cosmological interpretation—that this was indeed the radiation left over from the Big Bang. Within science this created almost as huge a sensation as the confirmation of the Big Bang itself. And it was this duo of papers, published in the *Astrophysical Journal*, which caused most physicists to finally accept the Big Bang theory. In 1978, Penzias and Wilson received the Nobel Prize in physics, even though they didn't know what they had discovered when they had first observed it.[10]

During the awards ceremony, the head of the Nobel physics committee said that the background radiation discovery had "turned cosmology [into] a science, open to verification and observation."[11] In fact, it was to be Peebles who did more than the others in the years that followed. He went on to calculate the amount of helium and deuterium that would have been produced in the Big Bang, and noted that the visible disc of stars in a galaxy such as our own Milky Way is extremely unstable, could barely survive a single rotation, and therefore can only be held together by a spherical halo of *dark matter*, with a mass of about ten times that of the bright stars in a typical galaxy. He also was one of the first to draw attention to the "flatness" problem of the universe, the extraordinary coincidence that the rate at which the universe is expanding sits exactly on the dividing line between allowing the expansion to continue indefinitely (this is known as an open universe) and allowing gravity to halt the expansion and eventually bring about a collapse in a Big Crunch (a closed universe). As one expert put it, "This is like finding a finely sharpened pencil balancing on its point for millions of years."[12] This observation was one of the factors that would lead to the idea of inflation in the early universe (see below).

The Ordered Life of Stars

But background radiation was not the only form of radio waves from deep space discovered in the 1960s. Astronomers had observed many other kinds of radioactivity unconnected with optical stars or galaxies. Then, in 1963, the moon passed in front of one of those sources, number 273 in the *Third Cambridge Catalogue of the Heavens*, and therefore known as 3C 273. Astronomers carefully tracked the exact moment when the edge of the moon cut off the radio noise from 3C 273—pinpointing the source in this way enabled them to identify the object as "star-like," but they also found that the source had a very large red shift, meaning it was well outside our Milky Way galaxy.

It was subsequently shown that these "quasi-stellar radio sources," or quasars, form the heart of distant galaxies that are so far away that such light as reaches us (often very dim) left them when the universe was very young, more than 10 billion years ago. What brightness there is, however, suggests that their energy emanates from an area roughly one light-day across, more or less the dimensions of the solar system. Calculations show that quasars must therefore radiate "about 1,000 times as much energy as all the stars in the Milky Way put together."[13]

In 1967 John Archibald Wheeler, an American physicist who had studied in Copenhagen and worked on the Manhattan Project, revived the eighteenth-century theory of black holes as the best explanation for quasars. Black holes had been regarded as mathematical curiosities until relativity theory suggested that they must actually exist. A black hole is an area where matter is so dense, and gravity so strong, that nothing, not even light, can escape. They may be equivalent to 100 million solar masses of material: "The energy we hear as radio noise comes from masses of material being swallowed at a fantastic rate." [14]

Pulsars were another form of astronomical object detected by radio waves. They were discovered—accidentally, like background radiation—in 1967 by Jocelyn Bell Burnell, a radio astronomer at Cambridge. Brought up in Northern Ireland, Jocelyn Bell Burnell is probably the only scientist to have produced work worthy of a Nobel Prize who failed her eleven-plus examination, meaning she was educated at a co-ed secondary school rather than an elite grammar. Fortunately her father—an architect—was commissioned to work on the observatory in Armagh. This kindled her interest in astronomy, which she studied at Glasgow University. She survived the intellectual rigors of the course there, which shrank from three hundred students in the first year to fifty in the third.

It took two years to build the Cambridge radio telescope, three miles out of the city on the site of the disused Cambridge-to-Oxford railway line. It had 2,048 aerials, held up by nine-foot poles, and it resembled nothing so much as a hop field. Burnell became adept at distinguishing between "twinkling quasars and passing cars." Then, on November 28, 1967, she noticed something that was neither a quasar nor a car—a completely unknown radio source. The pulses were extremely precise—so precise that at first the Cambridge astronomers thought they might be signals from a distant civilization. But the discovery of many more showed they must be a natural phenomenon.

The pulsing was so rapid (and so accurate, to one-millionth of a second per day) that two things suggested themselves: the sources were small, and they were spinning. [15] Only a small object spinning fast could produce such pulses, rather like a very rapid lighthouse beam coming round every so often. The small size of the pulsars told astronomers that they must be either white dwarfs (stars with the mass of the sun packed into the size of the earth) or neutron stars (with the mass of the sun packed into a sphere less than ten kilometers across). When it was shown that white dwarfs could not rotate fast enough to produce such pulses without falling apart, scientists finally had to accept the existence of neutron stars. These super-dense stars, midway between white dwarfs and black holes, have a solid crust of iron above a fluid inner core made of neutrons and, possibly, quarks. The density of neutron stars has been calculated by physicist John Gribbin as 1 million billion times greater than water, meaning that each cubic centimeter of such a star would weigh 100 million tons. Put another way by Brian Cox, "A single teaspoon of neutron star matter weighs more than a mountain."[16]

The significance of pulsars being identified as neutron stars was that it more or less completed the story of *stellar evolution*, allowing us to grasp the full biography of the universe.

Stars form as a cooling gas. As they contract they get hotter, so hot eventually that nuclear reactions take place; this is known as the "main sequence" of stars. After that, depending on their size and at what point a crucial temperature is reached, quantum processes trigger a slight expansion that is also fairly stable—and the star is now a red giant. Toward the end of its life, a star sheds its outer layers, leaving a dense core in which all nuclear reactions have stopped—it is now a white dwarf and will cool for millions of years, eventually becoming a black dwarf, unless it is very large, in which case it ends as a dramatic supernova explosion. It shines very brightly, very briefly, scattering heavy elements into space, out of which other heavenly bodies form, and without which life could not exist. It is these supernovae explosions that give rise to neutron stars and, in some cases, black holes.

And so the marriage of physics and astronomy—quasars and quarks, pulsars and particles, relativity, the formation of the elements, the lives of stars—was all synthesized into one consistent, coherent, unified story, to produce a detailed assessment about the origin and evolution of the universe.

The First Three Minutes and the Building Blocks of the Cosmos

The most famous summing up of these complex ideas was Steven Weinberg's book *The First Three Minutes*, published in 1977. Weinberg, a ruddy-faced, Bronx-born New Yorker, graduated from Cornell, studied at Copenhagen, and shared the 1979 Nobel Prize with Sheldon Lee Glashow and Abdus Salam. They had each come up with another aspect of unification— identifying (yet again independently) an amalgamation of the weak nuclear force and electromagnetic interaction between elementary particles.[17]

The first thing that may be said about the "singularity," as physicists call Time Zero, is that technically all the laws of physics break down. Therefore, we cannot know exactly what happened at the moment of the Big Bang, only nanoseconds later (a nanosecond is a millionth of a second). Weinberg gives the following chronology.

After 0.0001 (10^{-4}) seconds, this, the original "moment of creation," occurred approximately 13.8 billion years ago. The temperature of the universe at this near-original moment was 10^{12} K, or 1,000 billion degrees (written out, that is 1,000,000,000,000 degrees). The density of the universe at this stage was 10^{14}— 100,000,000,000,000—grams per cubic centimeter. (The density of water is 1 gram per cubic centimeter.) At this point, photons and particles were interchangeable.

After 13.8 seconds, the temperature was 3 billion K, and nuclei of deuterium were beginning to form. These consisted of one proton and one neutron, but they would have soon been knocked apart by collisions with other particles.

After three minutes two seconds, the temperature was 1 billion K (about seventy times as hot as the sun is now). Nuclei of deuterium and helium formed. After four minutes, the universe consisted of 25 percent helium with the rest "lone" protons, hydrogen nuclei.

After 300,000 years, the temperature was 6,000 K (roughly the same as the surface of the sun), when photons would be too weak to knock electrons off atoms. At this point the Big Bang could be said to be over. The universe expanded "relatively quickly," cooling all the while.

After one million years, stars and galaxies started to form, when nucleosynthesis took place and the heavy elements were formed, which gave rise to the sun and the earth.

At this point the whole process becomes more accessible to experimentation, because particle accelerators allow physicists to reproduce some of the conditions inside stars. These show that the building blocks of the elements are hydrogen, helium, and alpha particles, which are helium-4 nuclei. These are added to existing nuclei, so that the elements build up in steps of four atomic mass units: "Two helium-4 nuclei, for example, become beryllium-8, three helium-4 nuclei become carbon-12, which just happens to be stable." This is important: each carbon-12 nucleus contains slightly less mass than the three alpha particles that go to make it up. Therefore energy is released, in line with Einstein's famous equation, $E=mc^2$—providing E to produce more reactions and more elements.[18]

The building continued, in stars: oxygen-16, neon-20, magnesium-24, and eventually silicon-28. "The ultimate step," as Weinberg describes it, "occurs when pairs of silicon-28 nuclei combine to form iron-56 and related elements such as nickel-56 and cobalt-56. These are the most stable of all."[19] Liquid iron, remember, is the core of the earth.

But this is not all. All that exists now exists because of a minuscule imbalance in the laws of physics in the early universe. Had the forces been balanced equally, matter and antimatter would have annihilated each other, leaving only radiation.* As it was,

---

* The modern conception of antimatter was first formulated by Paul Dirac in the late 1920s, when he worked on an amalgamation of Einstein's theory of special relativity and its links to quantum theory. Dirac (whom some regard as second only to Einstein in terms of his physicist's imagination, and ahead

there was just one everyday particle (baryonic matter) left over for every billion photons of radiation remaining after the fireball. "The Universe today is made out of the one-in-a-billion particles (protons + neutrons) manufactured in this way in the Big Bang fireball."[20]

In our solar system, 90.8 percent of the atoms are hydrogen, 9.1 percent helium, and 0.1 percent everything else put together. But because hydrogen is the lightest element it comprises only 70.13 percent of the mass of the solar system, while helium comprises 27.87 percent, and oxygen, the third most common element by mass, is 0.91 percent.[21]

Dark matter. This is perhaps *the* outstanding mystery of the universe. If the "standard model" of the Big Bang is correct (the standard model being what has been described here), the average density of matter throughout the universe has been calculated to be about 5×10^{-27} kg per cubic meter. The amount of bright matter that we can detect is equal to about only one-hundredth of this critical density, and adding in the dark matter known to exist within galaxy clusters (because they are spinning too fast to be held in place by the gravity the bright material generates), we are still left with the fact that there is somewhere between thirty times and one hundred times more dark matter in the universe than bright, detectable matter.[22] We don't know why dark matter exists, or what it is.

That is not the only outstanding problem. We need twelve

---

of Bohr and the others) found that his calculations led to the surprising prediction of an electron with a positive charge. This became in time the positron and has been observed experimentally many times. All other particles have their antimatter equivalents, which are virtually identical but have opposite charges. Although antimatter particles and matter particles are so similar, their existence in the world differs enormously. All the world is made of matter, and the observed imbalance is a puzzle yet to be explained. Without this imbalance, matter and antimatter would no doubt have annihilated one another eons ago, at the beginning of time, and there would have been no universe. Small amounts of antimatter have been observed in galactic clusters but, essentially, the mystery persists. As the paragraph immediately overleaf shows, it is not the only one.

particles in the master equation of the standard model, but it is necessary to have only four to build a universe (up and down quarks, the electron, and the electron neutrino). The existence of the other eight is "a bit of a mystery."[23]

A Unified Universe

This narrative of the early universe and its marriage with particle physics was brilliant science, but also a great work of the imagination, the second evolutionary synthesis of the twentieth century. It was more even than that, for although imagination of a high order was required, it also needed to conform to the evidence (such evidence as there was, anyway). As an intellectual exercise it merited comparison with the great unifying ideas of Copernicus, Galileo, and Darwin.

Once one gets over the breathtaking numbers involved in anything to do with the universe, and accepts the sheer weirdness not only of particles but of heavenly bodies, one cannot escape the fact of how inhospitable much of the cosmos is—very hot, very cold, very radioactive, unimaginably dense. No life as we can conceive it could ever exist in these vast reaches of space. The heavens of the physicists and the cosmologists were as awesome as they had ever been, ever since man's observation of the sun and stars began. But heaven was no longer heaven, if by that was meant the same thing as paradise.

A BIOGRAPHY OF EARTH: THE UNIFIED CHRONOLOGY OF GEOLOGY, BOTANY, LINGUISTICS, AND ARCHAEOLOGY

B etween September and November 1965, the United States National Science Foundation RV (Research Vessel) *Eltanin* was cruising on the edge of the Pacific-Antarctic Ocean, collecting routine data about the seabed. The ship was essentially a laboratory belonging to the Lamont-Doherty Earth Observatory, part of New York's Columbia University. Oceanography had received a boost in World War II because of the need to understand U-boats and their environment, and since then with the arrival of deepwater nuclear submarines. The Lamont Institute was one of the most active outfits in this area.

On that 1965 voyage, *Eltanin* zigzagged back and forth over a deep-sea geological formation known as the Pacific-Antarctic Ridge, located at 51 degrees latitude south. Special equipment measured the magnetic qualities of the rocks on the seabed. It had been known for a while that for some reason the magnetism of rocks reverses itself regularly every million years or so, and that this pattern told geologists a great deal about the history of the earth's surface. The scientist in charge of legs 19, 20, and 21 of *Eltanin*'s journey that time was Walter C. Pitman

III, a Columbia-trained graduate student, still working on his PhD. While on board ship he was too busy to do more than double-check that the instruments were working properly, but as soon as he got back to Lamont, he laid out his charts to see what they showed.

What he had in front of him was a series of black-and-white stripes. These recorded the magnetic anomalies over a stretch of ocean floor. Each time the magnetic anomaly changed direction, the recording device changed from black to white to black, and so on. What was immediately obvious that November day was that one particular printout, which recorded the progress of *Eltanin* from five hundred kilometers east of the Pacific-Antarctic Ridge (aka East Pacific Rise) to five hundred kilometers west, was *completely symmetrical* around the ridge. That symmetry could be explained in only one way. The rocks either side of the ridge had been formed at exactly the same time as each other and "occupied the position they did because they had originated at the ridge and then spread out to occupy the seabed." In other words, the seabed was formed by rocks emerging from the depths of the earth, which then spread out across the seafloor and pushed the continents apart. This was a confirmation at last of continental drift, achieved by seafloor spreading. In the geological community, the printouts were called "Pitman's Magic Profile."[1]

These results, he said later, "hit me like a hammer," adding, "In retrospect we were lucky to strike a place where there are no hindrances to sea-floor spreading. There's no other place we get profiles that perfect. There were no irregularities to distract or deceive us. That was good, because people had been shot down an awful lot over sea-floor spreading. The symmetry was extraordinary."[2]

The "Coincident Patterns" of the Continents

Pitman was right. Seafloor spreading was a controversial subject, just as the mother theory—continental drift—was. Continental drift had been proposed by Alfred Wegener as long ago as 1912 as a way to explain the distribution—the order—of the landmasses of the world and the pattern of life forms.

Wegener was a German meteorologist. Nothing about his education prepared him for the great theory he was to propose— he had no background in geology or geophysics.[3] His *Die Entstehung der Kontinente und Ozeane* ("The Origin of Continents and Oceans"), which appeared in 1915, was not particularly original. Francis Bacon had pointed out the "coincident patterns" of continents in 1620. In the early nineteenth century the German geographer and explorer Alexander von Humboldt had commented—again in passing—on the similarities of mountain ranges in Brazil and West Africa.[4] Wegener's idea in the book— that the six continents of the world had begun life as one supercontinent—had been aired earlier by an American, Frank Bursley Taylor, in 1908. Taylor thought that the cause of continental movement was a combination of tidal action and an increase in the rate of the earth's rotation, forcing landmasses to move away from the poles by centrifugal force. This increased rotation, he said, was caused by "the capture of a comet that became the moon." Another theory, put forward in 1911 by Howard Baker, had it that Venus and the earth had been much closer together in the past, causing the moon to be "ripped from the Earth," the resulting chasm forming the Pacific Ocean.[5]

But Wegener collected much more evidence—and more impressive evidence—to support his claim than anyone had done before, rather like Lyell and Darwin in their respective fields. He set out his ideas at a meeting of the German Geological Association at Frankfurt am Main in January 1912. In fact, with the benefit of hindsight, one might ask why scientists had not reached Wegener's conclusions earlier.

By the end of the nineteenth century it was obvious that to make sense of the natural world, and its distribution around the globe, some sort of intellectual reconciliation was needed. The evidence of that distribution consisted mostly of fossils and the peculiar spread of related types of rocks. Darwin's *On the Origin of Species* had stimulated a new interest in fossils because it was realized that if they could be dated, they could throw light on the development of life in bygone epochs and maybe on the origin of life itself. At the same time, quite a lot was known about rocks and the way one type had separated from another as the earth had formed, condensing from a mass of gas, to a liquid, to a solid.

The central problem lay in the spread of some types of rocks across the globe and their links to fossils. For example, there is a mountain range that runs from Norway to north Britain and should cross in Ireland with other ridges that run through north Germany and south Britain. In fact, it looked to Wegener as though the crossover actually occurs near the coast of North America, as if the two seaboards of the North Atlantic were once contiguous. Similarly, plant and animal fossils are spread about the earth in a way that can only be explained if there were once land connections that linked, for example, Africa to South America, or Europe to North America. But if these land bridges had ever existed, where had they gone to? What had provided the energy by which the bridges had arisen and disappeared? What happened to the sea?

Wegener's answer was bold. There were no land bridges, he said. Instead, the six continents as they now exist—Africa, Australia, North and South America, Eurasia, and Antarctica—were once one huge continent, one enormous landmass, which he called Pangaea (from the Greek for "all" and "earth"). The continents had arrived at their present positions by "drifting," in effect floating like huge icebergs. His theory also explained the mid-continent mountain ridges, formed by the ancient colliding landmasses.

He provided a map that showed just how South America had adjoined Africa, how India had fitted between Africa and Antarctica, and how the concave base of Australia had fitted around Antarctica. Then, the distribution of four fossils—*Lystrosaurus*, a terrestrial reptile; *Cynognathus*, a second terrestrial reptile; *Mesosaurus*, a freshwater reptile; and the fern *Glossopteris*—could be explained.[6] Some of his calculations were a little exaggerated. For example, using lunar charts he calculated that Greenland had moved relative to Greenwich by 9 meters (30 feet) per year between 1823 and 1870, and by 32 meters (105 feet) per year between 1870 and 1907. Such rates would mean that the continents could have moved over the earth in as little as one million to five million years. This is three thousand times as fast as the rate accepted today, 4 centimeters (1.6 inches) per year.[7]

It was in any case an idea that took some getting used to. How could entire continents "float"? And on what? And if the continents had moved, what enormous force had moved them?

By Wegener's time, the earth's essential structure was known. Geologists had used analysis of earthquake waves to deduce that the earth consisted of a crust, a mantle, an outer core, and an inner core. The first basic discovery was that all the continents of the earth are made up of one form of rock—granite, or a granular igneous rock (formed under intense heat)—comprised of feldspar and quartz. Around the granite continents may be found a different form of rock—basalt, much denser and harder. Basalt exists in two forms, solid and molten (we know this because lava from volcanic eruptions is semi-molten basalt). This suggests that the relation between the outer structures and the inner structures of the earth was clearly linked to how the planet formed as a cooling mass of gas that became liquid and then solid.

The huge granite blocks that form the continents are believed to be about 50 kilometers (30 miles) thick, but below that, for about 3,000 kilometers (1,900 miles), the earth possesses the properties of an "elastic solid," of semi-molten basalt. And below that, to the center of the earth (the radius of which is about 6,000 kilometers, nearly 4,000 miles), there is liquid iron. Millions of years ago, of course, when the earth was much hotter than it is today, the basalt would have been less solid, and the overall situation of the continents would have resembled more closely the idea of icebergs floating on the oceans. On this view, the drifting of the continents becomes much more conceivable.[8]

Wegener's theory was tested when he and others began to work out how the actual landmasses would have been pieced together. The continents do not of course consist only of the land that we see above sea level at the present time. Sea levels have fallen and risen throughout geological time as ice ages have lowered the water table and warmer times raised it, so that the continental shelves—those areas of land currently below water but relatively shallow, before the contours fall off sharply by thousands of feet—are just as likely to make the "fit."

Various unusual geological features fall into place once this massive jigsaw is pieced together. For example, deposits from glaciation of the Permo-Carboniferous age (i.e., ancient forests, which were formed 200 million years ago and are now coalfields) exist in identical forms on the west coast of South Africa and the east coast of Argentina and Uruguay. Areas of similar Jurassic and Cretaceous rocks (roughly 100 to 200 million years old) exist around Niger in West Africa and around Recife in Brazil, exactly opposite, across the South Atlantic. And a geosyncline (a depression in the earth's surface) that extends across southern Africa also strikes through mid-Argentina, aligning neatly.[9]

How long was Pangaea in existence, and when and why did the breakup occur? What kept it going? These are the final questions in what is surely one of the most breathtaking ideas of modern times. (It took some time to catch on: in 1939, geology textbooks were still treating continental drift as "a hypothesis only.")

How Continental Drift and the Age of the Earth Fit Together

The theory of continental drift coincided with the other major advance made in geology in the early years of the century. This related to the age of the earth. In 1650, as is well known, James Ussher, archbishop of Armagh in Ireland, using the genealogies given in the Bible, had calculated that the earth was created at nightfall on October 22, 4004 BC.* In the late nineteenth century William Thomson, Lord Kelvin, using ideas about the earth's cooling, proposed that the crust formed between 20 million and 98 million years ago (chapter 1). All such calculations were over-taken by the discovery of radioactivity and radioactive decay. In 1907 Bertram Boltwood, at Yale, acting on a suggestion of Ernest Rutherford, established that lead was the final decay product of radioactive uranium, and realized that he could calculate the age of the rocks by measuring the relative constituents of uranium and lead, and relating it to the half-life of uranium. The old-est substances on earth, to date, are some zircon crystals from Australia dated in 1983 to 4.2 billion years old. The current best estimate for the age of the earth—linking physics, geology, and time—is 4.5 billion years.[10]

Wegener took the theory for granted, based on the evidence he had collected, but many geologists, especially in the United States, were not convinced. They were "fixists," who believed that the continents were rigid and immobile. In fact, geology was divided for years, at least until World War II. But with the advent of nuclear submarines the US Navy in particular needed far more information about the Pacific Ocean, the area of water that lay between it and its main enemy, Russia. The basic result to come out of this study was that the magnetic anomalies under the Pacific were shaped like enormous "planks" in roughly parallel lines, running predominantly north-south, each one 15 to 25 kilometers wide and hundreds of kilometers long.

Walter Pitman didn't know too much about all this. At the time of his Pacific voyage, he was unfamiliar with the classical arguments for continental drift and not over-familiar with the notion of seafloor spreading.[11] But the *Eltanin* findings produced a tantalizing piece of arithmetic: divide 25 kilometers by 1 million (the number of years after which, on average, the earth's polarity

---

* In some geology departments in modern universities, October 22 is still celebrated—ironically—as the earth's birthday.

changes), and you get 2.5 centimeters. Did that mean the Pacific was expanding at that rate each year? The broad answer was yes, though later research suggested that the Pacific has been expanding constantly by 1 centimeter per year for the past 10 million years. Even 1 centimeter multiplied by 10 million equals 100,000 meters, 100 kilometers, or 61 miles. (Today the width of the Pacific, at its widest, is 18,800 kilometers or 12,300 miles.)[12]

There was other evidence to support the mobilists. In 1953 the French seismologist Jean-Pierre Rothé produced a map at a meeting of the Royal Society in London that recorded earthquake epicenters for the Atlantic and Indian oceans. This was remarkably consistent, showing many earthquakes associated with the mid-ocean ridges. Moreover, the further the volcanoes were from the ridges, the older they were, and the less active. Yet another spin-off from the war was the analysis of the seismic shocks sent shuddering across the globe by atomic bomb explosions. These produced the surprising calculation that the ocean floor was barely four miles thick, whereas the continents were twenty miles thick. Just a year before the *Eltanin*'s voyage, Sir Edward Bullard, a British geophysicist, had reconstructed the Atlantic Ocean margins, using the latest underwater soundings, which enabled thousand-meter depth contours to be used, rather than sea-level contours. At that depth, the fit between the continents was even more complete. Despite these various pieces of evidence, it was not until *Eltanin*'s symmetrical picture came ashore that the "fixists" were finally defeated.[13]

Capitalizing on this, in 1968 William Jason Morgan, from Princeton, put forward an even more extreme "mobilist" view. His idea was that the continents were formed from a series of global, or "tectonic," plates, slowly inching their way across the surface of the earth. He proposed that the movement of these plates—each one about 100 kilometers (62 miles) thick—together accounts for the bulk of seismic activity on earth. His controversial idea soon received support when a number of "deep trenches" were discovered in the floor of the Pacific Ocean.

Labeled subduction zones, they could be up to 700 kilometers (435 miles) deep. It was here that the seafloor was absorbed back into the underlying mantle (one of these trenches ran from Japan to the Kamchatka Peninsula in Russia, 1,800 kilometers, 1,118 miles).[14]

The Order That Led to Life

Another way in which the earth has changed over time—in effect, aged, and in a most interesting way—is that, about 2,500 million years ago, we begin to see in the earth's rocks the accumulation of hematite, an oxidized form of iron. This appears to mean that, for the first time, oxygen was being produced but was "used up" by the minerals of the world. There is little doubt that, before then, the atmosphere of the earth contained little or no oxygen and that the first forms of life—bacterial organisms—were anaerobes, operating only in the absence of oxygen. So where was the oxygen coming from?

The best candidate for an oxygen producer is a blue-green bacterium that, in shallower reaches of water (where the light from the sun could act on its chlorophyll), broke carbon dioxide down into carbon (which it utilized for its own purposes) and oxygen—in other words, photosynthesis. For a time, the minerals of the earth soaked up what oxygen was going (limestone rocks captured oxygen as calcium carbonate, iron rusted, and so on), but eventually the mineral world became saturated and, after that, over a billion years, billions of bacteria poured out minuscule puffs of oxygen, transforming the earth's atmosphere.

According to Richard Fortey, in his history of the earth, the next advance was the formation of slimy communities of microbes, structured into "mats," almost two-dimensional layers. These are still found even today—on saline flats in the tropics where the absence of grazing animals allows their survival—though fossilized forms have also been found in South Africa and Australia in rocks dating to more than 3.5 billion years old. These structures are known as stromatolites. Resembling "layered cabbages," they could grow to immense lengths—thirty feet was normal, and one hundred meters not unknown. But they were made up of prokaryotes, or cells without nuclei, which reproduced simply by splitting.[15]

The advent of nuclei was the next advance. Freeman Dyson has described Lynn Margulis as one of the great bridge-builders in modern biology. As the Amherst professor has pointed out, "in all probability" one bacterium cannibalized another, which became an organelle within another organism, and eventually formed the nucleus. A prokaryote cell became a eukaryote cell. A chloroplast is another such organelle, a specialist subunit performing photosynthesis within a cell. The development of the nucleus and organelles was a crucial step, allowing more complex structures to be formed.

This, it is believed, was followed by the evolution of sex, which seems to have occurred about two billion years ago. Sex occurred because it allowed the possibility of genetic variation, giving a boost to evolution, which, at that time, would have speeded up. Cells became larger, more complex—and slimes appeared. Slimes can take on various forms, and can on occasion move over the surface of other objects. In other words, they are both animate and inanimate, showing the development of rudimentary specialized tissues, behaving in ways that faintly resemble animals.[16]

The Cambrian Explosion and Another Evolutionary Synthesis

By 700 million years ago, the Ediacara had appeared. These, the most primitive form of animal, have been discovered in various parts of the world, from Leicester, England, to the Flinders Ranges in South Australia. They take many exotic forms, but in general are characterized by radial symmetry, skin walls only two cells thick, and primitive stomachs and mouths—like primitive jellyfish in appearance, and therefore not unimaginably far from slime. The first truly multicellular organisms, the Ediacara for some reason became extinct, and this may have been ultimately because they lacked a skeleton. This seems to have been the next important moment in evolution. Paleontologists can say this with some confidence because, about 500 million years ago, there was a revolution in animal life on earth. This is what became known as the Cambrian explosion. Over the course of only 15 million years, animals with shells appeared, and in forms that are familiar even today. These were the trilobites—some with jointed legs and grasping claws, some with rudimentary dorsal nerves, some with early forms of eye, others with features so strange they are hard to describe.[17]

And so, by the mid to late 1980s, a new evolutionary synthesis began to emerge in biology, one that filled in the order of important developments and provided more accurate dating. Moving forward in geological time, we can leap ahead from the Cambrian explosion by more than 400 million years, to approximately 65 million years ago.

One of the effects of the landing on the moon in 1969, and the subsequent space probes, was that geology went from being a discipline with just a single planet to study to one where there was a much richer base of data. One of the ways that the moon and other planets differ from earth is that they seem to have a lot more craters on them, these craters being formed by impacts from asteroids or meteorites: bodies from space. This was important in geology because, by the 1970s, the discipline had become used to a slow-moving chronology, measured in millions of years. There was, however, one great exception to this rule, and that became known as the K–T boundary, the boundary between the Cretaceous and the Tertiary geological periods, occurring about 65 million years ago.* That is when the fossil records showed a huge and very sudden disruption, the chief feature of which was

---

* Cretaceous is derived from the Latin *creta*, for chalk. K is the traditional abbreviation, taken from the German word for "chalk": *Kreide*.

the very opposite of the Cambrian explosion—many forms of life on earth *disappeared.*

The most notable of these extinctions was that of the dinosaurs, dominant large animals for about 150 million years before that, and completely absent from the fossil record afterward. Traditionally, geologists and paleontologists considered that mass extinctions were due to climate change or a fall in sea level. For many, however, this process would have been too slow—plants and animals would have adjusted, whereas in fact about half the life forms on earth suddenly disappeared between the Cretaceous and the Tertiary. After the study of so many craters on other moons and planets, some paleontologists began to consider whether a similarly catastrophic event might have caused the mass extinctions seen on earth 65 million years ago. In this way there began an amazing scientific detective story that was not fully resolved until 1991.[18]

For a meteorite or an asteroid to cause such a devastating impact it needed to have been a certain minimum size, so the crater it caused ought to have been difficult to overlook. No immediate candidate suggested itself, but the first breakthrough came when scientists realized that meteorites have a different chemical structure to that of the earth, in particular with regard to the platinum group of elements. This is because these elements are absorbed by iron, and the earth has a huge iron core. Meteorite dust, on the other hand, would be rich in these elements, such as iridium. Sure enough, by testing rocky outcrops dating from the Cretaceous—Tertiary border, Luis and Walter Alvarez, from the University of California at Berkeley, discovered that iridium was present in quantities that were *ninety times* as rich as they should have been if no impact had taken place. It was this discovery, in June 1978, that set off this father-and-son (and subsequently daughter-in-law) team on the quest that took them more than a decade.

The second breakthrough came in 1981, in *Nature*, when

Jan Smit, a Dutch scientist, reported his discoveries at a K–T boundary site at Caravaca in Spain. He described some small round objects, the size of a sand grain, called spherules. These, he said, were common at such sites and on analysis were shown to have crystals of a "feathery" shape, made of sanidine, a form of potassium feldspar. These spherules, it was shown, had developed from earlier structures made of olivine—pyroxene and calcium-rich feldspar—and their significance lay in the fact that they are characteristic of basalt, the main rock that forms the earth's crust under the oceans. In other words, the meteorite had slammed into the earth not on land but in the ocean.[19]

This was both good and bad news. It was good news in that it confirmed there had been a massive impact 65 million years ago. It was bad news in the sense that it led scientists to look for a crater in the oceans, and also to look for evidence of the massive tsunami, or tidal wave, that must have followed. Calculations showed that such a wave would have been a kilometer high as it approached continental shorelines. Both of these searches proved fruitless, and although evidence for an impact began to accumulate throughout the 1980s, with more than one hundred areas located that showed iridium anomalies, the actual site of the impact remained elusive.

It was not until 1988, when Alan Hildebrand, a Canadian attached to the University of Arizona, first began studying the Brazos River in Texas, that the decade-long search moved into its final stage. It had been known for some time that in one place near Waco the Brazos passes over some rapids associated with a hard sandy bed, and this bed, it was recognized, was the remnant of a tsunami inundation. Hildebrand looked hard at Brazos and then went in search of evidence that would link it, in a circular fashion, with other features in the area. By examining maps and gravity anomalies, he finally found a circular structure, which might be an impact crater, on the floor of the Caribbean, north of Colombia, but also extending into the Yucatán Peninsula in

Mexico. Other paleontologists were skeptical at first, but when Hildebrand brought in help from geologists more familiar with Yucatán, they soon confirmed the area as the impact site.

The reason everyone had been so confused was that the crater—known as Chicxulub—was buried under more recent rocks. When Hildebrand and his colleagues published their paper in 1991, it caused a sensation, at least among geologists and paleontologists. They now had to revise their whole attitude: catastrophic events *could* have an impact on evolution.[20]

The disappearance of the dinosaurs also proved to have had a liberating effect on mammals. Until the K–T boundary, mammals were small creatures. This may have helped their survival after the impact—because they were so numerous—but in any event the larger mammals did not emerge until after the K–T, and in the absence of competition from *Tyrannosaurus rex*, *Triceratops*, and their brothers and sisters.

Ever-larger mammals started to appear all across earth, but Africa was now becoming distinctive, particularly in its abundance of primates, including the first true apes and ancestors of the Old World monkeys. Primates had vanished from Europe by the late Eocene period (56 to 34 million years ago) and were gone from much of Asia and North America by the same era. So Africa was their refuge and it was to be there that the apes and monkeys evolved.[21] By the next geological period, the Oligocene (34 to 23 million years ago), primates were restricted to Africa but were abundant there.[22]

We can say, then, that there would probably have been no humans unless the K–T meteorite had collided with earth. The earliest of the hominids, *Sahelanthropus tchadensis*, appeared in the region of the African Great Rift Valley about 7 to 6 million years ago (written 7 ma). Though its skull was small and chimp-like, with a small brain and large brow ridges, it already had very human-like features, such as a flattened face, reduced canine teeth, and an upright posture. So far, some twenty-one different

types of hominid have been identified—dated to between then and now.[23]

Interlocking Dates: The Romance of Coherence

Continental drift, tectonic plates, the Cambrian explosion, and the K–T boundary were initially of geological interest only. But geology is a form of history. One of the achievements of twentieth-century science has been to make accessible more and more remote areas of the past. Although these discoveries have arrived piecemeal, they have proved consistent—romantically consistent—in helping to provide the basis for one story, one narrative culminating (so far) in humankind. This is perhaps the crowning achievement of modern thought. Here are some of the other discoveries, presented piecemeal but chronologically, as they arrived, to emphasize their coherence.

In paleontology, and in classical archaeology, the traditional method for dating the past is stratigraphy. As common sense suggests, deeper layers are older than the layers above them. However, this only gives a relative chronology, helping to distinguish later from earlier. For absolute dates, some independent evidence is needed—like a list of kings with written dates, or coins with the date stamped on them, or references in writing to some heavenly event, like an eclipse, the date of which can be calculated back from modern astronomical knowledge. Such information can then be matched to stratigraphic levels. This is of course not entirely satisfactory. Sites can be damaged, accidentally or deliberately, by man or nature. Tombs can be reused. Archaeologists and paleontologists are therefore always on the lookout for other dating methods. The twentieth century was to offer several advances in this area.

The first came in 1929, and was covered in the Preface. This was Andrew Ellicott Douglass's identification of tree rings as

indicators of "fat" and "lean" years in history, which dovetailed nicely with sunspot activity and eventually allowed him to date accurately certain events in the pre-Columbian calendar.

Our understanding of early human life on earth was also revolutionized by the discovery of radiocarbon dating. This was first discovered by Willard Libby, in New York in 1949 (he won the Nobel Prize in chemistry in 1960 for his innovation).

Radiocarbon dating depends on the fact that plants take out of the air carbon dioxide, a small proportion of which is radioactive, having been bombarded by cosmic rays from space. Photosynthesis converts this CO_2 into radioactive plant tissue, which is maintained as a constant proportion until the plant (or the organism that has eaten the plant) dies, when radioactive carbon uptake is stopped. Radioactive carbon, ^{14}C (as opposed to the common isotope, ^{12}C, which comprises more than 98 percent of carbon), is known to have a half-life of roughly 5,700 years and so, if the proportion of radioactive carbon in an ancient object is compared with the proportion of radioactive carbon in contemporary objects, it is possible to calculate how long has elapsed since that organism's death.

Libby started by testing the dates he obtained against known historical dates—objects taken from ancient civilizations, such as: wood from a mummiform coffin of the Ptolemaic period in Egypt, with a known age of 2,280 years (Libby got 2,190 ± 450); wood from a funerary ship from the tomb of Sesostris III, king of Egypt, known age 3,750 (Libby got 3,621 ± 180); and an acacia wood beam from the tomb of Zoser, second king of the Third Dynasty in ancient Egypt, known age 4,650 (Libby got 3,979 ± 350).[24]

Later, he collected evidence from all over the world—Turkey, Iraq, Ireland, Denmark, and right across North and South America, 949 samples in total. The American ones began to answer crucial questions in archaeology there, such as when agriculture first began, when the great Mayan and Aztec temples were built, and what was contemporaneous with what. This really began the systematic study of the chronology of New World civilizations.

Some of Libby's dates were surprising, but the biggest and most important turnover came in a book by Colin Renfrew, then a professor at Southampton University, later at Cambridge, called *Before Civilisation: The Radiocarbon Revolution and Prehistoric Europe*, released in 1973.[25]

Before Civilisation had two core arguments. First, it revised the timing of the way the earth was populated. For example, from about 1960 on it was known that Australia had been occupied by humans as early as 4000 BC and maybe even as early as 17,000 BC. Maize, it was established, was gathered systematically in Mexico by about 5000 BC, and well before 3000 BC it was showing signs of domestication. The real significance of these dates was not just that they were earlier than anyone had hitherto thought, but that they killed off the vague theories then current that Meso-America had only developed civilization after it had been imported, in some indefinable way, from Europe. The Americas had been cut off from the world since 16,000–14,000 BC, in effect the last Ice Age, and had developed all the hallmarks of civilization— farming, building, metallurgy, religion—entirely separately.

This revision of chronology, and what it meant, was the second element in Renfrew's book, and here he concentrated on the area he knew best, Europe and the classical world of the Middle East. Sumer and Egypt, for example, were the mother civilizations, the first great collective achievements of humankind, giving rise to the Minoans on Crete, and the classical world of the Aegean: Athens, Mycenae, Troy. From there, civilization had spread north, to the Balkans and then Germany and Britain, and west to Italy and then France, and the Iberian Peninsula.

But after the ^{14}C revolution, there was suddenly a serious problem with this model. On the new dating, the huge megalithic sites of the Atlantic seaboard in Spain and Portugal, in Brittany and Britain, and in Denmark, were either contemporaneous with the civilizations of the Aegean or actually *preceded* them. This wasn't just a question of an isolated date here and there, but of many hundreds of revised datings, consistent with each other, and which in some cases placed the Atlantic megaliths up to a thousand years earlier than the Aegean cultures. The traditional model for Egypt, the Middle East, and the Aegean still held. But there was, as Renfrew put it, a sort of archaeological "fault line" around the Aegean. Beyond that, a new model was needed.[26]

The model Renfrew came up with started with a rejection of the old idea of "diffusion"—that there had been an area of mother civilizations in the Middle East from which ideas of farming, metallurgy, and, say, the domestication of plants and animals had started, and then spread to all other areas as people migrated. It seemed clear to Renfrew that up and down the Atlantic coasts of Europe, there had developed a series of chiefdoms, a level of social organization midway between hunter-gatherers and full-blown civilization as represented in Egypt, Sumer, and Crete, which had kings, elaborate palaces, a highly stratified society.

The sovereign areas of the chiefdoms were smaller (six on the Isle of Arran in Scotland, for example). And they were centered around large tombs and occasionally religious/astronomical sites, such as Stonehenge. Associated with these chiefdoms were rudimentary social stratification and early trade. Sufficient numbers were needed to build the impressive stoneworks—funerary religious monuments around which the clans cohered. The megaliths were always found associated with arable land, suggesting that chiefdoms were a natural stage in the evolution of society. When people settled with the first domesticated crops, chiefdoms and megaliths soon followed.[27]

Renfrew's analysis, now generally accepted, concentrated on sites in Britain, Spain, and the Balkans, which illustrated his argument. But it was his general thrust that counted. Although early people had no doubt spread out to populate the globe from an initial point (maybe East Africa), civilization, culture—call it what you will—had not developed in one place and then spread in the same way. Civilizations had grown up at different times and in different places of their own accord.

This had two important long-term intellectual consequences.

First, all cultures across the world were sui generis and did not owe their being to a mother culture, the ancestor of all. Combined with the findings of anthropologists, this made all cultures equally potent and equally original. And therefore the "classical" world was no longer the ultimate source.

At a deeper level, as Renfrew specifically pointed out, the discoveries of the new archaeology showed the dangers of succumbing too easily to Darwinian thinking. The old diffusionist theory was a form of evolutionism, but a form so general as to be almost meaningless. It suggested that civilization developed in an unbroken, single sequence. The new ^{14}C and tree-ring evidence showed that simply wasn't true. The new view was not any less "evolutionary" but it was very different.

The World's Earliest Man

Mary and Louis Leakey, archaeologists and paleontologists, had been excavating in Africa—Kenya and Tanganyika (later Tanzania)—since the 1930s without finding anything especially significant. In particular, they had dug at the Olduvai Gorge, a three-hundred-foot-deep, thirty-mile-long chasm that cut into the Serengeti plain, part of the so-called Great Rift Valley that runs north–south through the eastern half of Africa and is generally held to be the border between two massive tectonic plates or blocks.

For scientists the Olduvai Gorge had been of interest ever since it was discovered in 1911, when a German entomologist named Wilhelm Kattwinkel almost fell into it as he chased butterflies. Climbing down into the gorge, which cuts through many layers of sediment, he discovered innumerable fossil bones lying around, and these caused a stir when he got them back to Germany because they included parts of an extinct horse. Later expeditions found sections of a modern human skeleton, and this led some scientists to the conclusion that Olduvai was the perfect place for the study of extinct forms of life, including—perhaps—ancestors of humankind.[28]

It says a lot for the Leakeys' strength of character that they dug at Olduvai from the early 1930s to 1959 without making the earth-shattering discovery they always hoped for. (Until that time, it was believed that early humans originated in Asia.) During the 1930s, until most excavation was halted because of World War II, the Leakeys had dug at Olduvai more years than they had not. Their most notable achievement was to find a massive collection of early man-made tools. Louis and his second wife, Mary, were the first to realize that flint tools were not going to be found in that part of Africa, as they had been found all over Europe, say, because generally in East Africa flint is lacking. They did, however, find "pebble tools"—basalt and quartzite especially—in abundance. This convinced Leakey that he had found a "living floor," a sort of prehistoric living room where early man made tools in order to eat the carcasses of the several extinct species that by now had been discovered in or near Olduvai.

After the war neither he nor Mary visited Olduvai until 1951, but they dug there through most of the 1950s. Throughout the decade they found thousands of hand axes and, associated with them, fossilized bones of many extinct animals—pigs, buffaloes, antelopes, several of them much bigger than today's varieties—evoking a romantic image of an Africa inhabited by huge, primitive creatures. They renamed this living floor the "slaughter-house." At that stage, according to Virginia Morrell, the Leakey's biographer, they thought that the lowest bed in the gorge dated to about 400,000 years ago and that the highest bed was 15,000 years old. Louis had lost none of his enthusiasm, despite having reached middle age without finding any humans in more than twenty years of searching. The Leakeys were kept going by the occasional find of hominid teeth (being so hard, teeth tend to survive better than other parts of the human body), so Louis remained convinced that one day the all-important human skull would turn up.[29]

On the morning of July 17, 1959, he awoke with a slight fever. Mary insisted he stay in camp. They had recently discovered the skull of an extinct giraffe, so there was plenty to do. Mary drove off in the Land Rover, alone except for her two dogs, Sally and Victoria. That morning she searched a site in Bed I, the lowest and oldest, known as FLK (for Frieda Leakey's Korongo, Frieda Leakey being Louis's first wife and *korongo* being Swahili for "gully"). Around eleven o'clock, with the heat becoming uncomfortable, Mary chanced on a sliver of bone that "was not lying loose on the surface but projecting from beneath. It seemed to be part of a skull. . . . It had a hominid look, but the bones seemed enormously thick—too thick, surely," as she wrote later in her autobiography. Dusting off the topsoil, she observed "two large teeth set in the curve of a jaw."

At last, after decades. There could be no doubt. It was a hominid skull.

She jumped back in the Land Rover with the two dogs and rushed back to camp, shouting "I've got him! I've got him!" as she arrived.

Excitedly, she explained her find to Louis. He, as he put it later, became "magically well" in moments.[30]

When Louis saw the skull he could immediately tell from the teeth that it wasn't an early form of *Homo* but probably australopithecine—more apelike. But as they cleared away the surrounding soil, the skull revealed itself as enormous, with a strong jaw, flat face, and huge zygomatic arches—or cheekbones—to which great chewing muscles would have been attached. More important, it was the third australopithecine skull the Leakeys had found in association with a hoard of tools. Louis had always explained this by assuming that australopithecines were the victims of *Homo* killers, who then feasted on the more primitive form of ancestor. But now he began to change his mind—to ask himself if it wasn't the australopithecines who had made the tools. Tool-making had always been regarded as the hallmark of humanity—and now, perhaps, humanity should be seen as stretching back to the australopithecines.

Before long, however, Louis convinced himself that the new skull was actually midway between australopithecines and modern Homo sapiens and so he called the new find *Zinjanthropus boisei*—*Zinj* being the ancient Arabic word for the coast of East Africa, *anthropus* denoting the fossil's human-like qualities, and *boisei* after Charles Boise, the American who had funded so many of their expeditions.[31]

Because he was so complete, so old, and so strange, Zinj made the Leakeys famous. The discovery was front-page news across the world, and Louis became the star of conferences in Europe, North America, and Africa. At these conferences, Leakey's interpretation of Zinj met some resistance from other scholars who thought that his new skull, despite its great size, was not all that different from australopithecines found elsewhere. Time would prove these critics right and Leakey wrong. But while Leakey was arguing his case with others about what the huge, flat skull meant, two scientists elsewhere produced a completely unexpected twist on the whole matter.

A year after the discovery of Zinj, Leakey wrote an article for the *National Geographic* magazine, "Finding the World's Earliest Man," in which he put *Zinjanthropus* at 600,000 years old. As it turned out, he was way off.

Using the traditional archaeological device of stratigraphy (analyzing sedimentation layers) Leakey calculated that Olduvai dated from the early Pleistocene, generally believed to be the time when giant animals such as the mammoth lived on earth alongside man, extending from 600,000 years until around 10,000 years ago. Since 1947, as mentioned earlier, a new method of dating, the ^{14}C technique, had been introduced.

With its relatively short half-life, however, ^{14}C is useful only for artifacts of up to roughly 40,000 years old. Not long after Leakey's *National Geographic* article appeared, two geophysicists from the University of California at Berkeley, Jack Evernden and Garniss Curtis, announced that they had dated some volcanic ash from Bed I of Olduvai—where Zinj had been found—using the potassium-argon (K-Ar) method. In principle this is analogous to ^{14}C dating but uses the rate at which the unstable radioactive potassium isotope, potassium-40 (^{40}K), decays to stable argon-40 (^{40}Ar). This can be compared with the known abundance of ^{40}K in natural potassium, and an object's age calculated from the half-life. Because the half-life of ^{40}K is about 1.3 billion years, this method is much more suitable for geological material.

Using the new method, the Berkeley geophysicists came up with the startling news that Bed I at Olduvai was not 600,000 but 1.75 *million* years old. This was a revelation, the very first clue that early humans were much, much older than anyone suspected. This, as much as the actual discovery of Zinj, made Olduvai Gorge famous, eventually producing the breathtakingly audacious idea—almost exactly one hundred years after Darwin—that humankind originated in Africa and then spread out to populate the globe.[32]

The World's Earliest Woman

All that occurred in the 1960s. As we shall see in a later chapter, the discovery of the very great age of Zinj was one of the factors that helped to kick-start an interest in ancient man and primates in Africa. But it would be a quarter of a century before yet another form of dating fleshed out the picture in a fairly fundamental way. This time it was a piece of biological dating.

In a paper published in *Nature* in January 1987, Allan Wilson and Rebecca L. Cann, from Berkeley, California, revealed a groundbreaking analysis of mitochondrial DNA used in an archaeological/paleontological context. The study of DNA had come a long way since Watson and Crick's breakthrough. Mitochondria are organelles within cells that lie outside the nucleus and are, in effect, cell batteries—they produce a substance known as adenosine triphosphate, or ATP. The particular property of mitochondrial DNA that interested Wilson and Cann was that it is inherited only through the mother—it therefore does not change as nuclear DNA changes, through mating. Mitochondrial DNA can only change, much more slowly, through mutation. Wilson and Cann had the clever idea of comparing the mitochondrial DNA among people from different populations, on the reasoning that the more different they were, the longer ago they must have diverged from whatever common ancestor we all share. Mutations were known by then to occur at a fairly constant pace, so this change should also give an idea of how long ago various groups of people diverged.[33]

To begin with, Wilson and Cann found that the world is broken down into two major groups—Africans on the one hand, and everyone else on the other. Second, Africans had slightly more mutations than anyone else, confirming the paleontological evidence that humanity is older in Africa, very probably began there ("mitochondrial Eve"), and then spread from that continent to populate the rest of the world. Finally, by studying the rate of mutations and working backward, Wilson and Cann were able to show that humanity as we know it is no more than 200,000 years old, broadly confirming the evidence of the fossils. Biology and geology were in agreement.[34]

Mother Tongues, Superfamilies, and the Links Between Stones and Bones

One reason that the Wilson and Cann paper attracted the attention it did was because its results agreed well not only with what the paleontologists were discovering in Africa, but also with recent work on linguistics and archaeology. As long ago as 1786, Sir William Jones, a British judge serving in India at the High Court in Calcutta, had discovered that Sanskrit bore an unmistakable resemblance to both Greek and Latin. This observation gave him the idea of the "mother tongue," the notion that there was once, many years ago, a single language from which all other languages are derived.[35]

Beginning in 1956, Joseph Greenberg began to reexamine Sir William Jones's hypothesis as applied to the Americas. In 1987 he concluded a massive study of Native American languages, from southern South America to the Eskimos in the north, published as *Language in the Americas*, which concluded that, at base, the American languages could be divided into three. The first and earliest was Amerind, which covers South America and the southern states of the United States, and shows much more variation than other, more northern languages, suggesting that it is much older. The second group was Na-Dene, and the third, Eskimo-Aleut, covered Canada and Alaska. Na-Dene is more varied than Eskimo-Aleut, all of which, says Greenberg, points to three migrations into America, by groups speaking three different languages. He believes, on the basis of "mutations" in words, that Amerind speakers arrived on the continent before eleven thousand years ago, Na-Denes around nine thousand years ago, and that the Aleuts and Eskimos diverged about four thousand years ago.[36]

Greenberg's conclusions are deeply controversial but agree quite well with evidence from dental studies and surveys of genetic variation, in particular the highly original work of Luigi Luca Cavalli-Sforza of Stanford University. In a series of books— *Cultural Transmission and Evolution* (1981), *African Pygmies* (1986), *The Great Human Diasporas* (1993), and *The History and Geography of Human Genes* (1994)—Cavalli-Sforza and his colleagues have examined the variability of both blood, especially the rhesus factor, and genes around the world. This has led to fairly good agreement on the dates when early humans spread out across the globe.

It has also led to a number of extraordinary possibilities in our *longue durée* history. For example, it seems that the Na-Dene,

Sino-Tibetan, Caucasian, and Basque languages may be related in a very primitive way, and once belonged to a superfamily that was broken up by other peoples, shunting this superfamily into backwaters, and expelling Na-Dene speakers into the Americas. The evidence also shows great antiquity for Basque speakers, whose language and blood is quite different from those around them. Cavalli-Sforza notes the contiguity between the Basque nation and the early sites of cave art in Europe, and wonders whether this is evidence for an ancient people who recorded their hunter-gatherer techniques on cave walls and resisted the spread of farming peoples from the Middle East.

Finally, Cavalli-Sforza attempted to answer two of the most fascinating dating questions of all—when did language first appear, and was there ever a single ancestral language, a true mother tongue? At that time, some paleontologists believed that the Neanderthals had been wiped out because they did not have language. Against that, Cavalli-Sforza points out that the region in our brains responsible for language lies behind the eye, on the left side, making the cranium slightly asymmetrical. This asymmetry is absent in apes but present in the skulls of *Homo habilis* dated to 2 million years ago. Furthermore, our brain case ceased to grow about 300,000 years ago, and so on this basis it seems that language might be older than many paleontologists think. On the other hand, studies of the way languages change over time (a rate that is known, roughly) point back to between 20,000 and 40,000 years ago, when the main superfamilies split. This discrepancy has not been resolved.[37]

Regarding the mother tongue, Cavalli-Sforza relies on Greenberg, who claims that there is at least one word that seems to be common to all languages across the world. This is the root word *tik*.

Family or Language	Forms	Meaning
Nilo-Saharan	tok-tek-dik	one
Caucasian	titi, tito	finger, single
Uralic	ik-odik-itik	one
Indo-European	dik-deik	to point
Japanese	te	hand
Eskimo	tik	index finger
Sino-Tibetan	tik	one
Austroasiatic	ti	hand, arm
Indo-Pacific	tong-tang-ten	finger, hand
Na-Dene	tek-tiki-tak	one
Amerind	tik	finger

For the Indo-European languages, those stretching from Western Europe to India, Greenberg's approach has been taken further by Colin Renfrew, the Cambridge-based archaeologist who rationalized the effects of the ^{14}C revolution in dating. Renfrew's aim in *Archaeology and Language* (1987) was not simply to examine language origins but to compare those findings with others from archaeology, to see if a consistent picture could be arrived at and, most controversially, to identify the earliest homeland of the Indo-European peoples, to see what light this threw on human development overall. After introducing the idea of regular sound-shift, according to nation . . .

	French	Italian	Spanish
milk	lait	latte	leche
fact	fait	fatto	hecho

Renfrew went on to study the rates of change of language and to consider what the earliest vocabulary might have been. Comparing variations in the use of key words (like "eye," "rain," and "dry"), together with an analysis of early pottery and a knowledge of early farming methods, Renfrew examined the spread of farming through Europe and adjacent areas. He concluded that the central homeland for the Indo-Europeans, the place where the mother tongue, "*proto-Indo-European*," was located, was in central and eastern Anatolia about 6500 BC and that the distribution of this language was associated with the spread of farming.[38]

The intriguing thing about all this for us is the measure of agreement between archaeology, linguistics, and genetics. The spread of peoples around the globe, the demise of the Neanderthals, the arrival of humanity in the Americas, the rise of language, its spread associated with art and with agriculture, its link to pottery, and the different tongues we see about us today, all fall into a particular order, perhaps the beginnings of the last chapter in the evolutionary synthesis.

Joining the Hemispheres

The place of the Americas—the New World—in the general scheme of things took some time to figure out, but it makes a fitting end to this chapter.

The question as to whether America was part of Asia, or a landmass in its own right, was only settled in 1732, nearly a quarter of a millennium after Christopher Columbus had first set foot on Guanahani in what is now the Bahamas. Only in that year did Ivan Fedorov and Mikhail Grozdev finally discover Alaska. In 1778 Captain James Cook sailed through the Bering Strait, noting that only a short reach of sea sixty miles wide separated the continents, convincing many that this was the point of entry for the first Americans.

The man who first put forward the idea that there was once a land bridge between Russia and America was Fray José de Acosta, a Jesuit missionary, in 1590. By then he had lived in Mexico and Peru for nearly twenty years, and he took it as an article of faith that, since Adam and Eve had begun life in the Old World, man must have migrated to the Americas. Moreover, he thought transoceanic travel unlikely, preferring the idea that "the upper reaches" of North America were joined or "approach near" Russia, with a narrow enough water gap that migration would not be inhibited. He noticed too the spread of smaller animals, considering it unlikely that they had swum across even a short stretch of water. Overland travel was much more likely.[39]

Angelo Heilprin (1853–1907), a geologist, also drew inferences in 1887 from animal distribution, observing that Old and New World animal species were relatively dissimilar at southern latitudes, more similar in mid-latitudes, and "nearly identical" in the north. To him it was clear that "if species diversification was a function of distance from the north, then the species must have dispersed from that direction." Not long afterward, another geologist, the Canadian George Mercer Dawson (1849–1901), noted that the seas separating Alaska from Siberia were shallow and "must be considered physiographically as belonging to the continental plateau region as distinct from that of the ocean basins proper." Dawson added that "more than once and perhaps during prolonged periods [there existed] a wide terrestrial plain connecting North America and Asia." He had no idea of the ice ages but accepted that continental uplift had from time to time raised the seabed above the water level. In 1892 great excitement was caused when some mammoth bones were discovered on the Pribilof Islands, three hundred miles west of Alaska. "Either these giant hairy elephants were awfully good swimmers, or the islands were once high spots in a broad plain, conjunct . . . with the entire Alaskan and Siberian landmasses." W. A. Johnson, another Canadian geologist, added a final gloss. In 1934, he made a link between sea-level changes and the ice ages, the existence of which had been affirmed only in 1837. "During the Wisconsin stage of glaciation [~110,000–11,600 years ago]," he wrote, "the general level of the sea must have been lower owing to the accumulation of ice on the land. The amount of lowering is generally accepted to be at least 180 feet, so that a land bridge existed during the height of the last glaciation." This agreed well with the argument that the Swedish botanist Eric Hultén made about the same time, that the area of the Bering Strait had been a refuge for plants and animals during the Ice Age. It was Hultén who named the region Beringia in honor of Vitus Bering, a captain in the Russian Navy, who discovered the strait in the eighteenth century. Hultén argued that the Beringia refugium provided the terrestrial route by which ancient humans reached the New World.[40]

The Link Between Sea Levels and Life

The evolution of scholarship surrounding the Bering Land Bridge is a fascinating story in its own right. It comprises three aspects. First, the attempts to prove that there really *was* a land bridge, at various remote times in the past; second, inquiries as to what the land bridge was like physically, what its geography was like, what plants and animals it could support; and third, trying to ascertain what sort of people traveled across, and when.

The Pleistocene era, more popularly known as the Ice Age, began roughly 1.65 million years ago. Most scientists think that it ended some ten thousand years ago, though others argue we are still in it, "merely enjoying an interglacial reprieve." During the Pleistocene, warming trends followed cooling ones in great cycles that could last hundreds of thousands of years. The most recent cold cycle began about 28,000 years ago, with temperatures falling relentlessly until roughly 14,000 years ago—this is the Late Glacial Maximum, or LGM. Conditions were far harsher than anything known today, especially at the polar regions, and particularly so in the northern hemisphere. This was owing to the direction of the earth's rotation, which made ocean currents and the weather they affected worse there than anywhere else and, because there is more (heavy, dry) land in the northern hemisphere than in the southern one, producing irregularities in the earth's orbit.

All of which meant that more snow fell in winter than melted in summer, building up in layers that melted slightly in the short summers, then recrystallized. Each year's snowfall pressed down on the layer of the year before, creating great masses of ice—the

Laurentide Ice Sheet, the biggest in North America, built up to a height of nearly two miles. It was centered on what is now Hudson Bay, but it eventually smothered all of what would come to be called Canada, four thousand miles across. To the west, the Laurentide merged with North America's other great ice sheet, the Cordilleran, which stretched three thousand miles down the coastal mountains of western North America, from Puget Sound to the Aleutian Islands.[41]

Eventually, the glaciers held around one-twentieth of the world's water, and half of that was to be found in the Laurentide Ice Sheet. As a result, the seas eventually dropped by about 125 meters, or 400 feet (not 180, as Johnson said) below where they are now. As the waters receded, Asia and North America "began to reach for each other like the outstretched arms of God and Adam on the ceiling of the Sistine Chapel. When the finger tips touched, a charge of new life streamed into the Americas."[42] An imaginative analogy, but the contact was achieved with geological slowness. Gradually, the shelf widened until, at 18,000 to 14,000 years ago—the height of glacial activity, or LGM, to remind ourselves—the shelf between Alaska and Siberia was exposed as dry land for over nine hundred miles, north to south.

Geological studies, involving drilling cores of both land and ice, show that there have been some sixteen ice ages in the last million years alone, separated by "inter-glacials." The Bering Land Bridge would have connected the continents during most of these glacials, when animals—if not humans, who had yet to evolve or reach Siberia—could have switched between the New and Old worlds. The last land bridge is of most interest precisely because humans were around at that time. This period endured from, roughly speaking, 25,000 to 14,000 years ago.

Even without ice, it was a harsh landscape, dry and windy. Loess (windblown glacial silt) built up in great gray dunes. Vegetation was thin, the land little more than a polar desert, a drier version of today's tundra. Despite this, Beringia—as it came to be called—was home to a variety of animals. Or that is what research shows. Woolly mammoths were probably the biggest of the beasts that roamed the region. Their six-inch-thick hairy coats, hanging in ragged skirts, offered sufficient protection against the cold and the bitter winds. Ground sloths, weighing as much as six thousand pounds, and the long-horned steppe bison, were almost as big. Horses, which evolved in North America, migrated across the land bridge going the other way, *into* Asia, but they almost certainly had much thicker coats than today's breeds. Various forms of antelope, moose, caribou, and sheep—all these inhabited the land bridge during the Ice Age. And so too did huge saber-toothed tigers with their six-inch-long canines capable of piercing the thick hides of the mammoth and bison, plus giant lions and packs of timber wolves. An ancient form of bear, bigger even than today's Alaskan grizzly, completed this exotic bestiary.[43]

The identification of this plant and animal wildlife was itself an achievement of scholarship. Eric Hultén, the Swedish botanist referred to earlier, studied the plants of Siberia and Alaska in the 1930s to produce his *Flora of the Aleutian Islands*. Hultén was statistically minded, and as well as describing the plants, he noted their distribution, in particular their spread around Canada's Mackenzie River and Siberia's Lena River. Plotting these distributions on a map, he saw that they spread out in a series of ovals, stretching east–west. Moreover, these ovals, as well as being symmetrical, were also concentric, and the axis of symmetry was always located in a line drawn *through* the Bering Strait. A consistent picture therefore suggested itself. Hultén imagined there must have been at one time a dry landmass "stretching from mostly unglaciated Siberia into mostly unglaciated Alaska." This landmass was isolated from the much colder areas around it and therefore acted as "a great biological refugium, a place where northern plants and animals survived extinction and from whence they evidently spread when the glaciers receded."

Hultén's work was soon built on by J. Louis Giddings, a Texan archaeologist who, in the 1940s, at Cape Krusenstern, southeast of Point Hope, identified "no fewer than 114 beach ridges each parallel to the shore line and extending more than three miles inland." Not only that, each ridge furnished a series of archaeological discoveries in which the outer ridges (closer to today's coastline) were older than those further inland. The natural conclusion was that these seafaring cultures had transferred inland by stages as the seas rose and encroached on their dwellings, showing neatly how human habitation and sea level were intimately linked.[44]

But the individual who did more than anyone else to advance the scientific understanding of Beringia was David M. Hopkins, a graduate of the University of New Hampshire. Working together with William Oquilluk, a famous Iñupiat historian from north Alaska, Hopkins's first project was a study of fossil mollusc shells, because he understood that their distribution and sedimentation would reveal when the Bering Strait was open and when it was not. The basis for this study was the natural history of the giant snail *Neptunea*. This has existed in the north Pacific all the way back to the Tertiary geological epoch, some 65 million years ago. But there was no evidence of *Neptunea* in the Atlantic sediments until the early Pleistocene, about a million years ago. What this suggested was that a land bridge blocked marine migration for most of the Tertiary period and was only flooded out at the start of the Pleistocene, allowing *Neptunea* to move north and east and to reach the Atlantic.

Oquilluk introduced Hopkins to many shell deposits, and as a result they collaborated on a seminal paper, published in *Science* in 1959, dedicated to the Bering Land Bridge. Their conclusion was that the bridge had been in existence throughout most of the Tertiary era (from 65 to 2 million years ago), that the evidence from fossil fishes suggested that a waterway cut through the bridge in the middle of the Eocene (about 50 million years ago), and that it was fully submerged around a million years ago. Since then, the bridge had appeared and been submerged numerous times as the ice ages had come and gone. The land bridge disappeared for the last time, they said, some 9,500 years ago.[45]

The final confirmation of the Beringia landscape was discovered by Hopkins himself. In 1974, near a lake at Cape Espenberg on the northern Seward Peninsula, he came across a layer of tephra, or volcanic ash, that was fully one meter deep and had congealed within it masses of twigs and tufts of grass. The nearby lake was in fact a maar—a circular lake of the kind produced when a volcano erupts at ground level. Hopkins was aware of previous research, which had established that this maar, Devil Mountain, erupted eighteen thousand years ago, at a time when the land bridge existed. It therefore followed logically that the twigs, roots, and grasses congealed in the tephra *were land bridge vegetation.* Examination of this plant material showed it to be a dry meadow and herb-rich tundra, a mix of herbs and grasses, in particular the sedge *Kobresia*, plus the occasional willow, and a carpet of mosses. This amalgam of *Kobresia*-dominated vegetation has since been discovered in the stomachs of several preserved mammals and confirms the steppe-type landscape. Hopkins, who died in 2001, held firmly to the view that the Bering Land Bridge could support whole herds of grazing animals, and the predators who sought them out. Importantly, this environment meant that the land bridge would have supported humans.[46]

The Coherence of the Great Divide

The arrival of early humans in the Americas in, roughly speaking, 15,000 BC is important for our story. It is important because that arrival meant that, for approximately 16,500 years—15,000 BC to AD 1500, 640 generations—there were *two* populations of people in the world who, insofar as we know, were unaware of each other. Therefore, the development of these two populations over the intervening centuries, until Christopher Columbus arrived in October 1492, formed an extraordinary—and extraordinarily valuable—natural experiment. Did they develop differently or much the same?

Humanity—and civilization—developed very differently in

the two hemispheres. But the variation wasn't random. There were important, systematic differences between the Old World and the New in terms of, for example, climate, geography, geology, animal and plant biology, what animals existed where, the spread of certain kinds of plants, and much else. The important point is that when all this is taken into account, and collated, we find that Big History, though it differs between the hemispheres, is nonetheless derived from the same basic principles. Coherence is maintained.

13

THE OVERLAPS BETWEEN
NEW DISCIPLINES: ETHOLOGY,
SOCIOBIOLOGY, AND
BEHAVIORAL ECONOMICS

The Nobel Prizes, established in 1901—five of them, in physics, chemistry, medicine or physiology, peace, and literature—have played their own distinguished role in modern history. But there has been an interesting difference in the way the prizes have developed. In physics and chemistry, the prizes tend to have been uncontroversial. A list of winners of physics Nobels, for instance, would mark all the decisive twentieth- and twenty-first-century advances in the subject, with very few objections from specialists in the field. In peace and literature, it has been very different, the award often signifying someone the judges thought should be better known, or recognizing a body of work that was not always as popular as it might be. As a result there was much less agreement in the political and literary worlds about who merited the awards and who did not.

The award for medicine or physiology was different again. It was mainly presented, as you would expect, for mainstream developments in biology—Robert Koch on tuberculosis, Frederick Banting and John Macleod for the discovery of insulin, Thomas Morgan for discovering the role played by the chromosome in heredity—with a second string which honored those who developed new forms of treatment: Alexis Carrel, who devised methods to make transplants safer, Alexander Fleming for the discovery of penicillin, António Egas Moniz for the discovery of the therapeutic value of leucotomy in certain psychoses. But then, in 1973, the Nobel committee produced something entirely new.

In that year, the award of the Nobel Prize in medicine or physiology went to two Germans, Karl von Frisch and Konrad Lorenz, and to the Oxford-based Dutchman Nikolaas Tinbergen, for their work on ethology. This was doing to science what the other Nobel committees had been doing in peace and literature—recognition for a little-known area of study which, the committee thought, ought to be better understood. It was an interesting decision in several ways, for there was high drama implicit in the award. Lorenz had worked for the Nazis during the Third Reich. Von Frisch had been pursued by the Gestapo owing to the fact that his grandmother was "probably" Jewish. Tinbergen had been sequestered for two years in a camp in occupied Holland that contained hostages, held captive and liable to be executed in the event of German retaliation for sabotage and assassination attempts by the Dutch Resistance. Despite this, both von Frisch and Tinbergen, who had known Lorenz before the war, had forgiven him.

The award of the 1973 Nobel was recognition for a relatively new biological discipline in which each man had been a founding father. (Remarkably, Tinbergen's brother Jan had been awarded the Nobel Prize in economics four years earlier.)* The new discipline, ethology, was the study of animal behavior, with a strong comparative element. Ethologists are interested in animal

---

* In 1968 the Swedish Academy of Sciences, which administered the Nobel Prizes, according to Alfred Nobel's original will, was made responsible for administering the Sveriges Riksbank Prize in Economics in memory of Nobel, and this, now known as the Nobel Prize in economics, was first awarded in 1969.

behavior for what that might reveal about instinct, and what, if anything, separates man from all other forms of life.

Tinbergen's classic work, carried out since the war (and after he had moved from Leiden to Oxford), elaborated on Lorenz's idea of "fixed action patterns" and "innate releasing mechanisms" (IRMs). Experimenting with the male three-spined stickleback, Tinbergen showed the crucial importance of why at times the fish stood on its head to display its red belly to the female: this stimulated a mating response. Similarly, he showed the significance of the red spot on a herring gull's bill: it elicited begging from a chick. It was later shown that such IRMs were more complicated, but the elegance of Tinbergen's experiments caught the imagination of scientists and public alike.[1] The possibilities of revealing overlaps in the behavior of different species fascinated everyone.

Flo, Flint, and Flame: Almost Human

Three men won the 1973 Nobel, but far and away the most influential people in persuading a wider public that ethology could offer valuable insights were three extraordinary women in Africa, whose imaginative and brave forays into the bush proved remarkably successful, underlining above all that animals *could* be studied in the wild. These women were Joy Adamson, who worked with lions in Kenya, Jane Goodall, who investigated chimpanzees at Gombe Stream in Tanzania, and Dian Fossey, who spent several years working with gorillas in Rwanda.

Joy and her husband, George, also caught the imagination of the public with their exploits with a young lioness, Elsa. Dian Fossey became notorious when her gorilla studies were brought to a sudden end by her murder, possibly by a colleague who fled the country, fearing he could never receive a fair trial, and who was convicted in his absence. But the most important of the three women was Jane Goodall, whose book *In the Shadow of Man* was published just before Tinbergen and the others won their prize.

Goodall (like Fossey after her) was a protégé of Louis Leakey. Apart from his other talents, Leakey was a great womanizer, who had affairs with a number of female assistants. Goodall, born in London in 1934 and the daughter of a novelist, developed an interest in animals and dreamed of working in Africa from her earliest years. She had approached Leakey as early as 1959, the year of Zinj, begging to work for/with him. When he met her, Leakey noted that Goodall was very knowledgeable about animals, and so was born a project that had been simmering at the back of his mind for some time. He knew of a community of chimpanzees at Gombe Stream, near Kigoma, on the shore of Lake Tanganyika. Leakey's thinking was simple. Africa had a very rich ape population; man had evolved from the apes; and so the more we discovered about them, the more we were likely to understand how humankind had evolved. Goodall loved her assignment, and her official reports as well as her popular account, *In the Shadow of Man*, released in 1971, managed to be both scientifically important and moving at the same time.[2]

Goodall found that it took the chimpanzees some months to accept her, but once they did she was able to get close enough to observe their behavior in the wild *and* to distinguish one chimpanzee from another (Flo had "ragged ears and a bulbous nose"). This simple insight proved extremely important. She was later criticized by other, more academically grounded scientists, for giving her chimps names—David Greybeard, Satan, Sniff, Goliath, Flint, Flame, and Freud(!)—instead of more neutral numbers, and for reading motives into chimp actions. But these were lame criticisms when set against the richness of her material.[3]

Her first significant observation occurred when she saw a chimp insert a thin stick into a termite mound in order to catch termites that attached themselves to the stick. The chimp then raised the stick to its lips. Here was a chimp using a tool, hitherto understood to be the hallmark of humanity.[4]

As the months passed, the social/communal life of these

primates also began to reveal itself. Most notable was the hierarchy of males and the occasional displays of aggression that brought about this ranking, which by and large determined sexual privilege in the troupe, but not necessarily priority in food gathering. Goodall also recorded that many of the aggressive displays were just that—displays—and that once the less dominant male had made deferential gestures, the dominant animal would pat his rival in what, in a human, would have been a gesture of reassurance. Goodall also observed mother-offspring behavior, the importance of social grooming (picking unwanted matter out of each other's fur), and what appeared to be familial feeling. Young chimpanzees who for some reason lost their mothers shriveled physically and/or became nervous—what we would call neurotic.

She recorded that chimpanzees could tell *more* from *less*; that low-ranking chimps had to learn deception, to do things in secret if they wanted to get away with them; that Fifi was "enchanted" when she became sexually attractive to the big males and, on one occasion, tweaked a male's penis to get him interested (it worked). She described a four-year, three-way "war" between the chimp communities of Kasakala, Kahama, and Kalanda. She noted chimp psychopathology—females who preyed on other females' babies and even ate them. She felt that chimps sometimes showed depression, especially if a mother died, and she described a lot of "attachment behaviour." She witnessed some care for the sick, some sharing of food, and long, peaceful sessions of relaxed grooming behavior—one widely circulated photograph showed four chimps sitting in line, like a rowing crew, each one picking debris from the back of the one in front.[5]

Controversially, she thought that chimps had a rudimentary sense of self and that children learned much behavior from their mothers. In one celebrated instance, she observed a mother with diarrhea wipe herself with a handful of leaves. Immediately, her two-year-old infant did the same, although his bottom was clean. And, only slightly tongue-in-cheek, she underlined that chimps have a history, if only they could write it down. "As in human societies, certain individuals have played key roles in shaping the fortunes of their community. Some of the adult males who have demonstrated outstanding leadership qualities of determination, courage or intelligence would figure prominently in chimpanzee history books: Goliath Braveheart, Mike of the Cans, Brutal Humphry, Figan the Great, Goblin the Tempestuous."[6]

Goodall's books were a great success and she didn't need to labor in her accounts the overlaps between ape and human behavior—it was plain for all to see. Human behavior had evolved, just as the human form had.

The Emergence of Evolutionary Psychology and Our Original Nature

An American author, Robert Ardrey, drew yet more attention to Africa in a series of books, *African Genesis* (1961), *The Territorial Imperative* (1967), and *The Social Contract* (1970). Ardrey did much to familiarize the idea that all animals—from lions and baboons to lizards and jackdaws—had territories, which varied in size from a few feet for lizards to a hundred miles for wolf packs, and which they would go to extreme lengths to defend. He also drew attention to the rankings in animal societies and to the idea that there is a wide variety of sexual arrangements, even among primates, which, he thought, demolished Freud's ideas.[7]

In popularizing the idea that humans originated in Africa, Ardrey also emphasized his own belief that Homo sapiens is emotionally a wild animal who is domesticating himself only with difficulty. He thought that man was originally an ape of the forest who was defeated by the other great apes and forced into the bush. For Ardrey, mankind could only survive and prosper so long as he never forgot that he was at heart a wild animal. The fieldwork that lay at the center of Ardrey's book, *African Genesis*, also helped underline the idea that, contrary to the view prevailing before World War II, humanity originated not in Asia but in Africa and that, by and large, it emerged only once, somewhere along the Great Rift Valley, rather than several times in different places (his books appeared before the idea of "mitochondrial Eve" confirmed Africa as the place of origin of humankind).

This simplification of the paleontological/primate history was an important conceptual breakthrough, and as a result of the intellectual battle fought out in the early 1970s, we may say that agreement on a small but important number of conclusions was arrived at. In the first place, *evolutionary psychology* emerged as a separate discipline, though it was in fact a synthesis of anthropology, ethology, genetics, and psychology (and some mathematics). It is an approach that holds that we evolved as a species on the African savannah as hunter-gatherers and that it is a hunter-gatherer psychology and hunter-gatherer skills—skills both individual and social—that we are most adapted to. These skills, in being adapted and evolved over hundreds of thousands and even millions of years, are essentially part of our unconscious instinctive makeup. They are part of our innate biological human nature that we take for granted, that affect our actions and beliefs without our being aware of it. Anthropology shows that the hunter-gathering lifestyle has occupied 99.5 percent of the time humans have been on earth.

These ethological evolutionary strands all converge on two ruling principles. First, there *is* a *universal human nature* that arose, over aeons of time, through natural and sexual selection of gene traits that helped our prehuman, hominid, and human ancestors to first survive and then reproduce individually and inclusively. As a result, the actual behavior of human beings tends to revolve around a delimited set of universal themes in the domains of shelter and security, food procurement and nutrition, sexuality and sex roles, mating, parenting, and interaction within and between groups, all of which promote the probability of gene survival. These include the care and protection of children, peer bonding and peer play, status-seeking, competing for valuable resources, courtship, sexual bonding and marriage, sharing and storing food, seeking shelter, cooperating, reciprocal altruism, discriminating against strangers, the splitting of groups when they achieve a critical size, the expression of out-group hostility and in-group loyalty, grooming, teaching, subscribing to the beliefs and practices of myth, religion, and so on.

The second principle is that, in modern contexts, some environments are now toxic and deeply damaging and thwarting of our original nature. On this basis, psychopathological matters—sexual abuse, incest, infidelity, rejection, low self-esteem, repressed anger, depression, addiction, and the like—are seen as *species problems as well as individual ones*. This provides a meaningful and scientific frame of reference for making sense of our predicament.[8]

From now on, scientists began to extend the idea of evolution, and adaptation, to more and more aspects of human existence. In science outside medicine, evolutionary psychology would dominate the rest of the century, alongside molecular biology and particle physics.

The Biology of Psychology

As noted, the first account of Jane Goodall's work was published just before Tinbergen and the others received their Nobel Prize. Like many Nobels it was given for work carried out long before— in this case from before World War II. In fact, the ethological approach had been adopted and adapted well before 1973 and used in an amalgamation with psychology and pediatrics. This was seen most especially in the seminal work of John Bowlby and Mary Ainsworth discussed in the Preface.

Bowlby's work on attachment theory linked up neatly with Goodall's in Gombe Stream, Fossey's among the gorillas of Uganda, and with several other ethological studies on both sides of the Atlantic. The mother-infant bond became firmly established and not just among primates, but across a wide range of species. It soon formed a central plank in what became at times a vicious fight over our fundamental human nature.

For the plain fact is that, although ethology was linking up with paleontology and psychology in seemingly fruitful ways, at that time a major division was opening up between those who believed that human nature is a "blank slate," meaning that culture and experience create who we are, and those who are more receptive to the idea that evolution and genes account for much of our behavior. Bowlby's work on attachment (1969–73), and the award of the 1973 Nobel Prize to the three ethologists, therefore occurred in the middle of—and were themselves important ingredients in—a major paradigm shift in the way biology and psychology were being drawn together. It was in its way as important as the convergence of particle physics and quantum chemistry to create molecular biology.

Against "Forceless Generalities"

A major advocate of the blank-slate view was the influential American anthropologist Clifford Geertz (1926–2006), based at the Institute for Advanced Study in Princeton, New Jersey. He shared very strongly the view that the world is "a various place" and that we must confront this "uncomfortable truth" if we are to have any hope of understanding the "conditions" by which we live. In two books—one, *The Interpretation of Cultures*, published in the crucial year of 1973, and the other, *Local Knowledge*, a decade later—he detailed his view that subjectivity is *the* phenomenon for anthropologists like himself, and others in the biological sciences, to tackle. The basic unity of mankind, according to Geertz, is an empty phrase if we do not take on board that drawing a "line between what is natural, universal, and constant in man and what is convention, local and variable [is] extraordinarily difficult. In fact, it suggests that to draw such a line is to falsify the human situation, or at least to misrender it seriously."[9]

The hunt for universals began with the Enlightenment, says Geertz, and that aim directed most Western thought, and has been a paradigm of Western science, and the Western notion of "truth," ever since. Pursuing fieldwork in Java, Bali, and Morocco, Geertz dedicated his entire career to changing that view, to distinguishing between the "thin" and "thick" interpretations of cultures around the world. For Geertz, "thick" means to try to understand the signs and symbols and customs of another culture in its own terms, by assuming not, as Claude Lévi-Strauss did, for example, that all human experience across the globe can be reduced to structures, but instead that other cultures are just as "deep" as ours, just as well thought out and rich in meaning, but perhaps "strange," not easily fitted into our own way of thinking.

Geertz's starting point was paleontology. It was wrong in his view to assume that the brain of Homo sapiens evolved biologically and that cultural evolution followed. Surely, he argued, there would have been a period of overlap, of co-evolution. As humans developed fire and tools, our brains would have still been evolving—and have evolved to take into account fire and tools. This evolution may well have been slightly different in different parts of the world, so that to talk of one human nature, even biologically speaking, may be misleading. Geertz's own anthropology, therefore, involves the meticulous description of certain alien practices among non-Western peoples, where the examples are chosen precisely because they appear strange to "us."

The Balinese, for example, have five different ways of naming people. Some of these are rarely used, but among those still in use are names that convey, all at the same time, the region one is from, the respect one is held in, and one's relation to certain significant others. In another example, he shows how a Balinese man, whose wife has left him, tries to take (Balinese) law into his own hands, but ends up in a near-psychotic state since his actions cause him to be rejected by his society. These matters cannot be compared to their Western equivalent, says Geertz, because there *are* no Western equivalents.[10]

Cultural resources are, therefore, not so much accessory to thought as "ingredient" to it. For Geertz, an analysis of a Balinese cockfight can be as rich and rewarding about Bali thought and society as, say, an analysis of *King Lear* or *The Waste Land* is about Western thought and society. For him, the old division between sociology and psychology—whereby the sociology of geographically remote societies differed, but the psychology stayed the same—has now broken down. Geertz's own summing up of his work is that "every people has its own sort of depth." "The hallmark of modern consciousness . . . is its enormous multiplicity. For our time and forward, the image of a general orientation, perspective, *Weltanschauung*, growing out of humanistic studies (or for that matter out of scientific ones) and shaping the direction of cultures is a chimera. . . . Agreement on the foundations of scholarly authority, old books and older manners, has disappeared. . . . The concept of a 'new humanism,' of forging some general 'the best that is being thought and said' ideology and working it into the curriculum, [is] not merely implausible but utopian altogether. Possibly, indeed, a bit worrisome." Life will in future be made up of vivid vernaculars, rather than "forceless generalities."[11]

This view, of course, goes diametrically against the theme of this book, that there is an emerging order—a convergence, even a kind of unity—between the sciences. And that this order or unity is one of the most important (and satisfying) elements in scientific knowledge, which gives it an authority unrivaled among other forms.

Deep Structure and Surface Structure: Instinctive Knowledge?

One man who recognized this was Noam Chomsky. Born in Pennsylvania in 1928, Chomsky is the son of a Hebrew scholar who interested his son in language. The son's book, *Syntactic Structures*, published in 1957, and then his 1975 book on generative grammar, initiated what came to be called the Chomskyan revolution in psychology.

Chomsky, by then a professor at MIT, argued—and this

pitted him against Geertz and the likes of the behaviorist B. F. Skinner—that there are inside the brain universal, innate, grammatical structures. In other words, the "wiring" of the brain somehow governs the grammar of languages. He based much of his view on studies of children in different countries, which showed that whatever their form of upbringing (and whatever Geertz might say), they tended to develop their language skills in the same order and at the same pace everywhere. His point was that young children learn to speak spontaneously without any real training, and that the language they learn is governed by where they grow up. Moreover, they are very creative with language, using at a young age sentences that are entirely new to them and cannot have been related to experience. Such sentences cannot therefore have been learned in the way that behaviorists and the blank-slate advocates say.[12]

Chomsky argued that there is a basic structure to language, that this structure has two levels, surface structure and deep structure, and that different languages are more similar in their deep structure than in their surface structure. For example, when we learn a foreign language, we are learning the surface structure. This learning is in fact only possible because the deep structure is much the same. German and Dutch speakers may put the verb at the end of a sentence, which English and French speakers do not, but German, Dutch, French, and English *have* verbs, which exist in all languages in equivalent relationships to nouns, adjectives, and so on.

Chomsky went on to provide yet more examples of our uncanny language ability. For example, the meaning of a house is more intuitive than it looks. If someone says that *John is painting the house brown*, we know—apparently without instruction—that it is the external surface of the house that is being painted, rather than the inside. But the meaning of a house cannot be restricted to its external surface. "If two people are equidistant from the surface, one inside and one outside, only the one outside can be described as 'near' the house." Even very young children seem to know such facts, "suggesting that the knowledge is in some sense *antecedently available* to the organism." (Italics added.)[13]

Here Chomsky was on to something that brings us back to Bowlby. In saying that knowledge "is in some sense antecedently available" to an organism, he was hinting—more than hinting— at the idea of the unconscious, or preconscious.

This idea, the unconscious, had a very unusual—and uncomfortable—ride during the twentieth century. Made much of by Freud in his *The Interpretation of Dreams*, published in 1900, it had been dismissed by many as unscientific. At the same time, many ordinary people, and many artists—painters and writers, even choreographers—found it useful.

Chomsky had his idea of knowledge being antecedently available, Bowlby had his concept of "interior working model," and then, in 1972 and again in 1974, two celebrated papers by Robert Trivers, an anthropologist at Rutgers University, were published—exploring "reciprocal altruism," "parental investment theory," "parent-offspring conflict," and "female choice theory." Female choice theory argues that women—mothers—determine social evolution far more than males do because they have more "invested" in the outcome of families (and their genes) and this places the mother-infant attachment bond center stage. "Parent-offspring conflict theory" was also viewed in an evolutionary context, to show that the adaptive needs of the infant do not always coincide with those of the mother, and that this can give rise to such psychoanalytic-related ideas as repression, "oral manipulation of the mother," and Oedipal fixations. More and more, the idea of the unconscious, hitherto a psychological concept, was being reconceived as biological.[14]

A final element in this rash of developments around 1973 came from a quite different area, both geographically and conceptually. This was the publication, in 1974, of a paper, "Judgment Under Uncertainty: Heuristics and Biases," by Daniel Kahneman and Amos Tversky. This was a paper originally concerned with judgment and decision-making, especially so far as rational choice was related to economic behavior. But Kahneman and Tversky were psychologists, and in the late 1950s the former had done his national service in the Israeli Defense Forces. As a psychology graduate he was assigned to the army's psychology branch, where one of his duties was to help evaluate candidates for officer training. Ironically, the Israeli Army used methods devised by the British Army in World War II, conceived by psychiatrists at the Tavistock Clinic, including none other than John Bowlby.[15]

Not too much should be made of this coincidence, but the exercise did have a profound effect on Kahneman. He found that the British techniques he used *failed* to predict good leaders (though they had, by all accounts, helped win the war). "The evidence that we could not forecast success accurately was overwhelming." Time after time, the feedback was clear: "The story was always the same; our ability to predict performance was negligible." This was both worrying and disappointing, but much more interesting for Kahneman—and of longer-term importance—was that the failure to predict success in the tests *had no effect on his own behavior.* Despite the evidence, he and his colleagues went on giving the tests and went on being convinced that their judgments were correct. He had discovered, he said, his first "cognitive illusion." He was not behaving rationally but, seemingly, could do nothing about it. His unconscious was in the way.

And so, as a result, and after much more study, in 1974, he and Tversky published what has since become a classic paper. This was the first of a whole series of studies, which culminated in 2002 when Kahneman was awarded the Nobel Prize in economics for his work on behavioral psychology, again as it related primarily to economic decision-making. (Tversky died in 1996.) This work will be discussed more fully in chapter 16 but, to anticipate briefly, what it shows is that humans have a number of *inbuilt* cognitive biases—intuitive and instinctive ways of behaving, rapid shortcuts which are adaptations and have evolved over many years and generations. As adaptations, they have become instinctive and therefore resistant to change, as Kahneman had himself found—and which drew him to the phenomenon in the first place.

So, with Geertz and Skinner on one side, and Chomsky, Bowlby, and Kahneman on the other, the two approaches were at loggerheads over the basic understanding of human nature. At this point, however, a number of heavyweights came down on Chomsky and Bowlby's side, to support the new synthesis that was emerging between ethology, biology, and psychology—and sought to take it further.

The Watershed Moment

The first was *Chance and Necessity* by Jacques Monod, part of a three-man French team that had won the 1965 Nobel Prize for uncovering the mechanism by which genetic material synthesizes protein. In his book, published in 1970, Monod sought to use the latest biology, produced since Watson and Crick's discovery of the double helix, to define life—and in considering what life is, he went on to consider the implications that might have for ethics, politics, and philosophy.[16]

Although he was a biologist, Monod's underlying argument was that life is essentially a physical and even a mathematical phenomenon. His initial purpose was to show how entities in the universe can "transcend" the laws of that universe while nevertheless obeying them. For Monod, two of the great intellectual successes of the twentieth century—the free market and the transistor—shared an important characteristic with life itself: amplification. The rules allow for the constituent parts to spontaneously—naturally—produce *more* of whatever system they are part of. On this reasoning there is nothing in principle unique about life.

In the technical part of his book, Monod showed how proteins and nucleic acids, the two components which all life is made from, *spontaneously* adopt certain three-dimensional forms, and that it is these three-dimensional forms which predetermine so much else. This was taking up some of D'Arcy Thompson's ideas in *Growth and Form*, and the approach would become more and more influential as the decades passed. It is this spontaneous assembly that, for Monod, is the most important element of life. These substances, he says, are characterized by physical—and therefore geometric, mathematical—properties. "Great thinkers, Einstein among them, have often . . . wondered at the fact that mathematical entities created by man can so faithfully represent nature even though they owe nothing to experience." This is surely the greatest of mysteries, but Monod implies that there is nothing especially "wonder"-ful about this. Life is just as much about mathematics and physics as it is about biology.[17]

He felt that ideas, culture, and language are survival devices, that there is survival value in myth (he avoided the use of the term "religion"), but that they will in time be replaced. He thought Christianity and Judaism more "primitive" religions in this sense than, say, Hinduism, and implied that the latter would outlast Judaeo-Christianity. And he felt that the scientific approach, as epitomized in the theory of evolution, which is a "blind" process, not leading to any teleological conclusion, is the most "objective" view of the world, in that it does not involve any one set of individuals having greater access to the truth than any other group. In this sense, he thought that science disproves and replaces such ideas as animism, vitalism, and above all Marxism, which presents itself as a privileged scientific theory of the history of society. Monod therefore saw science not only as a way of approaching the world, but as an *ethical* stance from which other institutions of society could only benefit.[18]

This was an important moment. We began this chapter after recording the change from quantum chemistry to molecular biology. We can now see that, relatively quickly, this strong underpinning of the theory of evolution by what we might call the "hard" sciences had a major impact on thinking in a much wider realm.

It marks the watershed moment when the coming together—the convergence—of the sciences achieves such resonance that science itself becomes the basis for comprehending other forms of knowledge.

This had been hinted at often enough before, but Monod's book presented the evolving state of affairs quite starkly and, moreover, was followed by other, no less uncompromising, offerings, by other—no less distinguished—biologists.

The "Genes Hold Culture on a Leash"

The most impressive—and most uncompromising—of these biologists was the Harvard entomologist Edward O. Wilson, who, in 1975, released *Sociobiology: The New Synthesis.* Raised as a Southern Baptist in Alabama (where he read the Bible from cover to cover, twice), Wilson lost his faith suddenly on being introduced as a young man to evolution. ("It seemed to me that the Book of Revelation might be black magic hallucinated by an ancient primitive.") It also seemed to him, he said, that the biblical authors had missed the most important revelation of all—they had made no provision for evolution. "Could it be," he asked himself, "that they were not really privy to the thoughts of God? Might the pastors of my childhood, good and loving men though they were, be mistaken?" It was all too much, and he was a Baptist no more.[19]

Even so, he had no immediate desire to purge himself of his overall religious feelings. "I also retained a small measure of common sense. To wit, people must belong to a tribe; they yearn to have a purpose larger than themselves. We are obliged by the deepest drives of the human spirit to make ourselves more than animated dust, and we must have a story to tell about where we came from, and why we are here. Could Holy Writ be just the first literate attempt to explain the universe and make ourselves significant within it? Perhaps science is a continuation on new and better-tested ground to attain the same end?"[20]

Wilson had the distinction of coining three neologisms that were to prove influential and, again, as with Monod, extended the reach of science. The first of these neologisms was "sociobiology."

In this book, Wilson intended to show the extent to which social behavior—in all animals, including man—is governed by biology, by genes; that the "genes hold culture on a leash."[21] Widely read in every field of biology, and a world authority on insects, Wilson sought to demonstrate that all manner of social behavior in insects, birds, fish, and mammals could be accounted for either by the requirements of the organism's relationship to its environment or to some strictly biological factor—such as smell—that was clearly determined by genetics. He showed how territoriality, for example, was related to food requirements, and how population was related not only to food availability, but to sexual behavior, itself in turn linked to dominance patterns. He surveyed the copious evidence for birdsong, which showed that birds inherit a "skeleton" of their songs but are able to learn a limited "dialect" if moved geographically (not a million miles from what Chomsky was saying about human language). Wilson showed the importance of bombykol, a chemical substance that, in the male silkworm, stimulates the search for females, making the silkworm, according to him, little more than "a sensual guided missile." As little as one molecule of the substance is enough to set the silkworm off, he says, which shows how evolution might happen: a minute change in either bombykol or the receptor structure—equally fragile—could be enough to provoke a population of individuals sexually isolated from the parental stock.

Among the other findings he incorporated into his synthesis was the discovery that among honeybees, and ants of the genera *Formica* and *Pogonomyrmex*, some individuals are unusually active—they fulfill their share of "work"—while others are consistently "lazy." Although they are seemingly healthy and live long lives, the per-individual output of the "lazy" individuals is only a small fraction of that of the harder workers. He was of course inviting a human analogy, whether he intended it or not. In another case he showed that modern hunter-gatherer bands containing about twenty-five individuals occupy land territory that is much the same as carnivorous wolf packs, but a hundred times bigger than similar bands of vegetarian gorillas—it being known that a diet of animal food needs roughly ten times as much area to gain the same energy yield.[22] And in one table, of sixty-six agrarian societies, he showed that the more they subsisted on herding, the more likely they were to believe in an active, creator God—from 20 percent where they subsisted only 15 percent of the time on herding, to 92 percent when their subsistence level was 45 percent.[23]

Wilson surveyed many of the works referred to in this chapter—on gorillas, chimpanzees, lions—as well as the studies of *Australopithecus*, and provided at the end very contentious tables claiming to show how human societies, and human behavior, evolved. This produced a hierarchy with countries like the United States, Britain, and India at the top, Hawaii and New Guinea in the middle, and Aborigine and Eskimo regions at the bottom.

"Borderland" Disciplines

In an introduction to a later edition of the book, Wilson said that he had originally intended his synthesis to be only a union of entomology and population biology. He had been attacked, however, for his chapters on human behavior—for what was called "inappropriate reductionism," the proposal that human social behavior is ultimately reducible to biology. While he felt that "reductionism is the primary cutting tool of science," he said he was just as much an "interactionist" as the next man, though he did feel that there were emerging four borderland disciplines that were awaiting "cooperative exploration." These were: cognitive neuroscience, human genetics, evolutionary biology, and sociobiology itself.

Sociobiology proved as fertile as it was contentious. Much of it concerned the behavior—usually the mating behavior—of birds and insects, which were endlessly fascinating, but it was, perhaps inevitably, human behavior that became the focus of acrimony. Three examples will illustrate why this was so.

The first concerns "beauty," in particular: why do men have standards of beauty that they apply to women? Critics of the sociobiological approach thought it was "highly pernicious" because, in assigning value to women in a "vertical hierarchy" (as the sociobiological studies did), it imposed "cultural standards" that are an expression of "power relations" between men and women. The sociobiologists countered by arguing that men are drawn to women who display features that signal "high repro-ductive value" and, in the journals patronized by sociobiologists, one could find charts linking, for example, the body-mass index of women (kg/m^2) to men's ratings of them as attractive or not. In a wide series of societies, scattered across the globe, the charts showed that men found women attractive who, on average, were about three years younger than them, with large eyes, small noses, and full lips, and with high waist-to-hip ratios.[24]

This was crass, growled the critics.

But the sociobiologists would not be dissuaded. They pointed to the fact that in a variety of studies, it had been found that humans everywhere "share a host of attributes." These include "facial and other gestures when speaking." They share the same attitudes to frowning, and to time—as being divided into past, present, and future. Another example: "No societies exist in which newlyweds go to live with the wife's father's sister's fam-ily, whereas, in quite a few, newly married couples move to be near the husband's mother's brother."[25] Human behavior, across cultures, is stable.

Another study linked biology and economics. From their knowledge of history, Jennifer Billing and Paul Sherman noted that in the past people went to extraordinary lengths to obtain spices for cooking. These could easily be considered cultural practices, as indeed they were, but in England in the Middle Ages people used peppercorns as currency, even paying their rents in pepper. In 1265 one woman in Leicester paid twelve shillings for a pound of cloves, even though she could have bought a cow for less. Intrigued, Billing and Sherman asked themselves whether there was anything "Darwinian" about this economic behavior. This led them to the theory that, perhaps, and despite cultural practices (the proximate cause of using spices), there might be an evolutionary reason—the "ultimate" cause. When they looked into it they found that many spices contain chemicals that counter microbial action. Nutmeg, garlic, onion, and oregano, for example, block the effects of a wide range of bacterial infections.

To test their thesis, Billing and Sherman located cookbooks from thirty-six different countries ranging from Norway to Indonesia and examined 4,241 meat-based recipes. They found that the antimicrobial chilies, garlic, and onions were far more likely to appear in meals prepared in tropical countries than in cool, temperate countries—because they countered the actions of the microbes.[26]

Finally, there is Stephen Emlen's research, repeated by others, which shows that stepfamilies share similarities in both birds and humans. In bird species that form durable "marital" relationships, but where one partner dies and the survivor forms a second relationship, the stepparents are much less caring—and often downright violent—to the young who are not related to them, certainly as compared with the offspring that are born later in the second "marriage." This pattern is repeated more or less exactly with Homo sapiens: abuse of stepchildren among humans is about forty times as common as abuse of natural children. These "sociological" problems are, in effect, biologically—genetically—based.[27]

Sociobiology has taken some knocks (strangely, ethology hasn't, not to anywhere near the same degree). Jerry Coyne, writing in the *New Republic*, said: "In science's pecking order, evolutionary biology lurks somewhere near the bottom, far closer to phrenology than to physics." Someone else called sociobiologists "evo-psychos."[28] Still others saw the new discipline as a form of intellectual imperialism.

We shall return to this issue in chapter 18 when, for example, we shall explore more fully the biological basis of culture. For the moment, however, let us end this one with another quote from Chomsky, which is of particular interest because, to an extent, it goes against our theme of emerging order. In the book *Language and Problems of Knowledge*, published in 1988, Chomsky specifically argued that the unification of disciplines—which he conceded *was* such a feature of twentieth-century science— does not necessarily have to take the form of reduction. More important, and more problematic, for him was the assertion that the physical or the physiological has "some kind of priority." This, he insisted, is misconstrued. Linguistics is just as full of rich theories that lead to specific predictions across a wide domain as chemistry and biology. "Trying to reduce linguistics to neurology in the current state of our understanding is then unlikely to be productive."[29] A specific example that he gives is the association between "deviant" linguistic structure (departure from principles of grammar, for example) and electrical activity in the brain known as "event related potentials." Linguistics enables us to make sense of this deviancy, but there is no relevant electro- physiological theory in existence. "It is as impossible to express interesting generalizations about language in terms of the con- structs of cells or neurons, as it is to express generalizations about geology or embryology in terms of the constructs of particle physics. In both cases demands for reduction go too far."[30]

One would disagree with Chomsky only reluctantly. But, as he himself admitted, the unification of disciplines *was* a feature of twentieth-century science; and, in recognizing ethology as a new discipline, worthy of the Nobel its founding practitioners had won, the Swedish Academy of Sciences was itself acknowledging the convergence of biology, ecology, genetics, and psychology. The fact that sociobiology came along at much the same time as Kahneman and Tversky's amalgamation of psychology and economics shows that the mid-1970s was once again one of those times when, as on the eve of Darwin, "something was in the air." Despite the dramatic backstory to the 1973 Nobel award in medicine or physiology, the Swedish academy's choice of ethology was extremely apropos.

14

CLIMATOLOGY + OCEANOGRAPHY + ETHNOGRAPHY → MYTH = BIG HISTORY

It has been known for more than a century that the most wide-spread myth across the world—as well as the best known—is that referring to a vast flood, whose exact size was not calibrated, but which was reported not just in the Christian Bible, of course, but in the ancient legends of India, China, Southeast Asia, Australia, and the Americas. We shall be considering the flood myth(s) in some detail in just a moment, but, for reasons that will become clear, it suits us here to consider first the second most common myth on earth, that of the "watery creation" of the world.

The chief theme of this myth is separation, usually of the sky from the earth. This story is found in a band stretching from New Zealand to Greece (a significant distribution, as we shall see), and it invariably has a small number of common features. The first is the appearance of light. As it says in Genesis 1:3: "And God said, 'Let there be light': and there was light." Nearly all cosmogonies have this theme, where it is notable that *neither the sun nor the moon is the source of the first light at creation.* Rather, the first light is asso-ciated with the separation of heaven and earth. Only after heaven and earth have separated does the sun appear. In some traditions in the East the light is let in because the heavy substance of the clouds that envelop the earth sinks down to the ground, and the light, clearing the clouds, rises to become heaven. In other myths, the darkness is described as a "thick night."

Recent geological studies have identified a phenomenon known to scientists as the Toba volcanic explosion. Cores drilled in the Arabian seabed have shown that there was a volcanic eruption at Toba in Sumatra between 74,000 and 71,000 years ago. This is known to have been the biggest eruption on earth during the last 2 million years, a massive conflagration that released a vast plume of ash 30 kilometers high (an estimated 670 cubic miles, twice the volume of Mount Everest). It spread north and west, to cover Sri Lanka, India, Pakistan, and large areas of the Gulf region with a blanket six inches deep—though at one site in central India the ash layer is still twenty *feet* thick. Toba ash has recently been found in the Arabian Sea and in the South China Sea, 2,400 kilometers from Toba itself. The eruption left an immense caldera that now holds Indonesia's largest lake, Lake Toba—85 kilometers long, up to 25 kilometers wide, with cliffs 1,200 meters high and water 580 meters deep.[1]

A prolonged volcanic winter would have followed this eruption. Sea temperatures, according to geologist Michael Rampino, dropped by ten degrees Fahrenheit and a total darkness would have existed over large areas for weeks or months. The aerosol clouds of minute globules of sulphuric acid, now known to be produced by massive eruptions, could have reduced photosynthesis by 90 percent, or even shut it down completely, having a major effect on forest cover.[2]

Now if, as anthropologists and paleontologists believe, early humankind left Africa at about or sometime after eighty thousand years ago, and if the people followed a beachcombing route that took them around what is now Yemen and Aden, and on to the Iranian, Afghanistani, and Pakistani coasts, they would have arrived in the latter areas more or less on schedule to meet the

Toba eruption. This is in fact confirmed by excavations in India and Malaysia, which have found Paleolithic tools embedded in volcanic ash at this date. According to some estimates, the population of this vast area could have been reduced from an estimated one hundred thousand, to between two thousand and eight thousand (a similar population crash is known to have occurred among chimpanzees).[3]

At a conference in Oxford in February 2010 the "catastrophic" nature of the Toba eruption was queried, and new evidence presented to suggest that the temperature dropped by only 2.5°C. But no one is suggesting that Toba's effects were other than far-reaching, and the conference also heard fresh evidence that tools made by Homo sapiens straddled the ash layer.[4]

Two things follow. The volcanic winter may well have all but wiped out the early humans living in a wide swath centered on India, a mass-extinction event. This would have meant that certain specific survival strategies needed to be devised, which may have been memorized in myth form. Second, the area would have been recolonized later, both from the west and the east.

We shall come back to this, but for now the main point is that the "separation" myth is a not inaccurate description of what would have happened over large areas of the globe—in Southeast Asia—after the Toba eruption and the volcanic winter that would have followed.[5] Sunlight would have been cut out, the darkness would have been "thick" with ash, the ash would gradually have sunk to the ground, and, after a long, long time, the sky would gradually have got brighter, lighter, and clearer, but there would have been *no* sun or moon visible perhaps for generations. There would have been light but no sun, not for years, not until a magical day when, finally, the sun at last became visible. We take the sun for granted, but for early humankind it (and the moon, eventually) would have been a *new* entity in the ever-lightening sky. Mythologically, it makes sense for this event to be regarded as the beginning of time.[6]

The Deep Order in Myths

The discovery of the Toba eruption, therefore, was almost as important a breakthrough for mythology as it was for geology. And there are grounds for believing that many other ancient myths and legends, far from being the products of our deep unconscious—as Carl Jung, for one, insisted—are in fact based on real events that science has recently brought to light.

Next, we turn to the commonest myth of all. At one stage, while myths were interesting to anthropologists, they were treated as mainly fictional accounts, revealing more about early man's primitive beliefs than anything else. Sir James Frazer, the late-nineteenth-century anthropologist and author of *The Golden Bough*, recorded many of these myths in his book, *Folk-lore in the Old Testament*, published in London in 1918, and where he had this to say: "How are we to explain the numerous and striking similarities which obtain between the beliefs and customs of races inhabiting different parts of the world? Are such resemblances due to the transmission of the customs and beliefs from one race to another, either by immediate contact or through the medium of intervening peoples? Or have they arisen independently in many different races through the similar workings of the human mind under similar circumstances?"[7]

Attitudes evolved somewhat when, a few years later, in 1927, the British archaeologist Leonard Woolley began to dig at the biblical Ur of Chaldea, in Iraq, the alleged home of Abraham, founder of the Jews. Woolley was to make several important discoveries at Ur, two of them momentous. In the first place he found the royal tombs, in which the king and queen were buried, together with a company of soldiers and nine ladies of the royal court, still wearing their elaborate headdresses. However, no text had ever hinted at this collective sacrifice, from which he drew the important conclusion that the sacrifice had taken place *before writing had been invented* to record this extraordinary event, an inference that was subsequently substantiated. And second, when

Woolley dug down as far as forty feet he came upon nothing, nothing at all. For more than eight feet there was just clay, completely free from remains of any kind. For a deposit of clay eight feet thick to be laid down, he concluded that a tremendous flood must at some time have inundated the land of Sumer.[8] Was this then the Flood referred to in the Bible?

Many people—then and now—thought that it was. But just as many didn't. They didn't because the Bible text says that the flood covered mountaintops—i.e., it was rather more than eight feet deep—and because the flood was supposed to extend right across the world. An eight-foot flood of the Tigris and Euphrates in Mesopotamia did not suggest anything more than a local event. Or had the ancients exaggerated? Since in those days hardly anyone traveled far, perhaps a reference to a "worldwide flood" was just a manner of speaking.

That is more or less where matters remained for several decades. In recent years, however, new light has been cast on three events—or rather, three sets of events—deep in our past. The history of the years covering the transition from the Pleistocene to the Holocene—from the Ice Age to modern times—has undergone a major revision recently, and, to put the matter briefly, the latest scholarship of the period shows three things. It shows that the world suffered not one but *three* major floods, at (roughly speaking) ~14,000, ~11,500, and ~8,000 years ago, and that the last of these was especially catastrophic, changing life drastically for many of the people then on earth. Second, it shows that the area of the world that was most affected by the floods was not Mesopotamia but Southeast Asia, where a whole continent was drowned. If these floods did have most effect in Southeast Asia, it would mean that the inhabitants of that sunken continent would have been forced to migrate all over the world—north to China and then to the New World, east to the Pacific islands and Australia, and back west to India, and possibly as far as Asia Minor, Africa, and Europe, taking their skills with them. The third aspect of this new chronology is that many of the early skills of civilization, such as agriculture—which have always been understood as being invented in the Middle East—were actually first developed much further east, in Southeast Asia and in India.[9]

There is much more to the story than this of course, but these are the main bones. The story overturns dramatically the understanding that went before, not least because it may help to explain the arguments of those archaeologists who assert that there was a great linguistic split between the Old World and the Pacific peoples, who played an important role in the peopling of the New World.

A New Order of Civilization

There is now substantial evidence to suggest that the rise in sea levels after the last Ice Age was neither slow nor uniform. Instead, three sudden ice-melts, the last only eight thousand years ago (6000 BC), had a devastating effect on certain tropical coastlines, which had extensive flat continental shelves. These changes were accompanied by massive earthquakes, caused as the weight of the great ice sheets was removed from the land and transferred to the seas.[10] These massive earthquakes would have generated super-waves, tsunamis. Geologically, the earth was much more violent then than it is now.

The overall oceanographic record between 20,000 and 5,000 years ago reveals that sea levels rose at least 120 meters (~400 feet) and affected human activity in three profound ways. In the first place, in Southeast Asia and China, which have a large flat continental shelf, all examples of coastal and lowland settlement were inundated and for all time. Those settlements have been underwater for thousands of years and will most likely remain so. Second, during the final rise in sea level, 8,000 years ago, the water did not retreat for about 2,500 years, with the result that many areas there that are now above water are nevertheless covered with a layer of silt that is many feet thick. Third, as already mentioned, the floods that devastated Southeast Asia required the inhabitants to move out.[11]

This picture is supported by the curious dating pattern of the Neolithic Revolution in eastern Eurasia. According to such sites as have been found, the Pacific Rim cultures seem to have begun their development well *before* those in the West but then, apparently, stopped. For example, pottery appeared for the first time in southern Japan around 12,500 years ago; 1,500 years later it had spread to both China and Indochina. It is important to say that these examples predate any of the sites in Mesopotamia, India, or the Mediterranean region by as much as 3,500 to 2,500 years. To give another example, stones for grinding wild cereals are found in the Solomon Islands of the southwest Pacific dated to 26,000 years ago, but are not found in Egypt until about 14,000 years ago and Palestine 2,000 years after that. These early signs of civilization occurred much, much earlier in Southeast Asia than anywhere else.

In addition to the early beginnings of pottery in Japan and Indochina, around 12,000 to 11,000 years ago, a wide range of Neolithic tools have been found in East Asia—choppers, scrapers, awls, and grinding stones, as well as hearths and kitchen waste—but these finds tend to be located in *inland* caves. There are almost no Neolithic sites in lowland coastal areas dating to between 10,000 and 5000 BC.[12]

Two explanations have been put forward to account for this anomaly. One view has it that in island Southeast Asia the Neolithic period only started 4,000 years ago, with migrants coming down through Taiwan and the Philippines and introducing new skills and artifacts. The other view is more ambitious. People were living in Southeast Asia at the end of the Ice Age and had developed their agricultural (and sailing) skills much earlier than people elsewhere (in the Near East, for example) but were forced into long-distance migration, east, north, and west, as a result of flooding brought about by the melting glaciers.[13] And, as well as forcing these people out, the associated silt covered up many sites.

The Rhythms of the Floods

These are clearly important assertions, and so the floods need to be fully described if we are to be able to judge the merit of these new theories. A consistent picture will emerge, the importance of which suggests that the people who entered the New World first did so after a distinct set of experiences that separates them from those they left behind in the Old World.

We now know that three catastrophic floods referred to above occurred because of three interlocking astronomical cycles, each different and each affecting the warmth transmitted by the sun to various parts of the earth. Stephen Oppenheimer calls these the 100,000-year "stretch," the 41,000-year "tilt," and the 23,000-year "wobble." [14] The first arises from the earth's orbit around the sun, which is elliptical and means that the distance from earth to sun varies by as much as 18.26 million miles, producing marked variations in the force of gravity. The second cycle relates to the tilt that the earth presents to the sun as it rotates. This varies— over 41,000 years—between 21.5 and 24.5 degrees and affects the seasonal imbalance in heat delivered from the sun. Third, the earth rotates on its own axis, in a so-called axial precession, every 22,000 to 23,000 years. These three cycles perform an elaborate dance that produces an infinite array of combinations but which, when they come together in a "perfect storm," can provoke very dramatic and very sudden climate change on earth. It is these complex rhythms which triggered not one but three floods in the ancient world.

The glaciers that melted to cause these floods were massive, the largest covering huge areas such as Canada, and were several miles thick. One has been estimated as being 84,000 cubic kilometers. They could take hundreds of years to melt completely but eventually raised sea levels by as much as forty-four feet.

One of the interesting consequences of the changes that followed the second catastrophe (after 11,000 years ago) was that, as sea levels rose, river gradients were lowered and, after 9,500 years ago, river deltas began to form all around the world. The importance of this lay in the fact that these deltas formed very fertile alluvial plains—in Mesopotamia, the Ganges, the Chao Phraya in Thailand, the Mahakam in Borneo, and the Chiang Jiang (Yangtze) in China. Overall, more than forty such deltas have been identified as forming at that time on all continents. Many of these alluvial plains/deltas played a role in the growth of agriculture and the subsequent birth of civilization (chapter 15).[15]

But it was the most recent flood, at 8,000 years ago, that had the greatest effect. The sudden removal of the ice sheets from the North American and European continents, releasing massive amounts of ice and water into the world's great ocean basins, meant that a sudden change in the spread of weight across the earth occurred, and this would have caused great earthquakes, increased volcanism and massive tsunamis crashing ashore on all continents. It was an epic period of natural disasters that had a profound effect on the mental life of ancient men and women.

This flood, so recently established, had several important consequences. One was that a flood and tsunamis of such dimensions would have deposited layer upon layer of silt many feet thick across huge areas, layers that must have covered crucial examples of early human development between, say, 8,000 years ago and when the seas receded again many hundreds if not thousands of years later. This "silt curtain," as Stephen Oppenheimer calls it, must in turn affect our understanding of world chronology.[16] A second consequence arises from the natural geography of the world, where the largest landmass that was inundated by the flood was almost certainly Southeast Asia, where there could be found the largest shallow continental shelf, stretching out into the South China Sea for 1,600 kilometers. Crucially, for an understanding of early chronology, and perhaps for a full grasp of the emergence of civilization, these two consequences can be put together.

The starting point for this synthesis comes from the fact that this area has the highest concentration of flood myths in the world.[17] Does this prove that the flood had the most devastating impact here? No, but the possibility is tantalizing and it fits exactly with what William Meacham, a Hong Kong–based prehistorian, noted in 1985—that the most important gap in the Neolithic record now "is the total absence of open sites in lowland areas [in Southeast Asia] dating from 10,000 BC to 5000 BC." Moreover, after sea levels started to fall again, from 6,000 years ago, pot-making maritime settlements began to occupy sites all the way down from Taiwan to central Vietnam. Charles Higham, a New Zealand archaeologist based mainly in Thailand, argues that these settlements were actually *re*locations of maritime people who had always lived in these areas but had been flooded out much earlier on. Inland sites, on the other hand, were continuously inhabited from more or less the end of the Ice Age.[18]

And so, the largest amount of low-lying land in the world, which would have been most affected by any rise in sea levels, where a flood would have been most catastrophic, was the Sunda Shelf, on the southeast "corner" of Southeast Asia. It stretched 5,400 kilometers east to west and 2,700 kilometers north to south. This may well account for why flood myths are more prevalent in that region than anywhere else. Such a flood would have provoked large-scale migration—east, west, and north.

Astronomy and the Great Cycles of Nature

We now need to consider, briefly, one other relatively new finding about ancient Asian history, before moving on to compare Old World and New World myths. It concerns the Vedas, the sacred writings of the Hindus, which envisage a "Yuga" theory of historical and cosmic development—great cycles of humanity and of nature, disrupted by enormous natural cataclysms. One of these cycles is said to last 24,000 years, not so very different from the 23,000-year "wobble," as Stephen Oppenheimer calls it, but more relevant, perhaps, is the newly discovered fact of three great floods in recent geological history, at ~14,000, ~11,500, and

~8,000 years ago. Is this not in effect cyclical history, broken by great catastrophes?

More specifically, however, the Vedantic literature refers to a land of seven rivers—identified as the Indus, Ravi, Sutlej, Sarasvati, Yamuna, Ganga, and Sarayu—in which the Sarasvati was the most important for Vedic people, both spiritually and culturally, irrigating their central land and place of origin and supporting a large maritime culture. One verse of the Vedas describes the Sarasvati as "the best of mothers, the best river, the best goddess," and another places it between the Sutlej and the Yamuna.[19]

The problem is—or was—that today there is no major river flowing between the Yamuna and the Sutlej, and the area is well known as the Punjab (*panca-ap* in Sanskrit), or the Land of *Five* Rivers or Waters. This discrepancy led some scholars for many years to dismiss the Sarasvati as a "celestial" river, or an imaginary construct.

Beginning just after World War II, however, archaeological excavations started to uncover more and more settlements which *seemed* related to the well-known Mohenjo-daro and Harappa Indus civilizations but, paradoxically, were up to 140 kilometers distant from the Indus River itself, at sites where there is today no obvious source of water. It was only in 1978 that a number of satellite images from the spacecraft launched by NASA and the Indian Space Research Organisation began to identify traces of ancient river courses that lay along locations where the Vedas said the Sarasvati had been. Gradually, these images revealed more details about the channel, including the fact that it had been six to eight kilometers wide for much of its course, and no fewer than fourteen kilometers wide at one point. It also had a major tributary and between them the channel and its tributary converted the Land of Five Rivers (of today) into the Land of Seven Rivers (*sapta-saindhava*) in the Vedas. Moreover, the Rig Veda describes the Sarasvati as flowing from the "mountains to the sea," which geology shows it would have done only between 10,000 and 7,000 years ago, as the Himalayan glaciers were melting. Over time, the rivers feeding the Sarasvati changed their course four times as a result of earthquakes, feeding the Ganges instead, and the Sarasvati dried up.[20]

So the Veda myths were right all along and the rediscovery of the Sarasvati underlines two things. One, the basic skills of civilization—notably domestication, pottery, long-distance trade, sailing—were in place in South Asia (India) and island Southeast Asia by 5000 BC. And second, the great myths which we find spread across the world are almost certainly based on real catastrophic events that actually occurred and devastated early humankind. They form a powerful collective memory to warn us that such terrible events may one day recur.

The Order Beneath the Myths

Now that we have done our groundwork, what, we may ask, do myths tell us about the early experiences of mankind, and how do the historical sciences add to the picture, to reveal what order?

The basic difference we shall examine is that between Old World and New World myths, where we find that there is a broad overlap between genetics, linguistics, and popular legend. The genetic evidence shows that the Chukchi in Siberia and the first human groups to enter the Americas reached Beringia by central Eurasia and arrived sometime between 20,000 and 12,000 years ago at the latest. The linguistic evidence in particular suggests that a second, later group of ancient peoples traveled up the western coast of the Pacific Ocean—Malaysia, China, Russia. If the earliest peoples reached the New World at any time between 43,000 and 29,500 years ago, as some of the genetic evidence suggests, they may well have had a memory of the Toba earthquake, but none of the great floods had yet occurred. On the other hand, the second group—the Na-Dene speakers, who have a distinctive genetic marker (M130), whose bearers migrated up the Pacific Rim and into the Americas at about 8,000 to 6,000 years ago—should have had fairly recent experience of flood. What do we find?

In the first place, and by way of generalization, we may say that there is an extensive constellation of myths that occurs in both the Old World *and* the New, far too many for them to have all been jointly conceived by coincidence. Allied to this, there are some important myths that occur only in the Old World and in Oceania but do *not* appear in the New World. At the same time, there are a few myths—of origin, creation—that appear in the New World and not in the Old. This is all what you would expect if early humankind originated in the Old World and migrated to the New.

A good starting place is the myths to be found on both sides of the Bering Strait, where we can examine the systematic ways in which they vary.

As has been said, many myths describe a "watery chaos" flood, out of which land gradually emerges. In the subarctic regions of North America, however, the most common myth is that of the "land diver." In these myths, following the flood, land doesn't emerge gradually but is created by raising it up from the floor of the ocean bed. A common procedure to ensure this happens is the use of what have become known as "land divers." These are animals, often diving birds, who are sent down to the bottom of the ocean (by either the creator or earth's first inhabitants) to pick up a scrap of earth on the ocean bed. Typically, after a few unsuccessful attempts, one diver returns with earth or clay in its claws or beak, and this small amount is transformed into the growing earth.

The Huron, of Ontario, for example, have a myth in which a turtle sends various animals diving for earth, all of whom drown except the toad, who returns with a few scraps of land in its mouth. These are placed on the back of the turtle by the female creatrix, who has descended from heaven for this purpose, and the scraps of earth grow into the land. The Iroquois and Athapascan tribes (on the northwestern Pacific coast of what is now the United States) also have this myth, which is in fact confined to two linguistic groups, Amerind speakers and Na-Dene speakers.

The motif is not found in Eskimo flood myths or in Central or South America.[21]

Two other aspects need to be pointed out. First, the distribution of the land-diver stories overlaps with a characteristic genetic marker in subarctic North America. Certain population groups (not Eskimos or Aleuts, for example) have what is known as the "Asian 9-base-pair deletion": nine pairs of proteins are missing from their DNA. This marker, this pattern of absence, is shared with certain clans in New Guinea, and also with peoples in Vietnam and Taiwan. Not only does this further confirm the Southeast Asian origin of at least some native Americans (and underline the distinction between Eskimos and Na-Dene speakers), but the sheer size and diversity of the 9-bp deletion on both sides of the Pacific suggests a very old origin.

Second, these land-raiser myths lend themselves to two phenomena recognized by geographers and oceanographers. The first is "coastline emergence." This is something that happened on a grand scale, especially in North America. It is a phenomenon that occurs because, after the Ice Age, as the glaciers melted, they grew lighter and, with less weight on it, the continental crust lifted up. Moreover, the change in weight brought about a rise in the land that was *more* than the rise in sea level. Since the land had hitherto been crushed below sea level, it would at that time have *risen* out of the sea. Photographs of Bear Lake in Canada show several shorelines that have risen hundreds if not a few thousand feet above sea level.[22] People alive at the time would, over the generations, have noticed that the shoreline had moved and, we may assume, incorporated this strange phenomenon into their myths, explaining it as best they could.

The second phenomenon involves the well-established fact that the Pacific Rim is known as the "Ring of Fire," because that is where the world's most active volcanoes are located. All we need to add here is that many volcanoes in the Ring of Fire are offshore, underwater volcanoes, forming part of the seabed. During underwater offshore eruptions (of which there were more than fifty in 2001–2, with the latest being documented off Oregon in April 2015) solid matter—"land"—would have been propelled forcefully to the surface.

The rest of the Americas lack the "land diver" and "land raiser" myths, though they have a rich stock of flood myths, including those with birds who fly out to seek land.

The Unifying East-West Corridor

According to Stephen Oppenheimer, there are very few myth motifs that are totally absent in the New World, but there *are* systematic variations and these form a consistent and ordered picture. The most substantial systematic difference is a constellation of ten linked motifs not generally found in Africa, the Americas, or Central and Northeast Asia. Instead, this constellation occurs in a distinctive swath (referred to earlier) from Polynesia across China, South Asia, and then the Middle East, ending in northern Europe (as far as Finland).

For example, in the Americas there is a relative dearth of "watery chaos myths" beyond the northwest Pacific coast. Another difference is that New World myths lack almost any references to sea monsters or dragons (the one exception is an Aztec myth). In any one of the three floods referred to above, far more areas—and far more populated areas—would have been accessible to crocodiles, whose main range of activity was in Indochina. This is confirmed by the fact that, in the myths, most dragons and serpents attack coastal peoples, not fishermen. On this reading, the dragon and sea monster stories are perhaps a deep folk memory of a plague of crocodiles that occurred when shallow coastal areas were flooded at one stage in the distant past.[23]

Besides "watery chaos," first light, and the separation of heaven and earth, other elements in this constellation of myths include incest, parricide, and use of the deity's body parts and fluids as building materials for the cosmos. None of this occurs in New World myths, though it is common across Eurasia, where we often find a divine couple who are bound together, and separate to create heaven and earth, who are then mutilated and torn apart by their offspring, who use the parts of the parent deity to create the landscape. (Blood is used for rivers, for example, or the skull for the dome of the sky.) Many myths in this constellation contain episodes of post-flood incest, usually between a brother and a sister. Sometimes the participants are aware of the taboo, at others it isn't mentioned. This would appear to be a forceful way for primitive peoples to reinforce the memory that, after the flood and/or the Toba eruption, the race almost died out and/or was isolated (from other islands?), the population reduced to such an extent that brothers were forced to mate with sisters. Again, such myths are not in general found in the New World.[24]

This overall pattern is amplified by a second group of myths that are also notable by their absence in the New World. These include the "dying and rising tree god," and the myth of the warring brothers. The dying and rising tree god or spirit had a very wide distribution across the earth, from the Norse myth of Odin to the Egyptian myth of Osiris, to the Christian story of Jesus, to the Moluccan myth of Maapitz, to the New Britain myth of To Kabinana. Moreover, this myth overlaps in certain locations with the theme of warring brothers, or sibling rivalry: Set/Isis in Egypt; Bangor/Sisi in Papua New Guinea; Wangki/Sky in Sulawesi; and, of course, Cain and Abel in the Bible. This conflict is generally taken to reflect the different lifestyles of agriculture and either foraging or nomadism—in other words, it is *post*-agriculture.[25]

What matters for us with these myths is less their meaning (for the moment) than their distribution which, broadly speaking, once again extends from Indonesia and Borneo, up through the Malay Peninsula, India, the Arabian Gulf, Mesopotamia, the Mediterranean civilizations to Western and Northern Europe. This range of locations has in common that it occupies and overlaps with the great "East–West Corridor," a broad swath of coastline running from the tip of Malaysia at Singapore as far west as Pointe Saint-Mathieu, near Brest in Brittany, in France. This continuous coastline would have enabled the conduct of early coastal trade and the ideas that accompanied such movement.

The new synthesis of cosmology, geology, genetics, and mythology is exciting, but we have taken it about as far as it will go. From it we may conclude (and repeating the proviso that this is all very speculative) that one group of early humans who first peopled the Americas arrived no later than 14,000 years ago and very probably 16,500–15,000 years ago. They shared with everyone else an experience of a global seaborne flood, but they showed no awareness of either agriculture or navigation, having reached the cold and limiting region of Siberia, and then Beringia, before these skills were invented (or needed) somewhere on the Sunda Shelf. Similarly, they showed a very rudimentary awareness of a great global catastrophe, other than flood, in which the skies darkened for generations and only slowly cleared.

These myths concur with the genetic and linguistic evidence, that there was a later migration, possibly 11,000 but more likely 8,000–6,000 years ago. It follows from this that such people had no awareness of the cultural conflict that gave rise to the "warring brothers" myths, or the dying and rising tree god myth, which originated in Southeast Asia too late for it to be incorporated into New World mythology. This too suggests these people left the Sunda Shelf before agriculture was invented. These two constellations of myth motifs were arguably the most important ideas of ancient times in the Old World, shaping most of the religions and traditional histories from Europe to Southeast Asia.[26]

This all suggests (and still speculating) that the period between 11,000 and 8,000 years ago on the Sunda Shelf was very problematical, entailing several catastrophes which both expelled many people and gave rise to powerful myths among those who remained. There was a major rupture between the people who headed north, eventually to colonize the New World, and those who remained, or headed back west, to form part of the civilizations of Eurasia.

In general, two broad conclusions may be drawn from this

brief survey of myths, one concerning the New World, the other the Old World. In the New World the very ancient myths (such as the watery creation of the world) tended to be superseded by those that forced themselves on people by their experiences in the American continent itself—land divers, land raisers, violent tsunamis. This is an early indication of a trend or theme we shall see more of in the next chapter: the role that extreme weather—storms, hurricanes, volcanoes, and earthquakes—plays in New World ideology. In the Old World, what draws our attention is the distinctive distribution of the watery creation myth, the creation of light, and the dying and rising tree god set of myths. Dying and rising refers to fertility, a dominant issue in Old World ideology that, as we shall see, did not have quite the same resonance in the New World.

Patterns of Venus

Apart from the separation of earth from sky myth, and a near-universal flood myth, the widespread depiction of the female form in very early Paleolithic art also needs some explanation and comment. These so-called Venus figurines are found in a shallow arc stretching from France to Siberia, the majority of which belong to the Gravettian period—around 25,000 years ago.[27] Many of them (but by no means all) are buxom, with large breasts and bellies, possibly indicating they are pregnant. Many (but not all) have distended vulvas, indicating they are about to give birth. Many (but not all) are naked. Many (but not all) lack faces but show elaborate coiffures. Many (but not all) are incomplete, lacking feet or arms, as if the creator had been intent on rendering only the sexual characteristics of these figures. Some (but not all) were originally covered in red ocher—was that meant to symbolize (menstrual) blood? Some figures have lines scored down the backs of their thighs, perhaps indicating the breaking of the waters during the birth process.

Some critics have argued that we should be careful about

reading too much sex into these figures, saying that it tells us more about modern paleontologists than it does about ancient humans. Nevertheless, other early artworks do suggest sexual themes. Among the images found in 1980 in the Ignateva cave in the southern Urals of Russia is a female figure with twenty-eight red dots between her legs, very possibly a reference to the menstrual cycle. At Mal'ta, in Siberia, Soviet archaeologists discovered houses divided into two halves. In one half only objects of masculine use were found, in the other half female statuettes were located. Were they ritually divided according to gender? [28]

Whether these early "sexual images" have been over-interpreted, it remains true that sex is one of the main images in early art, and that female sex organs are far more often depicted than male organs. In fact, there are no depictions of males in the Gravettian period, and this would therefore seem to support the claims of the distinguished Lithuanian archaeologist Marija Gimbutas that early humans worshipped a "Great Goddess" rather than a male god. The development of such beliefs, she says, possibly had something to do with what at that time would have been the great mystery of birth, the wonder of breast-feeding, and the disturbing occurrence of menstruation. Randall White, professor of anthropology at New York University, adds the intriguing thought that these figures date from a time—and such a time must surely have existed—when early people had yet to make the link between sexual intercourse and birth. At that time, birth would have been truly miraculous, and early humans may have thought that, in order to give birth, women received some spirit, say from animals. Until the link was made between sexual intercourse and birth, women would have seemed mysterious and miraculous creatures, far more so than men.

Anne Baring and Jules Cashford, in their book, *The Myth of the Goddess: Evolution of an Image*, describe these early Venus figurines as the Paleolithic Mother Goddess and this goes straight to the heart of the matter. [29] Randall White is surely correct in his argument that there must have been a time when ancient humans had not made the link between sexual intercourse and birth, 280 days of human pregnancy (on average, from the end of menstruation to parturition) being just too long a delay for such a link to be observed. This may help explain why the Venus figurines were only ever carved—in ivory or stone—and hardly ever painted on cave walls. And why, often, they lacked heads and feet, or these features were highly stylized. The figurines were carved because they had to be portable. The clan or tribe took the statuettes with them as they followed the herds, and the figurines were carved in such a way that only the important, practical features were included.

Elizabeth Wayland Barber and Paul Barber, in their book *When they Severed Earth from Sky: How the Human Mind Shapes Myth*, have convincingly shown how many common myths are based on fairly accurate observations by ancient peoples of phenomena they thought important and yet used mental devices still familiar to us to remember these phenomena and warn their descendants.[30] The Barbers show how giants are to be understood as volcanoes or the remains of mammoth bones, how volcanic eruptions may be understood as gigantic "pillars of stone," how their eruptions are impressive evidence of an underworld with powerful forces, how storm gods became winged horses and tsunamis "bulls from the sea." On this basis, we may conclude that ancient people were perfectly well aware of the miraculous nature of birth but they were no less aware, too, that it could be a sensitive, even perilous time, and so they recorded on the figurines the elaborate and detailed configurations of the female body that indicated that birth was imminent. Without being aware of the link between sexual union and birth, they could have had no conception of the biological rhythms governing gestation, and so the bodily signs that birth was imminent would have been the only practical way they could know when to organize their lives so that the chances of a danger-free birth were maximized, perhaps by secluding the soon-to-be mother in a favorite or familiar cave where predators could be kept at bay.

The discovery of a small Venus statuette, carved from mammoth ivory, was announced in May 2009, having been excavated in the Hohle Fels caves in Germany and dated to 35,000 years ago. Nicholas Conard, part of the excavating team at the University of Tübingen, noted that the figure had exaggerated sexual features and a small loop where the head would be.[31] He thought the statuette was "hung on a string and worn as a pendant." This fits the general picture yet tends to confirm that the statuettes were carried around, supporting the interpretation offered here.

The earliest Venus figurines are recorded at 40,000–35,000 years ago, and they die out (the Venus of Monruz, Switzerland) about 11,000 years ago. This dating is suggestive too, being close to the time when mammals were first domesticated. The gestation period of cows is much the same as for humans (285 days), and the horse is even longer (340–342 days). But the dog is a mere sixty-three days, barely two months, so it may have been through observation of the (newly) domesticated dog's behavior that early peoples first spotted the all-important link between coitus and gestation.[32] Findings reported in March 2010, by Bridgett M. vonHoldt and Robert K. Wayne, of the University of California at Los Angeles, using DNA evidence, puts the domestication of dogs somewhere in the Middle East at 12,000 years ago.[33] Intuitively, this seems very late for humankind to have made the discovery of the link between sexual union and birth. And yet, according to Malcolm Potts and Roger Short, Australian aborigines did not associate intercourse with pregnancy until they domesticated the dingo—with a similarly short gestation period (sixty-four days).[34]

Nor is the idea inconsistent with the fact that male gods do not appear to have evolved until the seventh millennium BC, unless we include the (predominantly male) shaman himself. If we do allow the shaman, then this would mean that humankind's earliest ideology would have involved the worship of two principles: the Great Goddess and the mystery of fertility, and the drama of the hunt, highlighting the problem of survival. (Shamans drew their power in hunter-gatherer societies from their facility to communicate with animals.) An image of a wounded bear at Les Trois Frères cave in France shows it covered in darts and spears, with blood pouring from its mouth and nostrils.

This theme of the Great Goddess was to become the dominant ideological motif in the Old World, in contrast to the New, where different images prevailed. We need therefore to ask ourselves why this image, this motif, occurred where it did, when it did, and why its range was limited. We can go some way towards an understanding of the phenomenon.

The first thing to say is that, as the Ice Age came to an end, between 40,000 and 20,000 BC, when the glaciers and permafrost retreated, and as grassland spread, the woolly mammoth, woolly rhinoceros, and reindeer gave way to great herds of bison, horse, and cattle. Later still, the grassland itself gave way to thick forests, and so the herds moved east, with the hunters following them. A third of cave paintings are of horses, while bison and wild ox make up another third. Reindeer and mammoth hardly appear, though many bones of such animals have been found.[35]

If the "new" animals were being followed across the "new" grassland habitat, by people carrying the figurines with them, the distribution of the steppe would also explain the spread of Venus statuettes, which end where the steppe ends, around Lake Baikal. A look at the map will show that Lake Baikal and the Lena River, to the north, mark a kind of natural boundary.

Therefore, if, as was outlined earlier, ancient humans reached the New World via the Bering Strait from Mongolia and/or beachcombed their way north from Southeast Asia, deriving their protein chiefly from fish, they would not have incorporated the Eurasian post-glacial ideology into their psychological makeup. They would have presumably discovered independently the link between coitus and birth but, for one reason or another, perhaps the relative absence of herding mammals, did not elevate it into a general principle to be worshipped. Maritime peoples and beachcombers are unlikely to have had any conception of reproduction among fish or sea mammals. Similarly, rain forest inhabitants, sharing their habitat with only wild animals, would have had much less opportunity to observe mating and almost no opportunity to witness births, since the process is so dangerous in the wild.

Once the link between sexual union and birth was understood, and that it applied generally, in regard to other mammals that might serve as food, the idea of controlling fertility—a crucial aspect of domesticating animals—would have become a possibility. This too is therefore consistent with the discovery of the link between coitus and birth at around only 12,000 years ago.

The Link Between Climate Change and Agriculture

The domestication of plants and animals took place across the Old World some time between 14,000 and 6,500 years ago, and it is one of the most heavily studied topics in prehistory. It is safe to say that while we are now fairly clear about where agriculture began, how it began, and with what plants and animals, there is no general agreement, even today, about why this momentous change occurred.

The domestication of plants and animals (in that order) occurred independently in two areas of the world that we can be certain about, and perhaps in seven. These areas are: first, southwest Asia—the Middle East—in particular the "Fertile Crescent" that stretches from the Jordan Valley in Israel up into Lebanon and Syria, taking in a corner of southeast Turkey, and round via the Zagros Mountains into modern Iraq and Iran, the area known in antiquity as Mesopotamia. The second area of undoubted independent domestication lies in Mesoamerica, between what is now Panama and the northern reaches of Mexico. In addition, there are five other areas of the world where domestication also occurred but where we cannot be certain if it was independent or derived from earlier developments in the Middle East and Mesoamerica. These areas are: Southeast Asia, as we have seen, together with the highlands of New Guinea and China, where the domestication of rice seems to have had its own history; a narrow band of sub-Saharan Africa running from what is now the Ivory Coast, Ghana, and Nigeria across to the Sudan and Ethiopia; the Andes/Amazon region, where the unusual geography may have prompted domestication independently; and the eastern United States, where it very likely derived from Mesoamerica.[36]

In the case of animal domestication, the evidence is somewhat different. One or more of three criteria are generally taken as evidence of domestication: a change in species abundance (a sudden increase in the proportion of a species within the sequence of one site); a change in size (most wild species are larger than their domestic relatives, because humans found it easier to control smaller animals); and a change in population structure (in a domestic herd or flock, the age and sex structure is manipulated by its owners to maximize outputs, usually by the conservation of females and the selection of sub-adult males). Using these criteria, the chronology of animal domestication appears to begin shortly after 9000 BP (before the present)—that is, about 1,000 years after plant domestication. The sites where these processes occurred are all in the Middle East, indeed in the Fertile Crescent, at locations that are not identical to, but overlap with, those for plant domestication.[37]

Much more controversial, however, are the reasons for why agriculture developed, why it developed then, and why it developed where it did. It is also a more interesting question than it looks when you consider the fact that the hunter-gathering mode is actually quite an efficient way of leading life. Ethnographic evidence among hunter-gatherer tribes still in existence shows that they typically need to "work" only three or four or five hours a day in order to provide for themselves and their kin. Why, therefore, would one change such a set of circumstances for something different, where one has to work far harder?

The most basic of the economic arguments stems from the fact that, as has already been mentioned, some time between 14,000 and 10,000 BP, the world suffered a major climatic change. This was partly a result of the end of the Ice Age, which had the twin effects of raising sea levels and, in the warmer climate, encouraging the spread of forests. These two factors ensured that the amount of open land shrank quite dramatically, segmenting formerly open ranges into smaller units. The reduction of open ranges encouraged territoriality, and people began to protect and propagate local fields and herds. A further aspect of this set of changes was that the climate became increasingly arid, and the seasons became more pronounced, a circumstance that encouraged the spread of wild cereal grasses and the movement of peoples from one environment to the next, in search of both plants and animal flesh. There was more climatic variety in areas that had mountains, coastal plains, higher plains, and rivers. This accounts for the importance of the Fertile Crescent.

Added to the climatic changes occurring at that time are contrasting arguments that the world was either over- or under-populated, driving people to abandon the hunter-gatherer lifestyle.

One thing that recommends these theories is that they divorce sedentism from agriculture. This discovery is one of the more important insights to have been gained since World War II. In 1941, when the archaeologist V. Gordon Childe coined the phrase "The Neolithic Revolution," he argued that the invention of agriculture had brought about the development of the first villages and that this new sedentary way of life had in turn led to the invention of pottery, metallurgy, and, in the course of only a few thousand years, the blossoming of the first civilizations. This neat idea has now been repeatedly overturned, for it is quite clear that sedentism, the transfer from a hunter-gathering lifestyle to villages, was already well under way by the time the agricultural revolution took place. This has transformed our understanding of early man and his thinking.

Breakthrough: Discovering the Link Between Coitus and Birth

In particular, the fact that sedentism preceded agriculture stimulated the French archaeologist Jacques Cauvin to produce a
wide-ranging review of the archaeology of the Middle East. He
started from a detailed examination of the preagricultural villages.
Between 12,500 BC and 10,000 BC, the so-called Natufian culture
extended over almost all of the Levant, from the Euphrates to
Sinai (the Natufian takes its name from a site at Wadi an-Natuf in
Israel). Excavations at Ain Mallaha, in the Jordan Valley, north of
the Sea of Galilee, identified the presence of storage pits, suggesting "that these villages should be defined not only as the first sedentary communities in the Levant, but as 'harvesters of cereals.'"[38]
The Natufian culture boasted houses, grouped together (about
six in number) as villages, and were semi-subterranean, built in
shallow circular pits "whose sides were supported by dry-stone
retaining walls." Most importantly for us, single or collective
burials were interred under the houses or grouped in communal
cemeteries. Some burials, including those of dogs, may have been
ceremonial, since they were decorated with shells and polished
stones (note the early presence of dogs). Mainly bone artworks
were found in these villages, usually depicting animals.

Cauvin next turned to the so-called Khiamian phase. This,
named after the Khiam site, west of the northern end of the
Dead Sea, was significant for its "revolution in symbols." Natufian art was essentially zoomorphic, whereas in the Khiamian
period female human figurines (but not "Venus"-shaped) begin
to appear. Around 10,000 BC the skulls and horns of aurochs
(a now-extinct form of wild ox or bison) are found buried in
houses, with the horns sometimes embedded in the walls, an
arrangement which suggests they already have some symbolic
function. Then, around 9500 BC, we see dawning in the Levant,
"in a *still unchanged* economic context of hunting and gathering"
(italics added), the development of two dominant symbolic figures, the Woman and the Bull. The Woman was the supreme

figure, Cauvin says, and was often shown giving birth to a Bull.[39]

His main point is that this is the first time humans have been represented as gods, that the female and male principle are both represented, and that this marked a change in mentality *before* the domestication of plants and animals took place.

It is easier to see why the female figure in Khiamian art should be chosen than the male. The mystery of birth had conferred on the female form a sacred aura, easily adapted by analogy as a symbol of general fertility. For Cauvin, therefore, the bull conveniently symbolized the untameability of nature, the cosmic forces unleashed in storms, for example. Moreover, Cauvin discerns in the Middle East a clear-cut evolution. "The first bucrania [cattle skulls, with horns] of the Khiamian . . . remained buried within the thickness of the walls of buildings, not visible therefore to their occupants." Was this because they wanted to incorporate the power of the bull into the very fabric of their buildings, so that these structures would withstand the hostile forces of nature?

But there is surely more to the story—the transition—than this. As many prehistorians have noted, at certain stages in the moon's cycle it resembles the horns of a bull. This would have been recognized by ancient peoples, who would also have noted the menstrual cycle, linked to the phases of the moon and therefore, by implication, to the bull (recall the Venus figurine with twenty-eight red dots painted on it, mentioned above). Early peoples would have observed that menstruation stopped immediately before gestation and may therefore have linked the "bull-shaped" moon with human gestation and birth. Is this why so many early Neolithic images show women giving birth to bulls?

It is important to say that bulls are represented by their bucrania far more than by their penises, a discrepancy that is difficult to understand on straightforward "fertility symbol" reasoning. Furthermore, no one can actually have seen a woman giving birth to a bull, so what did this imagery mean? In shamanistic religions, shamans undergo soul flight to other realms and can take the form of animals.[40] On such a system of understanding, the bull/shaman of the heavens could visit earth and enter inside a woman. Did early peoples think that women were made pregnant by one or other of the mysterious forces of nature?

That is a plausible theory, made more so by the fact that, in the Middle Eastern sites of Çatalhöyk and Jericho, which followed the Natufian and Khiamian cultures, at ~11,000 to 9,500 years ago, yet more changes are observable—two in particular. First, as Brian Fagan and Michael Balter have each pointed out, at both Çatalhöyk, in Turkey, and Jericho, in Palestine, "there was a new preoccupation with ancestors and with the fertility of animal and human life."[41] And there was a second change, with people being buried not in communal graves (as Cauvin noted) but under the houses where they had lived. These are the very houses where, previously, bucrania had been sequestered, in the walls and under the floors.

Now recall that the dog is present in these Middle Eastern sites—this is precisely the area where, vonHoldt and Wayne say, domestication of the dog occurred (which fits with more recent evidence that dogs evolved from wolves sometime between 32,000 and 19,000 years ago). If people had only recently discovered the link between coitus and birth, a number of things would have followed. For example, not only would the discovery have revealed new relationships in Neolithic society between men and women, and between parents and children, but it would have transformed ideas about ancestors as well. Until that point, the understanding of "ancestors" would have been general, communal, tribal—ancestors were "the people who had gone before." After the breakthrough, however, "ancestorship" would have become a much more individual, personal phenomenon, which may be why ancestors at Çatalhöyk and Jericho were now buried under the floors of the houses where they had lived, replacing bucrania.

Is the disappearance of the Venus figurines across Eurasia, at about 12,000 years ago, and the appearance of the female figures at these Middle Eastern sites, about 11,000 years ago, giving birth rather than nearing birth, coincidental, or yet more evidence of the change in understanding that we are considering?[42]

There is one other factor that has only recently come to light. We now know that, as people transferred from hunter-foraging to a cereal-based, more sedentary diet, changes occurred in the pelvic canal of females. The pelvic canal, recent science shows, is very susceptible to nutrition and the change in diet caused it to narrow. Even today, the human pelvic canal has not regained its Paleolithic dimensions.[43] And so we may conclude that at the very time sedentism and a change in diet were taking place, and a new understanding of reproduction—and what it meant for family/religious life—was occurring, the act of birth itself was getting more traumatic and dangerous.

At this point we may well ask ourselves whether, with all that was happening, more or less at once, and with some of the changes being shocking—even moving—then, like other powerful events, would they have been remembered and rendered coherent in myth form?

A Genetic Code in Genesis?

There is one piece of evidence, one myth, which suggests there was just such a powerful change in human consciousness.

Could it be that this all-important change in mentality is in fact contained in the very first book of the Bible? Is this why the Bible begins as it does? Genesis is known partly for its account of the Creation of the world ("Let there be light"), and of humankind, but also for the otherwise rather strange episode of the expulsion of Adam and Eve from the Garden of Eden because they had eaten from the Tree of Knowledge against God's explicit instructions.

Some parts of this story are easier to understand/decipher than others. The expulsion itself, for instance, would seem to represent the end of horticulture, or the end of humankind's hunter-gatherer lifestyle and its transfer to agriculture, and the recognition, discussed earlier, that the hunter-gatherer lifestyle was easier, more enjoyable, more harmonious, than farming. The Bible is not alone in making this observation. In more or less contemporaneous traditions (Elysian Fields, Isles of the Blessed, in Hesiod or Plato) human beings are understood to have hitherto lived free from toil, on a fruitful earth, "without help from agriculture," and "untouched by hoe or ploughshare." [44] This is one reason why the transition is felt as a Fall.

The central drama of Genesis, however, is of course that Eve acts on the serpent's advice and induces Adam so that they both eat from the Tree of Knowledge, after which they discover that they are naked. This is utterly incomprehensible unless we acknowledge that both "knowledge" and nakedness here refer to sexuality, or sexual awareness in some form. And, indeed, as the biblical scholar Elaine Pagels tells us, the Hebrew verb "to know" (*yada*) "connotes sexual intercourse." (As in, "He knew his wife.")[45] Once Adam and Eve have eaten of the Tree of Knowledge, they discover they are naked. What can this mean other than that they become aware of their bodies, how they differ, how and why that difference matters, and that the "knowledge" they now have is of how sexual reproduction works? This knowledge is shocking and moving because it shows that reproduction is "natural"; humans are not made by some miraculous divine force but by sexual intercourse. This is a second reason why it is felt as a Fall.

There are other clues to this change once we look for them. As Potts and Short again observe, hunter-gatherers are polygynous, but it is now, Elaine Pagels says, that marriage becomes monogamous and "indissoluble." People understood the nature of paternity for the first time and it became important to them. It also becomes relevant, as again Elaine Pagels points out, that in Genesis 3:16 the text reads: "To the woman he [God] said, 'I will greatly multiply your pain in childbearing; in pain you shall bring forth children.'"*

---

* This is the wording of the Revised Standard Version (1952). The King James Version (1611) of the same passage reads: "I will greatly multiply thy sorrow and thy conception; in sorrow thou shalt bring forth children."

Are we not seeing here a mythical account of the transition from hunter-gathering to farming and some of its associated consequences? More than that, is it not accompanied by an attitude that is not entirely at variance with modern scholarship? That the transition was not wholly good, that harmony with nature had been lost, and that even childbirth had become more painful and dangerous? (In addition to the changes in the dimensions of the birth canal, other contemporary research has shown that birth intervals of fewer than four years—more likely in sedentary peoples than hunter-gatherer nomads—are more perilous than longer intervals.)

Genesis does not "date" when humanity linked sex and birth, not directly. But it does associate the link with the transition to farming. Therefore, is Genesis, and the Fall it records, in fact reporting a very great, shocking breakthrough that humanity made around 12,000 to 10,000 years ago—the insight that coitus and gestation are related? Timothy Taylor, in *The Prehistory of Sex*, reports that around that time, among the Inuit of Alaska, they had what archaeologist Lewis Binford recognized as "lovers' camps"—places where new couples could "get away from it all" to cement their relationships. Taylor makes the further point that, around 10,000 years ago, the caves where the Ice Age art proliferated seem to have been "forgotten."[46] Whatever was going on was pretty important. Venus figurines were no longer needed, cave art was no longer needed. Mammal reproduction was understood and domestication was in place.

These are tentative arguments, but their main strength lies in the consistent picture they paint. Around 12,000 to 10,000 years ago, as well as a transition to sedentism and domestication, people discovered the link between coitus and birth and this produced a seminal change in attitudes to ancestry, the male role, monogamy, children, privacy, property. It was above all a momentous psychological change as much as anything else, and that is why it was recorded, in coded form, in Genesis.

·   ·   ·

One more brief example will underline the point of this chapter: the scientific order underlying myths. It also extends the argument to another continent: Africa.

The Lemba are a tribe from Zimbabwe who have an origin myth that is, on the face of it, hard to believe. The legend, carried down through many generations, relates how, 3,000 years ago, a man called Buba led the tribe out of the lands that now make up Israel and journeyed southward, through Yemen, through Somalia, down the eastern coast of Africa, finally settling in Zimbabwe. They were, in the words of their legend, one of the lost tribes of Judea, and, like Jews, they observed the Sabbath, eschewed pork, and circumcised their sons.

There are many peoples across the world who claim to be one of the lost tribes of Judea, yet in 1997 scientists found a distinctive genetic marker among Lemba males.

According to Jewish tradition, the priestly caste, the *cohanim*, were all descended from a single male: Aaron, the brother of Moses. Now, the Y chromosome is the packet of genes that gives a male child his maleness.[47] It is passed down from father to son to grandson, just like the status of belonging to the priestly caste. And, indeed, as stated, in 1997 geneticists uncovered a marker of Jewish priesthood on the Y chromosomes of Lemba males. These markers show that the Lemba priests "were of the same stock as the other Jewish priests around the world."[48]

And, as we have come to expect, there are parallels in their language. Though they speak a Bantu language, a group of tongues that includes Swahili and Zulu, some of the Lemba words suggest a wider heritage. For example, some of their clans bear "Semitic-sounding" names, such as "*Sadique*." *Sadiq* means "righteous" in Hebrew, and in the Jewish regions of Yemen the name "Sadique" is not unknown. Here too, then, genetics and language concur and underline the veracity of ancient myth.[49*]

---

* In yet another recent case, modern astronomy has been used to explain the spread of early Christianity. William K. Hartmann, cofounder of the

• • •

In his book on sociobiology, Edward O. Wilson said he had asked himself—as a young man but not entirely rhetorically—if "Holy Writ" had been early man's first literate attempt to explain the universe. Little did he know how right he was. More than that, though, by now—the second decade of the twenty-first century—the genetic mosaic around the world is well known and is widely accepted as showing how different groups are related to one another. The overlaps between archaeology, mythology, genetics, and the various dating techniques are many and consistent. Together, they have transformed our understanding of the peopling of the world, not just between the Old World and the New, but the spread of tribes into Europe from the Middle East, migrations throughout Africa, and across the Pacific from Southeast Asia into Australia. With the rates of mutation more or less known, genetic markers have become a reliable form of history, taking us back into the distant past—and ever more accurately. Big History and science have come together.

Planetary Science Institute in Tucson, Arizona, has argued that Saul's conversion on the road to Damascus, thought to have taken place in AD 35, when he saw a bright light in the sky and heard the voice of Jesus, closely parallels "the sequence you see with a fireball," and neatly matches accounts of the fireball meteor seen above Chelyabinsk, Russia, in 2013. Among the parallels are that the Chelyabinsk fireball shone three times as powerfully as the sun, as Saul/Paul proclaimed, that he and his companions fell to the ground (people were knocked over when the Russian meteor exploded in the sky and generated a shockwave across Chelyabinsk), and that he was blinded (in Chelyabinsk people suffered photokeratitis, temporary blindness caused by intense ultraviolet radiation).

15

CIVILIZATION = THE ORCHESTRATION OF GEOGRAPHY, METEOROLOGY, ANTHROPOLOGY, AND GENETICS

When the Spanish first arrived on mainland America, there were two civilizations that were prominent—even dominant—in the New World. These were the Aztecs in what is now Mexico, and the Incas in Peru. Both were flourishing at the time—each had elaborate capital cities, organized religion with ritual calendars and associated artworks. Both societies were rigidly divided into social classes and each had successful methods of food production. But there was more to the Aztec and the Inca than met the eye.

The Aztecs were reached first, in 1519, thirteen years before the Incas. When the Spaniards crossed the ring of mountains which surrounded Tenochtitlán and descended into the Valley of Mexico, and beheld the astonishing cities which formed the core of the Aztec Empire, with its network of shallow lakes surrounded by active volcanoes, they could scarcely believe their eyes. So elaborate were these cities, and of such a size, that some of Cortés's soldiers could not be sure whether what was before them was real or a hallucination. But what the conquistadors discovered soon enough was that, despite being themselves capable of a very practical and calculating brutality, the Mexica, as the Aztecs were also known, "presided over a city of pyramids and sacred temples that reeked with the blood of human sacrifice."

This also took some getting used to. "The dismal drum sounded again," wrote Bernal Díaz, one of Cortés's disaffected aides, in his *The True History of the Conquest of New Spain.*

> [It was] accompanied by conches, horns, and trumpet-like instruments. It was a terrifying sound, and we saw [the captives] being dragged up the pyramid steps to be sacrificed. When they had hauled them up to a small platform in front of the shrine where they kept their accursed idols we saw them put plumes on the heads of many of them; and then they made them dance with a sort of fan. . . . Then after they had danced the [priests] laid them on their backs on some narrow stones of sacrifice and, cutting open their chests, drew out their palpitating hearts which they offered to the idols before them. Then they kicked the bodies down the steps, and the Indian butchers who were waiting below cut off their arms and legs and flayed their faces.

Bernal Díaz, who served as a *rodelero* (a "shield" man) under Cortés and claimed to have been in 119 battles, was appalled as much as his colleagues by the extent of human sacrifice among the Aztec, and he recorded that the priests had many ways of carrying out sacrifice: by shooting the victims with arrows, by burning or beheading them, by drowning them, by throwing them from a great height onto a bed of stones, by skinning them alive or crushing their heads. But the most common method was to rip out the victim's heart.

In the actual sacrifice itself, the priests painted the victims with red-and-white stripes, then reddened their mouths and glued white feather-down to their heads. Dressed in this way, the victims were lined up at the foot of the pyramid steps, before being led up one by one, symbolic of the rising sun. Four priests held the victim down over the sacrificial stone, while a fifth pressed hard on his (or her) neck, causing the chest to stand out. The leading priest thrust his obsidian knife swiftly through the rib cage and tore out the heart while it was still beating. Figures are hard to be sure of. Two conquistadors say they saw 136,000 skulls on a rack, but most scholars believe this to be a gross exaggeration.

On the other hand, in 1487, the forces of Ahuitzotl (1486–1501), the eighth Aztec ruler, put down a rebellion among the Huaxtecs (who occupied the coast of the Gulf of Mexico), the annexation coinciding with the completion of the Great Temple in the Aztec capital, Tenochtitlán. The joint celebrations that followed involved lavish gift-giving and the sacrifice of no fewer than twenty thousand captives, some bound together by ropes through their noses, who were formed into four lines running down the steps of the temple "and out along the four causeways of the island city." Every one of them had his heart cut out and the entire ceremony lasted four days. The city walls were plastered everywhere with the blood of victims.

The point of these grisly stories is to highlight the fact that, by that time—the European conquest of the Americas—blood sacrifice had long since died out in Eurasia, and so this throws up an extraordinary contrast between the civilizations of the two hemispheres that needs to be explained. This chapter shows that the very different behavior in the two hemispheres can nonetheless be explained by a coherent set of factors that marks in each case the same principle: the coming together of geography, weather, plant and animal distribution, anthropology, and genetics.

Why East-West is Best

As was introduced in chapter 12, between 15,000 BC and AD 1500—speaking in round numbers—two populations existed on earth, each unaware of the other. Comparisons of the trajectories of these populations should be uniquely educative. The systematic differences between the Old World and New World civilizations—the Great Divide, as it has been called—form one of the largest and most instructive "natural laboratories" in the history of humankind. What follows is a synthesis of widespread research—in geology, geography, climatology, ethnography, botany, zoology, agriculture, and genetics—which come together coherently to show how the earth's features have exerted a profound influence on the development and shape of civilizations, to provide perhaps the ultimate form of scientific understanding of Big History—how we have arrived at the modern world.

To begin with, the Americas are a much smaller landmass than Eurasia, even without Africa added on. Moreover the New World is, as Hegel, Jared Diamond, and others have pointed out, oriented in a north–south direction rather than east–west, as Eurasia is. This orientation in itself impeded development in the New World, relatively speaking, by slowing down the rate at which plants—and therefore animals and civilization—could spread.[1]

Alongside the general geographical alignment of the continents went an associated long-term climatic variation, of which the most important elements were, in the mainly *temperate* Old World civilizations, the monsoon, and, in the mainly *tropical* New World civilizations, the El Niño–Southern Oscillation (ENSO), reinforced by the more widespread violent activity caused by volcanoes, earthquakes, winds, and storms.

The importance of the monsoon lies in the fact that, for the last eight thousand years, since the time of the last great flood, glaciers have been released into the North Atlantic in larger numbers than before, causing massive evaporation. This has affected the amount of snow forming in Central Asia, which in turn has consumed solar energy in unprecedented amounts. The main result of this is that the monsoon (which has its origins in a vast pool of warm water thousands of kilometers long in the southwest corner of the Pacific, held in place by the Malay Archipelago) has been decreasing in strength.[2] This has affected the weather from the eastern end of the Mediterranean and northern/eastern Africa right across to China (the "monsoon civilizations"). As a consequence, approximately two-thirds of the world's farmers have found water progressively harder to come by.[3] The varying strength of the monsoon, and its relation to the emergence (and subsequent collapse) of Old World civilizations, means that the major environmental/ideological issue in Eurasia over that time has been *fertility*. The landmass, bit by bit, has been drying.

In the New World, on the other hand, the major factor affecting weather has been the *increasing* frequency of ENSO, the El Niño storms that sweep across the Pacific from west to east. These occurred a few times a century about six thousand years ago, but now occur every few years. Besides the occurrences of ENSO itself, its relationship with volcanic activity, given the makeup of the Pacific Ocean (an enormous body of water over a relatively thin crust), also appears to have been important. Meso- and South America are the most volcanically active mainland areas of the world where major civilizations have formed.[4] Put all this together, and the most important environmental issue in the Americas over the past few thousand years, which has had fundamental ideological consequences, has been the increasing frequency of *destructive weather*. These systematic differences in climate across the hemispheres dovetail plausibly with the historical patterns that are observable between the New World and the Old World.

The Link Between Biology and the Gods

After the geographical and climatological factors that determined basic and long-term differences between the two hemispheres, the next most important factors lay in the realm of biology—plants and animals. In the plant realm, we may say that, again, there were two main differences between the hemispheres. The first concerned cereals: grain. In the Old World there was a naturally occurring range of grasses—wheat, barley, rye, millet, sorghum, rice—susceptible of domestication, and because of the east–west configuration of the landmass, they were able to spread relatively rapidly once domestication was achieved. Surpluses were therefore built up relatively quickly, and it was on this basis that civilizations were able to form. Moreover, these grasses grew well on the estuaries that formed around the world at about 9000 BC thanks to rising sea levels caused by glacial melting, which slowed the rate at which rivers ran into the sea.[5]

In the New World, on the other hand, what turned out to be the most useful grain there, maize, was evolved from teosinte, which, in the wild, was, morphologically speaking, far more distant from the domesticated form than was the case with the Old World grasses. Furthermore, as we now know, because of its high sugar content (as a tropical rather than a temperate plant) maize was first used for its psychoactive properties rather than as a foodstuff—it was developed as beer.[6] On top of everything else, maize—even when it did become a foodstuff—found it harder to spread in the New World because of the north–south config-uration of the landmass, which meant that mean temperatures, rainfall, and sunlight varied far more than they did in Eurasia. For this reason, the development of maize *surpluses* was much harder—and slower—to build up. This slowed the development of civilization.

The second area where plants differed between the Old World and the New was in the realm of hallucinogens. The influence of these plants on history has not been appreciated before to this extent, but post–World War II research now makes it clear that the distribution of psychoactive plants across the world is very anomalous.[7] The figures confirm that between eighty and one hundred hallucinogenic species occur naturally in the New World, compared with not more than eight or ten in the Old World.

One result of this is that hallucinogens played a large and vital role in the religious thought of the New World but especially so in Central and South America, where the most advanced civili-zations evolved.

The role of hallucinogens was essentially twofold. First, they made the religious experience in the Americas much more *vivid* than in the Old World. Second, because of their psycho-active properties, hallucinogens fostered ideas of *transformation*, between humans and other forms of life, and of travel, or "soul flight," between the "middle" world and the "upper" and "lower" realms of the cosmos. Combined with a society in which, because of the lack of wheeled transport, or riding (see below), and the *north–south* configuration, people found it relatively difficult to travel far, the journeys to the upper and lower realms were all the more important. The sheer vivid-ness, and the fearsome nature, of some of the transformations experienced by shamans in hallucinogenic-induced trance, the overwhelming psychological *intensity* of altered states of con-sciousness brought about by hallucinogens, would among other things have made New World religious experiences far more *convincing* and therefore more resistant to change than those in the Old World, where, as we shall see, horse riding and wheeled transport—carts and chariots—meant that different groups, with different beliefs, came into contact with each other far more.[8]

This is not to say that there were no hallucinogens in the Old World, or that they were not important. Opium, cannabis (hemp), and soma were all widely used ritual substances in various regions of Eurasia. For a variety of reasons, however, the more powerful psychoactive substances gave way relatively early on to milder alcoholic beverages (see below).[9]

Nature and Big History in the Old World

In the Old World what was worshipped instead was two aspects of fertility—the Great Goddess and the Bull. Though the Bull was worshipped as an aspect of fertility, we do well to remember that this animal was more often represented by his distinctive bucrania—his head and horns—than his sexual organs. Probably this was due—as was outlined in the previous chapter—to the similarity between the bull's horns and the shape of the new

moon, added to which was the link between the phases of the
moon and the menstrual cycle, especially the cessation of menstruation, which would have been noted. Whatever the exact
beliefs, the essential thing is that throughout the Neolithic period
in the Old World—whether the object of worship was the Great
Goddess, the Bull, the cow, rivers, or streams—the central issue
was *fertility*.[10]

On top of all that was a developing interaction between humans
and domesticated mammals that had immense consequences and
didn't happen in the New World. Put succinctly and chronologically, those developments were as follows:

1. The domestication of cattle, sheep, and goats enabled the
   exploitation of less good land. This brought about the development of pastoralism, as a result of which these kinds of
   farmers spread beyond village life and became more dispersed.
   This dispersal in turn had an effect on religious ideology—a
   move beyond shamanism. Among pastoralists the calendar
   was less important, because domesticated mammals give
   birth at different times of the year, unlike plants which, in
   temperate zones, are more directly linked to the cycle of
   the seasons. (Cattle can give birth at any time of the year,
   goats in winter or spring. In temperate zones, sheep lamb
   in spring, but in warmer climates the lambing season can
   extend throughout the year. For horses the natural breeding
   season is May to August.) A further aspect of domesticated
   mammals is that their whole life takes place—as it were—
   aboveground. Unlike plants, which need to be sown in the
   soil and spend some time out of sight before reemerging in
   a different form, animals are less mysterious. In a pastoral
   society the Underworld is less important, less necessary, less
   ever-present. Together with the relative absence of hallucinogens, this development made the "netherworld" far less
   of an issue in the Old World than in the New.

This may have had other consequences. Although people in the New World never developed the wheel, for good reasons (their domesticated mammals were not much stronger than humans and were confined to the mountains, not in grain-growing areas), they did have the concept of roundness. They had rubber balls for their famous ball games; they sometimes formed balls out of human heads or captives' bodies, which they rolled down pyramid steps, and the combatants in boxing games fought with purely spherical carved stones in their fists.[11] Furthermore, New World peoples witnessed the sun and moon in the day and night sky, and eclipses of both, without ever appearing to consider that the earth itself was spherical. This is surely because, in a predominantly vegetal world, with the experience of the netherworld so vivid (and other "realms" so accessible via hallucinogens), "flatness" and layers were much more obvious than roundness. Not traveling great distances, particularly across the sea, aided by useful winds, they had less chance—and were therefore less prepared—to experience the world as a spherical object.

2. The domestication of the horse (which did not exist in the New World prior to the Conquest) had a number of different consequences. It accompanied the development of the wheel and the chariot and led to riding. These were enormous advances, adding to the mobility of men and women in the Old World, aiding in particular the creation of palace states, far larger than most of the New World states because the horse and chariot allowed larger territories to be conquered *and then held.*[12] In the same way, the wheel and cart meant that more goods could be carried further, boosting trade and the prosperity and exchange of ideas that went with it. The Old World was mobile in a way that the New World was not.

3. Horses and cattle in particular are large mammals, valued for their power. That power, however, meant that, as well as being useful, they were potentially dangerous. In such a context, the regular and frequent use of mind-altering substances was hazardous. A shaman in trance could not have handled a horse or a cow, let alone a bull. On top of that, as dispersed pastoral populations developed the habit of coming together for spouse selection, marriage, and to resist threats from outside (now greater, because wealth in the form of domestic mammals could be stolen as land couldn't), people were driven away from hallucinogens, which offered powerful, vivid (and at times threatening) but *private* experiences, and were led instead towards alcohol, which offered milder, euphoriant *communal* social-bonding experiences.[13]

4. In this way, pastoral nomadism—which never developed in the New World—emerged as one of the "motors" of Old World history. This is because of its inherent instability as a way of life. Because the weakening monsoon caused the drying of the steppes—the natural home of pastoral nomads—they could no longer subsist as easily in their traditional fashion, and had to disperse still further and invade the settled societies at the edges of the great grass-lands. The predominantly east–west nature of the Central Asian steppes ensured that peoples and ideas traveled right across Eurasia. Since weather was more important to the nomads than vegetal fertility, and because they lived on milk, blood, and meat, their gods were sky gods—storms and winds—and horses.[14] Their religious ideology was very different from those of the more settled societies, and the endemic conflict between nomads and settled peoples was both destructive and, in the long run, creative. This endemic conflict between lifestyles never existed in the New World.

5. The virtually continuous conflict introduced into Eurasian history over 2,700 years, from 1200 BC to AD 1500, by the

fact that highly mobile pastoral nomads were at all times more or less threatened by climatic factors (the weakening monsoon and the drying steppe) was one element in bringing about the end of the Bronze Age, the destruction of the great palace states created on the back of the horse-drawn chariot, and eventually provoking the great spiritual change known as the Axial Age. This was the great turning away from (man-made) violence and the epochal "turning in," which produced a new ideology, or morality, and culminated— this time among the pastoral nomadic Hebrews—in the idea of monotheism. It was the fact that the nomadic Hebrews wandered between so many different ecological habitats that gave them the idea of one overarching God that governed *all* environments.[15] This too never happened in the New World.

6. Greek rationalism, and Greek science, in particular the Greek concept of nature, partly brought about by a close examination of domesticated mammals and how their nature compared with human nature (whether they had souls, whether they had morals, whether they had language, whether they could suffer), when adapted to the Hebrew idea of a single, abstract God, eventually gave rise to the Christian idea of a rational God. This was a God whose own nature could be gradually uncovered at some point in the future *because He favored order.* And this idea, of the possibility of progress, helped to create many of the innovations that would enable humankind to explore the earth via its great oceans.[16]

7. The many and varied tribes of pastoral nomads emerging on and then escaping from the Central Asian steppes continued in the millennium and a half after Jesus Christ to both assist and hinder the east–west movement of goods and ideas. But above all it maintained Eurasia as a landmass across which there was much rapid movement. The horse also proved to be a vector for the transmission of disease (the plague)

across the same landmass, which had the dual effect—again in the long run—of promoting the wool industry in the north of Europe. Sheep provided the substance of the first great industry in the world, but also forced the inhabitants of the western Mediterranean to look for alternative routes to the east, where so many spices, silk, and other luxuries came from. Together, these factors helped open up the Atlantic. The very latest research has identified the steppes people as the Yamnaya, who clung to their pastoral nomadism until 5,500 years ago, when they were driven out by the drying climate.

In a sense this is a science-based (evidence-based) meta-narrative of history.

A further point is that most of this activity in the Old World took place in temperate zones (between seven degrees north and fifty degrees north). That is to say where the seasons were pronounced, where the planting and growing periods were carefully delineated. The seasonally rhythmic nature of what was essentially fertility worship contributed to the organized nature of early religious life, but it also had a far more important ideological corollary: *it worked.* The simple biology underlying a religion where fertility was the central issue, in temperate zones, was that—sooner or later—vegetation started to grow again. The cycles of planting and growth didn't invariably work, of course, when drought or rain bringing floods or other factors interfered with the rhythm (the Bible's "fat" and "lean" years). But, essentially, far more times than it failed, fertility worship worked.

Nature and Big History in the New World

Ideological life in the Americas was very different, although there were some similarities. For example, David Carrasco's research among the Aztec shows that their view of history was not so different from the Hindus, in the Vedas, in that they too viewed the past as a series of cycles, interspersed by great catastrophes. But the differences are more widespread. For a start, there were no domesticated mammals in the New World, save for the llama, vicuña, and guanaco in South America, and the guinea pig. (The llama had wool, but neither it nor the other mammals could carry much above one hundred pounds, hardly more than a human could, and so they were not in any way revolutionary as new forms of energy.) One effect of this absence of domesticated mammals was to make plants far more important in the New World than they were in the Old, and this brought with it certain ideas.

The simple, the most obvious, and the most powerful, was that plants need to be planted underground, where they undergo a transformation from seed into shoot. This, combined with hallucinogenic experiences, helps explain why, for the ancient Americans, the cosmos was divided vividly into three zones—the Upper, Middle, and Underworld. It was convincingly reinforced by the experiences of the shaman who, in trance, underwent soul flight, in order to consult the gods and/or the ancestors, and who used hallucinogenic plants to achieve these feats in these other worlds.[17] Fertility was an issue in the New World but, in the tropical rain forests, teeming with life and growing in profusion all through the year (as manioc did, for example), and where the seasons hardly varied, it was never the *overwhelming* issue it was in the temperate Old World. Manioc, the staple food of many early people in the New World, did not like the waterlogged soil of estuaries, preferring conditions upstream, which had the effect of keeping people away from the coasts, and the opportunities for travel and mixing with others that it offered.

Much more important in the New World mind-set were the feared and admired jaguar, and the weather gods—gods of lightning, rain, and hail or violent winds, of thunderstorms, erupting volcanoes, earthquakes, and tsunamis—"dangerous weather." Moreover, volcanic activity, "tectonic religion," comprises evidence of "humanity's closeness to the underworld in general" and its sheer power in particular.[18]

And we now know that, just as the monsoon has been weakening over the past 8,000 years, so the ENSO has been increasing in frequency, at least during the last 5,800 years. Which means that violent, catastrophic storms, with great tsunamis and their associated earthquakes, volcanic eruptions, together with hurricanes, tornadoes, floods, and famine, are stronger, more common, and have been getting considerably more frequent in the New World. In other words, the gods, far from smiling on humankind in the New World, have been getting angrier.

Put all this together and you have a crucial difference between New World and Old World gods, leading to very different civilizations.

If you worship angry gods, whether they be tsunamis or earthquakes, volcanoes or jaguars, your worship essentially takes the form of propitiation, of asking—petitioning—those gods *not* to do something: not to erupt if the god is a volcano, not to produce destructive tsunamis and winds if the god is an ENSO event, not to attack humans if the god is a jaguar. In the New World—in Central and South America certainly—the predominant form of worship was directed toward making unpleasant things *not* happen.

And here is the crucial point: that form of worship doesn't work. That is to say, it didn't/doesn't work all the time, or to anything like the extent that fertility worship works. It no doubt works for *some* of the time. No one in the village is carried off by a jaguar for a certain number of weeks. There is no tsunami for a few years, or even decades. A volcano dies down, as the Icelandic one did in early 2010. But, and it is an important "but," the angry gods are never totally appeased. Sooner or later, their wrath recurs.

We also know that, in the case of ENSO episodes, they have been getting quite a lot more common. Looked at from the point of view of an Olmec or Mayan or Toltec or Aztec shaman, with their extremely accurate calendar, it would have seemed to them that worship wasn't working, that whatever traditional level of ritual had been practiced in the past, it wasn't enough.

In such circumstances, religious specialists would have decided that, if the current level of worship wasn't working, they must redouble their efforts. And this is why the most profound and revealing difference between the Old World and the New occurs in the realm of human sacrifice. In the Old World, thanks to the proximity of domestic mammals, human sacrifice was gradually replaced by animal sacrifice and then, after AD 70 (when the Romans sacked Jerusalem and destroyed the Temple), thanks again to the close similarity of domestic mammals to humans, blood sacrifice was abandoned altogether.[19] In the New World, however, far from being abolished, human sacrifice became more and more widespread until, in the fifteenth century, tens of thousands of Aztec victims were being sacrificed each year. Inca sacrifice was not quite so numerous, but there were still hundreds of mountain *huacas* where people were sacrificed and, according to some accounts, hundreds of children were killed at a time.

The fact that (animal) sacrifice effectively ended in the Old World in AD 70, while human sacrifice (and many other forms of painful, religion-related violence, such as sacred sports where the losers were sacrificed) continued to grow in frequency in the Americas, is a salutary reminder of how environment and ideology can interact to produce marked differences in human behavior—in the very meaning of humanity. But all these ideologies and differences were based on events and processes that can be scientifically explained.

Science likes to depict itself as a forward-looking activity. But part of its strength in recent years—as chapters 11, 12, 14, and this one have shown—has lain in how its technological and theoretical advances enable us now to look back with ever-greater accuracy and precision.

The biography of the earth is not complete yet, but it is far fuller—and far more ordered—than we could have ever imagined at, say, the end of World War II. In putting together this picture of how the two hemispheres developed so differently side by side, almost every science you can think of has been brought into play. So that we can now say: Civilizations don't just happen. They arise where they do, when they do, for reasons that can be reconstructed with the aid of interlocking sciences. Everywhere you look—the Americas, Eurasia, Africa, Australia—ancient history is now an interlaced, interdisciplinary branch of science.

16

THE HARDENING OF PSYCHOLOGY AND ITS INTEGRATION WITH ECONOMICS

In the wake of World War II, it fell to a philosopher to advance the course of "emerging order" and convergence in a defining way. Gilbert Ryle was the Waynflete professor of metaphysical philosophy at Oxford University, and in 1949, in his book *The Concept of Mind*, he effectively laid metaphysics to rest, delivering a withering attack on the traditional, Cartesian concept of duality, which claimed that there is an essential difference between mental and physical events.

Using a careful analysis of language, Ryle gave what he himself conceded was a largely behaviorist view of man. There is no inner life, Ryle said, in the sense that a "mind" exists independently of our actions, thoughts, and behaviors. When we "itch" to do something, we don't really itch in the sense that we itch if a mosquito bites us. When we "see" things "in our mind's eye," we don't see them in the way that we see a green leaf or a tiger (try counting the stripes of a tiger in your mind's eye). This is all a sloppy use of language, Ryle says, and most of his book was devoted to going beyond this sloppiness. To be conscious, to have a sense of self, is not a by-product of the mind; it *is* the mind in action. The mind does not, as it were, "overhear" us having our thoughts; having the thoughts *is* the mind in action. In short, there is no ghost in the machine—only the machine. Ryle considered the will, imagination, intellect, and emotions in this way, demolishing at every turn the traditional Cartesian duality, ending with a short chapter on psychology and behaviorism.[1]

In making his argument, Ryle was effectively replacing the mind with the brain, and he found rapid indirect support from a very different direction: the hard sciences. One development in particular in the 1950s helped to discredit the traditional concept of mind. This was medical drugs that influenced the working of the brain.

As the century had worn on, one "mental" condition after another had turned out to have a physical basis: cretinism, general paralysis of the insane, pellagra (nervous disorder caused by niacin deficiency)—all had been explained in biochemical or physiological terms and, as a result, shown themselves to be amenable to medication.

Until the 1950s, however, the "hard core" of insanity—schizophrenia and the manic-depressive psychoses—lacked any physical basis. But, beginning in that decade, even these illnesses began to come within the scope of the harder sciences—biochemistry in particular—three avenues of inquiry joining together to form one coherent view. From the study of nerve cells and the substances that governed the transmission of the nerve impulse from one cell to another, specific chemicals were isolated. This implied that modification of these chemicals could perhaps help in treatment by either speeding up or inhibiting transmission. And, during the 1950s, six entirely new types of psychoactive drug were introduced and remain—more or less—the backbone of psychiatric treatment today.[2] We can call this the beginnings of a more or less successful union of chemistry and psychology.

Starting in 1949, the same year as Ryle's book came out, a French naval surgeon, Henri Laborit, working in Tunisia, examined several victims of shock. Shock can have several causes, but Laborit was interested in shock caused by major surgical procedures, which produces what can be a dangerous fall in blood pressure. His idea was that the fall in pressure was caused by the release of chemicals known as histamines and that therefore if the release of these chemicals could be somehow blocked, shock would be prevented. Some further background is needed here. The antihistamines were developed in the 1940s as remedies for motion sickness and were found to have the side effect of making people drowsy—i.e., they exerted an effect on the brain. It was also known that the Indian plant *Rauwolfia serpentina*, extracts of which were used in the West for treatment of high blood pressure, was also used in India to control "overexcitement and mania." In other words, the Indian drug acted like the antihistamines. Its most active substance was promethazine, commercially known as Phenergan. And so Laborit treated his patients with a cocktail of drugs that included promethazine. He claimed in the article he published in 1949 that this treatment worked, but, more important in the long run, he also added the observation that "antihistamines produce a *euphoric quietude*. . . . Our patients are calm, with a restful and relaxed face."[3] It did indeed control "overexcitement."

Alerted to this observation, in the following year the drug company Rhône-Poulenc began a major research initiative to test the psychiatric effects of promethazine. The group of drugs to which promethazine belongs is called the phenothiazines, and Paul Charpentier, the company's chief chemist, thought it worthwhile to synthesize as many variations of its molecular structure as possible, in the hope that one might be even more effective than the promethazine he started with. These compounds were tested in an interesting way. They were given to rats that had been conditioned to climb a rope to avoid an electric shock signaled by a ringing bell. "One compound in particular, chlorpromazine, left the rats unmoved when the bell rang."[4]

Alerted to this, two psychiatrists in Paris, Jean Delay and Pierre Deniker, tried out chlorpromazine on a schizophrenic patient,

Giovanni A, who had been admitted to hospital for "making improvised speeches in cafés, becoming involved in fights with strangers, and walking around the street with a pot of flowers on his head proclaiming his love of liberty." After nine days on chlorpromazine, Giovanni A was able to have a normal conversation and after three weeks he was discharged.[5] The news spread to Britain and the United States where again, in a growing number of patients, remission from schizophrenia was seen within weeks once chlorpromazine was administered.

Chlorpromazine was, as James Le Fanu put it, the first swallow. In rapid succession over the following few years, four other major drug groups were introduced, which applied across a wide psychiatric spectrum. These included lithium for depression, and tranquilizers, which appeared to work by inhibiting neurotransmitter substances, like acetylcholine or noradrenaline. It was only natural to ask what effect might be achieved by substances that worked in the opposite way—might *they*, for instance, help relieve depression?

Administering the new anti-tuberculosis drug isoniazid, doctors found there was a marked improvement in the well-being of patients. Their appetites returned, they put on weight, and they cheered up. Psychiatrists quickly discovered that isoniazid and related compounds were similar to neurotransmitters, in particular the amines found in the brain. These amines, it was already known, were decomposed by a substance called monoamine oxidase. So did ioniazid achieve its effect by inhibiting monoamine oxidase, preventing it from decomposing the neurotransmitters? The monoamine oxidase inhibitors, though they worked well enough in relieving depression, had too many toxic side effects to be lasting as a family of drugs. Shortly afterward, however, another relative of chlorpromazine, imipramine, the first of several tricyclic antidepressants,* was found to be effective

---

* So called because of their three-ringed chemical structure, which differs by only two atoms from chlorpromazine.

as an antidepressant, as well as increasing people's desire for social contact. This entered widespread use as Tofranil.[6]

The Link Between Chemistry and Behavior

More interesting results still have come from research into impulsivity/aggression. That aggressive behavior is heritable in humans (and therefore physically based) is strongly suggested by twin studies, which have distinguished four aggression-related behavior clusters: direct (physical) assault = a heritability of 47 percent; verbal assault = 28 percent; indirect assault (temper tantrums; malicious gossip) = 40 percent; and irritability = 37 percent. Repeated mating of aggressive or non-aggressive mice produced totally different phenotypes (very aggressive or very mild mice) within four or five generations. The chemical basis of aggression is clearly shown by the fact that castration of such mice and rats specifically bred for aggression virtually eliminates aggressive behavior, which can be fully restored by administration of testosterone.[7]

The brain, including the human brain, contains numerous cells with testosterone receptors, particularly in the hypothalamus, which is a site of synthesis of many other hormones. A role for testosterone in mediating aggressive behavior in humans is recognized, but in practice it has proved difficult to follow through, as it seems naturally occurring levels fluctuate with the time of day and day of the week, among other things. On the other hand, the role in aggression of the neurotransmitter serotonin led to the development of a class of drugs known as "serenics," which greatly reduce aggression in mice, making them more "serene."[8]

The role of serotonin in human personality disorders involving aggression has been under intense study for some time. Decreased concentrations of serotonin have been detected in a wide range of individuals displaying aggressive behavior: for example, in children showing outbursts of aggressive behavior at home and at school, in young men discharged from military service for repeated crimes of violence, and several others. Various drugs are available that block the serotonin transporter in humans, chief among them the so-called SSRIs (selective serotonin reuptake inhibitors), substances such as Prozac, Paxil, and Zoloft, primarily used for the treatment of depression and mood disorders. In a recent clinical trial, involving forty patients, the effect of Prozac on impulsive/aggressive behavior was assessed. After ten weeks of taking the drug, there was a substantial decrease in impulsive/aggressive behavior, compared with those receiving placebos.[9]

Neurotransmitters have also been found to be involved in appetite control (bulimia, anorexia, obesity), substance abuse, memory and mental function, Alzheimer's, and sexual behavior (promiscuity, for example). These are all fairly new fields, and while they are promising, the same caveat is due here as with impulsivity/aggression. Not everyone responds in the same way, and some people do not respond at all. Not everyone who is at risk succumbs, and not everyone who becomes ill responds to treatment.

Neuroplastic studies have also shown that short-term memory becomes long-term memory when a chemical in the neuron, called protein kinase A, moves from the body of the neuron into its nucleus, where genes are stored. And that the longer people are depressed, the smaller their hippocampus gets. The hippocampus of depressed adults who suffered prepubertal childhood trauma is on average 18 percent smaller than that of depressed adults without childhood trauma. More, if the stress experienced is brief, this decrease in size is temporary; if it is too prolonged, the damage is permanent.

Antidepressant medications increase the number of stem cells that become new neurons in the hippocampus. Rats given Prozac for three weeks had a 70 percent increase in the number of cells in their hippocampi. Other studies show that new cells in the hippocampus, in adult rats, are associated with an increase in learning, while findings reported to the Society for Neuroscience in October 2015 suggested that perineuronal nets, a scaffolding of linked proteins and sugars that resembles cartilage, could be the storage sites for long-term memories. Psychological discoveries are getting more specific all the time.[10]

While this is a "good news" story, and while it supports the idea that many "mental" illnesses are in fact physiological ones, and while it was later shown that chlorpromazine "interferes" with the action of the neurotransmitter dopamine, and that imipramine acts on adrenaline, and while many neurotransmitters (substances which occupy the synapses, the junctions between nerve cells in the brain) have been isolated, it is also true that neither schizophrenia nor bipolar disease is by any means eradicated yet. The intensity of the symptoms is undoubtedly lessened, making patients more "manageable," but there is no sense in which the underlying nature of either disorder is understood.

Nevertheless, the links between chemistry and some of the psychological functions of the brain are now well established.

The Physics of Psychology

Alongside the advances in brain chemistry has been the development of a whole raft of physical techniques exploring brain function. Brain mapping has been a triumph of technology, of physics. The achievements of CAT, PET, MRI, and fMRI are nothing short of extraordinary, distinguishing more and more structures of the brain, and narrowing down to ever-smaller areas which are responsible for (or associated with) this or that particular aspect of behavior.

The best-known and oldest technique, electroencephalography, measures the electrical activity of the brain through electrodes placed on the scalp, and is capable of detecting changes in electrical activity on a millisecond level. Functional magnetic resonance imaging (fMRI) works by detecting changes in blood oxygenation and flow that occur in response to neural activity (when a brain area is more active it consumes more oxygen and to meet this demand the blood flow increases to that area). The technique can therefore be used to show which parts of the brain are being used in particular mental processes. Computed tomography scanning (CAT) builds up a picture of the brain based on the differential absorption of X-rays. Positron emission tomography (PET) uses trace amounts of short-lived radioactive material to map functional processes in the brain. When the material undergoes radioactive decay a positron is emitted, which can be detected, so that areas of high radioactivity are associated with brain activity.

This does not exhaust the number of techniques now available, but it is enough to show the ever-expanding link between physical processes and brain activity, and psychology. The latest research at Oxford shows that the brain contains specific neurons to identify faces, direction, even eyes, noses, and mouths.

The most recent development in brain mapping is connectomics. "Connectome" is the new word for the way the brain is organized into small- and large-scale sets of pathways. The approach here is that illnesses like schizophrenia are disorders of brain *organization* rather than something more specific in the way of cellular pathology or neurotransmitter abnormality. The National Institutes of Health in the United States has backed this approach with a multimillion-dollar grant to the Human Connectome Project. Some new techniques have been developed for staining neurons and neurological pathways, and some very beautiful images have been obtained (individual neurons can now be stained with over one hundred distinct colors). However, since the human cerebral cortex alone contains some 10^{10} neurons, linked by 10^{14} synapses, whereas the number of base pairs in the human genome is 3×10^9 the size of the project becomes clear. While no one seems to doubt that brain pathology and psychological functioning will eventually be explained by connectomes, and while some new techniques are being developed to recognize patterns of statistical probability, no one seems to believe that such patterns will ever represent a "one-to-one" type of map of functions. Instead there will only ever be *probabilistic* representation of connectivity. And while the networks high-lighted in this way underline the apparent importance of such areas as the precuneus, the insula, the superior parietal, and the superior frontal cortex, and while seventy-eight cortical regions have been discerned, nothing has yet shown how the brain is divided into *functional* parcels. At the time of writing, the jury is still out on the Human Connectome Project. It is predicated on a reductionist model but has still to deliver the goods.

A new beginning of sorts was announced in May 2014 when the National Institutes of Mental Health published a list of twenty-three core brain functions and their associated neural circuitry, neuro-transmitters, and genes, and the behavior and emotions that go with them. The basic idea here is that there are circuits—the fear circuit being the best known—that may be related to such symp-toms as anxiety and post-traumatic stress disorder (PTSD).

At the same time, we may mention results reported exactly one year later, in May 2015, and referred to in the Preface. This was the experiment with a tetraplegic individual, whose brain was fitted with microscopic electrodes—relatively few in number—which recorded the activity of an equally small number of nerve cells. Based on the firing pattern of these few cells, as recorded in the electrodes, not only could the researchers predict where the individual intended to move his arm, but they could also predict whether he wanted to move his left arm or his right. Separately, they observed that different nerve cells were activated according to whether he imagined rotating his shoulder or touching his nose.

This discovery was important for at least two reasons. Hitherto, no one had imagined that brain location and function could be so specific, in general researchers believing that it was brain *organiza-tion* that lay at the root of most of our behavior. At the same time, the observation that the firing of brain cells also appeared related to the man's *intentions* was something of a surprise, to say the least. How can *intention* be represented in the brain? Do we even know how to talk about such a phenomenon? The experiment would appear to show an intimate link between a physical state of nerve cells and a complicated psychological state. These are clearly early days, but on the face of it, the experiment takes reductionism into new territory.[11]*

Evolutionary and Ethological Psychiatry

Evolutionary psychiatry is the latest development in our expanding understanding of the concept of the unconscious. This idea, made famous by Sigmund Freud in the late nineteenth century, was dismissed as unscientific throughout most of the twentieth, but gradually has been brought more and more into line with modern biology (see above, chapter 13).

Evolutionary psychiatry envisages psychopathological symptoms as maladaptive strategies, unsuccessful attempts to adapt to circumstances. This has given rise to a new way of understanding psychopathology and a new vocabulary. On this account, mental illnesses are understood as disorders of *attachment* and *rank*, as *spatial* or *reproductive* disorders—in other words, these are ethological pathologies, aberrations of the natural world.

On this new understanding, depression for example is linked with the "ubiquitous mammalian tactic of submission," a depressive reaction to the loss of status, and evolutionary psychiatrists now have a new vocabulary to distinguish various forms of the illness, as follows:

- *Deprivation depression*: caused by the loss of affiliative opportunities.

---

* In October 2015 it was reported that a twenty-six-year-old man who was paralyzed in both legs had walked for the first time in five years "just by thinking about it." Scientists at the University of California, Irvine, developed an electrode cap that "detects when a person is thinking about walking or standing still." They tailored it to pick up brain signals from their volunteer. He walked 3.8 meters to a traffic cone.

- *Defeat depression*: caused by failure to achieve desired goals.
- *Anaclitic depression*: caused by failure to form lasting attachment bonds.
- *Introjective depression*: caused by people failing to live up to the high standards they set themselves.
- *Horizontal depression*: which results from dissatisfactions in interactions involving "closeness" and "separateness."
- *Vertical depression*: which results from dissatisfactions in "upper-to-lower" (rank) interactions.

Evolutionary psychiatry is therefore based now on attachment theory, rank theory, self-assessment theory, and such phenomena as RHP (resource holding power), RAB (ritual agonistic behavior), SAHP (social attention holding power). SAHP, for example, is revealed by, among other things, *avoidant personality disorder* (uncontrollable blushing or sweating in social situations), *dysmorphophobia* (fear of looking or smelling abnormal), and *sphincteric phobias* (fear that others might hear one urinating or defecating). Here again the important point is the new concepts, the new vocabulary they bring with them, and their biological base.[12]

The Genetics of Behavior

All this is underlined by the field of behavioral genetics. This is a relatively new field—the first gene was not mapped with a specific chromosome until 1986. But, since then, behavioral genetics has become a very sophisticated ("hard") subject. Many experiments have been carried out, some on animals—ranging from insects to mammals such as mice and rats—and many others on identical human twins (raised together or apart), non-identical twins, parents, brothers and sisters, and longitudinal adoptive studies.

Their main results are, for the most part, clear-cut. The studies show that *most* human behaviors are heritable to *some* extent, and of these we might begin with the following:

- The results from five twin studies show that schizophrenia is 82 to 84 percent heritable. About a fifth of people with schizophrenia "recover or markedly improve," though in the majority of instances the disorder still runs a chronic or a recurrent course.

- The studies for bipolar disorder are not statistically strong, but from the six twin studies that have been done, heritability rates range from 36 to 75 percent. Family studies show that heritability is about 70 percent. Bipolar affective disorder is a serious *recurrent* condition.

- Autism spectrum disorders. It is now generally accepted that this is a neurodevelopmental disorder with characteristic cognitive deficits. Many studies show the heritability to be of the order of 90 percent. The very latest research, in mice, shows that animals which later develop symptoms similar to human autism could be identified by abnormalities in the placenta.

- ADHD: attention deficit hyperactivity disorder. A dozen reliable twin studies show the heritability of this disorder to be "well above 70 percent."

- Antisocial behavior. This can be difficult to define but "early-onset" physical aggression (i.e., before puberty) and "oppositional/defiant behavior" have been found, in many studies, to have a heritability in the 40 to 50 percent range.

- Unipolar depression. This is very common and, indeed, it can be argued it is part of the normal human condition to feel depressed at times. But it is important to stress that unipolar depression is not like ordinary sadness, and, over the course of a lifetime, probably about one in four females and one in ten males suffers from major depressive disorder. These episodes are associated with an increased risk of suicide and an increase in mortality from other disorders. Six studies of twins show a heritability of between 48 and 75 percent.

- Substance abuse. Heritability 25 to 50 percent, but much higher for drug dependency.

- Impulsivity is emerging as an important (inherited) behavioral trait, related to a wide range of disorders, such as aggression, pyromania, gambling, kleptomania, voyeurism, exhibitionism, and trichotillomania (a compulsive pulling out of one's own hair).[13]

Resilience: New Light on the Links Between Environment and Genes

In the realm of genetics—and not just behavioral genetics—two other phenomena are emerging as important. The first is the concept of *resilience*. One of the very consistent findings with respect to all types of environmental hazard is that there is a huge heterogeneity in individual responses. Some with disorders succumb, but others come through relatively unscathed. The latter phenomenon has been termed resilience.

For example, in a study of cardiovascular disease, known as the Framingham Heart Study, subjects who had high dietary fat intake developed abnormal HDL cholesterol concentrations, *or did not*, depending on their genotype on the polymorphic hepatic lipase (HL) gene promoter. A separate study showed that tobacco smokers developed coronary heart disease, *or did not*, depending on their lipoprotein lipase genotype and their apolipoprotein E4 (APOE4) genotype. In a study of stroke-prone hypertension, rats exposed to a high-salt diet developed elevated systolic blood pressure, *or did not*, depending on their genotype on the polymorphic angiotensin-converting enzyme (ACE) gene. In a study of low infant birth weight, women who smoked tobacco during pregnancy gave birth to underweight infants, *or did not*, depending on their genotype with respect to two polymorphic metabolic genes, CYP1A1 and GSTT1.[14]

Are some illnesses the result of a special link between an unfavorable environment and specific forms of genes? Does this genetic presence have consequences for treatment and the success of treatment?

The Great Paradigm Shift: Epigenetics

The other new phenomenon in genetics research is *epigenetics*. It is described in a recent issue of *Science* as "the greatest paradigm shift in science in recent history."

Epigenetics helps explain why identical twins, with identical genes, don't always turn out exactly the same. Epigenetics helps explain why some deprived children respond positively to better treatment later in life, and why some don't, why some are resilient and others aren't.

The science itself is fairly complicated, but the two main mechanisms are methylation and acetylation. In methylation a methyl group (H_3C) is added to DNA, as a result of various environmental experiences, usually at the cytosine location (one of the four bases of DNA that code for proteins—chapter 9), and this has the all-important effect of *turning the gene off*. That gene, in the jargon, is "unexpressed." As noted, methylation occurs as a result of certain life experiences and, in general, but not always, is not heritable. The crucial factor to understand, however, is that if one twin (say) has a gene that is methylated and the other twin hasn't, that gene will only be "expressed" in the twin whose gene is not methylated. This is why identical twins can show different traits.

The second epigenetic process is called acetylation and it affects proteins that are associated with DNA, called histones. (This all considerably simplified of course.) When histones are modified by acetylation—the addition of an acetyl group (H_3CO), which attaches to the lysine amino acid on the tail of the histone—the expression of that gene is *increased*.[15]

To begin with, then, the picture was simple: methylation turned genes off, acetylation turned genes on. That was in 1996. Since then, however, more than fifty epigenetic modifications of histone proteins have been identified.

Despite the complex nature of the exact mechanism, for our purposes the picture is simpler. The epigenetic mechanisms are a way for the environment to affect the expression—or non-expression—of genes, and mostly in a way that it is not inherited. For example, several large studies have shown that if people carry two copies of a gene variant (one from each parent) that controls the transport of a key brain transmitter, serotonin (also known as 5-HTT), they have twice the risk of major depressive episodes. *But this is only true if they also suffer a major life crisis.* This may be put together with studies in young macaque monkeys, which showed that in those who had the key 5-HTT gene variant, it was methylated (and so inactivated) when they were stressed. Therefore, this would appear to be the mechanism by which stress produces its psychological effects: stress (if it occurs early in life) methylates the gene that would help us later on to cope with stress, by preventing that gene's expression. This effect is not inherited but it does last a lifetime.[16]

The Biology of Thought

A whole raft of research has now been conducted following on from Amos Tversky and Daniel Kahneman's original paper on "Judgment Under Uncertainty: Heuristics and Biases" (chapter 13), reaching a high point in 2002 when Kahneman was awarded the Nobel Prize. The broad conclusion of these studies is that human beings have a large (and largely unconscious) group of rapid, intuitive biases in our everyday actions. This approach has spawned new expressions to explain our unconscious psychology: ego depletion, miswanting, cognitive ease, anchoring, availability cascades, duration neglect, priming, optimism bias, loss aversion, negativity neglect, self-inflation, false personal narrative.

Most important of all, perhaps, is that Daniel Kahneman finds that in many ways we are two selves, an "experiencing" self and a "remembering" self, and that these combine to produce our psychology. He also divides thinking into "fast" and "slow," which, he argues, stem from two systems in the brain. System 1 thinking is fast, automatic, intuitive, not readily educable, and therefore resistant to change. We might call this the unconscious, to distinguish it from System 2—conscious thought, which is slow, deliberate, reflective, long-term in its aims.[17]

Michael Gazzaniga, director of the SAGE Center for the Study of the Mind at the University of California, Santa Barbara, and president of the Cognitive Neuroscience Institute, has covered much the same ground, and he also finds that thinking can be divided into slow conscious thoughts and rapid, automatic thoughts, which are driven, he says, by natural selection, and that we have within our brains, specifically in the left hemisphere, an "interpreter mechanism," which imposes a coherence, an order, on the world that we experience. Furthermore, this coherence, brought about non-consciously, is by no means always correct.[18]

This realm of research began when a psychologist and an economist looked at judgments under uncertainty but ended up underlining the biological basis of much of our behavior, our psychology. The converging order of the sciences finds support from unexpected directions.

Behavioral Economics

It is probably too much to argue that psychology is disappearing. Yet it could also be said that, at one margin, it is dissolving into chemistry and physics, at another into genetics—which is also chemistry and physics—and at a third into ethology and sociology.

Yet in another way psychology is expanding. This is in its association with economics. For example, Richard Thaler is a colleague of Kahneman and Tversky but has been even more assiduous in describing how the economic profession has been transformed by the experimental discoveries of behavioral science. In his 2015 book, *Misbehaving: The Making of Behavioral Economics*, he charts advances in behavioral economics over a forty-year period, from the wilderness to the point when he himself became (in 2015) the president of the American Economic

Association.[19] To begin with, as he makes clear, the economic profession was dominated by classical rationalists, economists who argued that people are rational beings, who always behave in rational ways, with the result that markets invariably reflect the most efficient state of affairs, financially speaking.

After pointing out a number of anomalies in the economic sphere, which were inconsistent with rational choice, and even writing a column in an economics journal on anomalies, Thaler showed, increasingly, that heuristics produce *predictable* errors and that rationality is, in the jargon, "bounded." What this means is that people are classically rational only up to a point, and that many SIFs (supposedly irrelevant factors) are anything but. As a result he came up with his concept of "mental accounting," which is very different from classical accounting, in which "loss aversion" is the single most powerful idea: our aversion to loss is twice as strong as our desire for gain. This explains why we are not always as classically rational as traditional economists claim.

Thaler also shows for instance how in poker, and in so much else in life, people behave differently with money they have won compared with their original stake—they now take far more risks. He shows that people have an instinctive understanding of fairness and will punish people who exploit them financially if given the opportunity. He shows that teachers produce better results for their pupils if they are given a bonus at the *beginning* of the school year and then have it progressively taken away if they fail to meet a prescribed performance level than if they are given a bonus at the end of the year.[20] He shows that we are "present-biased" and that people are more likely to commit to a pension scheme at some point in the near future than they are right now. He shows that parents who were texted on the eve of their children's school exams liked receiving such information and as a result prevailed on their children to work harder.[21]

And so, from a time when it took authors as long as six years to get behavioral economics articles published, we are now in a situation where there exist the *Journal of Economic Behavior*

& *Organization*, the *Journal of Economic Psychology*, the *Journal of Behavioral Decision Making*, and *The Oxford Handbook of Behavioral Economics and the Law*. As that latest title shows, behavioral psychology is now making inroads into the legal profession, and Thaler and others have recently worked in both Britain's 10 Downing Street and in America's White House.[22] Building on Daniel Kahneman's Nobel Prize in 2002, the integration of psychology and economics is maturing steadily.

The very latest developments are summed up in the titles of two recent investigations: "Computational Rationality: A Converging Paradigm for Intelligence in Brains, Minds and Machines" and "Economic Reasoning and Artificial Intelligence." The latter paper considers "a future synthesis of economics and Artificial Intelligence engendered by the emergence of *machine economicus.*"[23]

Psychology, as it now stands, has become—in some ways—the epitome of this book's argument. Behavior—especially human behavior—has become a point on which many sciences now converge: physics, chemistry, genetics, evolutionary theory, ethology, sociobiology. The experiment reported in May 2015 in which a patient's intentions could be anticipated by the pattern of activity of certain brain cells, if supported by replication, takes us into new territory scientifically and even philosophically (the physics of "intention"?). The overlap between psychology and economics has both produced new ways to understand—and control—certain aspects of human behavior and introduced a whole new vocabulary that helps explain ourselves to ourselves. This is no small achievement.

The fact is that these are developments on several different fronts, so that synthesis is some way off. Nonetheless, the link from brain mapping, which has identified more and more specific brain areas, may well lead in time to an entirely new understanding of the chemistry/physics of the brain and its arrangement in an evolutionary context. These advances in psychology, if they occur, will be made in the language of physics, chemistry, and evolutionary theory. This convergence, which has already begun, may produce in time the most exciting emerging order there is to be had.

17

DREAMS OF A FINAL UNIFICATION: PHYSICS, MATHEMATICS, INFORMATION, AND THE UNIVERSE

I t was not until the Russians surprised the world with the launch of the *Sputnik 1* in October 1957 that the first faltering steps were taken toward the "Net" as we now know it. The launch of the satellite galvanized America, and among the research projects introduced as a result of this change in the rules of engagement was one to explore how the United States' command-and-control systems—military and political—could be dispersed around the country so that, should she be attacked in one area, America would still be able to function elsewhere.

Several new agencies were set up to consider different aspects of the situation, including the National Aeronautics and Space Administration (NASA) and the Advanced Research Projects Agency, or ARPA. It was this latter outfit that was charged with investigating the safety of command-and-control structures after a nuclear strike.

The origins of the digital universe, as it has been called, have been the subject of several recent detailed accounts, one by Walter Isaacson and another by George Dyson, son of Freeman Dyson, a British physicist working at the Institute for Advanced

Study at Princeton in the wake of World War II. Different people place the origins of the Internet at different times, and one of the earliest accounts put it in the mind of Vannevar Bush, the same man as was involved in convincing President Roosevelt that America should pursue the construction of an atomic bomb. As early as 1945, Bush envisaged a machine that would allow the entire compendium of human knowledge to be "accessed." And it was a student of his, Claude Shannon, who worked on a mechanical differential-equation solver that had been developed by Bush. This would eventually lead Shannon to conceive the science of information.[1]

It was Shannon who, after he had finished his PhD under Bush, and had transferred to the Bell Laboratories, embarked on a project that would create what has been called the third great revolution in physics in the twentieth century: "As did relativity and quantum theory, information theory radically changed the way scientists look at the universe."[2] It was in Shannon's 1948 paper, "A Mathematical Theory of Communication," that the term "binary digit," or "bit," first appeared (though Shannon himself said that the term was actually coined by a colleague, John Tukey). This, too, may be regarded as the birth of the digital age.

At much the same time, the community that was gathered together at the IAS comprised a close-knit social network of unusual talents that helped to kick-start more than computers. It was intimately associated with the design and manufacture of the hydrogen bomb, leading on to microprocessors, the Internet, and intercontinental ballistic missiles. The individuals who gathered at the IAS were possibly the greatest group of mathematical minds ever brought together in one place—this was a real but rarely reported center of excellence in intellectual history. John von Neumann and Alan Turing were the guiding spirits, but the group included Kurt Gödel, Oswald Veblen, Stanley Ulam, Norbert Wiener, and Nils Barricelli.

And in these—what we might call the twin births of the digital age—we have the theme of this chapter. That we are, in effect, living in a golden age of mathematics; when physics, as we shall see, has reached a point where direct observation is ever more difficult, even impossible, and so the only way forward is by mathematical extrapolation from the physics we *do* know, to see where it takes us. When put alongside the allied theme of the next chapter—the increasingly close relationship between mathematics and biology—questions are raised about the ultimate nature of reality, and to what extent that reality is mathematical. Whether order, as defined by mathematical equations, is not just an organizing principle of reality, but reality itself.[3]

In the narrative that follows, keep in mind the overall theme, of the ever-present—and ever-growing presence in other sciences of—mathematical calculation.

The Link Between Numbers and Biology

Princeton, New Jersey, in summer, was often described at the time as being "like the inside of a dog's mouth." But the IAS was also described, often by envious colleagues who didn't know what (classified) research was being done there (in fact, calculations for "Ivy Mike," the hydrogen bomb test, carried out in 1952), as the "Institute for Advanced Salaries," as an "intellectual hotel," or the "Princetitute."

Von Neumann, head of this greatest group of mathematical minds, was born in Budapest in 1903, the son of a lawyer and investment banker. He was prepared for his Hungarian gymnasium by not one but two private governesses (one German, the other French) and by private tutors in Italian, fencing, and chess. He eventually attained fluency in Latin, Greek, German, English, and French. Always formally dressed in suit and tie, von Neumann had an earthy sense of humor and a tireless social life—he adored fast cars and was always being ticketed. He began serious training in mathematics at age thirteen.

He studied math at Berlin (where he passed his exams without attending class) and, at the same time, chemistry in Zurich. He took his doctorate in 1926 at Göttingen, where David Hilbert, the most distinguished mathematician of his day, is alleged to have asked just one question: "In all my years, I have never seen such beautiful evening clothes: pray, who is the candidate's tailor?" Edward Teller, "father of the hydrogen bomb" and an early friend and colleague, confided that "if a mentally superhuman race ever develops, its members will resemble Johnny von Neumann."

Following his doctorate, von Neumann published more than a score of papers in the next three years, including one on the theory of games, and a book, *Mathematical Foundations of Quantum Mechanics*, which was still in print eighty years later. And so, when Oswald Veblen was recruiting for the Mathematics Department of Princeton, he offered visiting lectureships to Eugene Wigner and von Neumann. The cable offering the positions mentioned salaries about eight times what the Europeans were earning. With Nazism on the rise, von Neumann didn't think twice about emigrating.[4]

Colossus and ENIAC emerged during the war as prototype calculating machines, honed on code-breaking. Afterward, the main activity of IAS was to develop ever more sophisticated and rapid-acting calculating machines that would, in the first place, enable the calculations needed to manufacture a hydrogen bomb. This was therefore a matter which the military regarded as their own, though at the same time there were a number of commercial companies interested in the future possibilities of what were coming to be called computers.

The personnel at the IAS fully realized they were in on the ground floor of something new, something that could radically change the world. One of the people already mentioned, Stanley Ulam—yet another refugee, from Poland, via Cambridge University and Wisconsin—was responsible for conceiving what George Dyson calls four of the twentieth century's most imaginative ideas for "levering our intelligence." The "Monte Carlo method" was the (secret) name given to Ulam's idea to use random numbers to test levels of uncertainty in a system—the name was chosen because of the method's similarity to the gambling techniques applied by Ulam's raffish uncle in the casinos of Monte Carlo. The "Teller-Ulam design" was Ulam's proposal to use the reaction (explosion) in one system to set off a further one in a bigger system. (This led to the first hydrogen bomb.) "Self-reproducing cellular automata" was the theory that explored the mathematics/logic/algorithms of reproduction in, say, biochemical organisms. "Nuclear pulse propulsion" explored the use of nuclear explosions for thrust in, say, a rocket or intercontinental missile.

Described as being on the dividing line between a true original and a crank, Nils Barricelli's main interest—and the reason von Neumann invited him to the IAS—was the mathematics of life, the link between numbers and biology (considered more fully in the next chapter). One part of Barricelli's nonconformity was his suspicion that DNA was "molecular shaped numbers," that polynucleotide chains were basically digital. This line of reasoning did not, at the time, go very far, but it underlined—and still underlines—the fact that, in 1953, three technological revolutions dawned, and the IAS was part of all three: thermonuclear weapons, stored-program computers, and the elucidation of how life stores its own instructions as strings of DNA.[5]

The next breakthrough designed to increase the speed of calculation came in the early 1960s, with the idea of "packet switching," developed by Paul Baran. A Jewish immigrant from a part of Poland that is now Belarus, Baran took his idea from the brain, which can sometimes recover from disease by switching the messages it sends to new routes. Baran's idea was to divide a message into smaller packets and then send them by different pathways to their destination. The same idea occurred almost simultaneously to Donald Davies, a Welsh mathematician and engineer, then working at the National Physical Laboratory in Britain together with Alan Turing. The new hardware was accompanied by new software, a brand-new branch of mathematics known as queuing theory, designed to prevent the buildup of packets at intermediate nodes by finding the most suitable alternatives.[6]

Progressive Joining of the Nets

In 1968 the first "network" was set up, consisting of just four sites: UCLA, Stanford Research Institute (SRI), the University of Utah, and the University of California, Santa Barbara. The technological breakthrough that enabled this to proceed was the conception of the so-called Interface Message Processor, or IMP, whose task it was to send bits of information to a specific location. In other words, instead of "host" computers being interconnected, the IMPs would be instead, and each IMP would be connected to a host. The computers might be different pieces of hardware, using different software, but the IMPs spoke a common language and could recognize destinations. By the end of 1970 there were fifteen "nodes," all at universities or think tanks.

And by the end of 1972 there were three cross-country lines in operation and clusters of IMPs in four geographic areas—Boston, Washington, DC, San Francisco, and Los Angeles—with, in all, more than forty nodes. By now ARPANET, as it was at first called, was usually known as just the Net, and although its role was still strictly defense-oriented, more informal uses had also been found: chess games, quizzes, the Associated Press wire service. It wasn't far from there to personal messages, and one day in 1972, email was born when Ray Tomlinson, an engineer at BBN, one of the commercial nodes, located in Boston, devised a program for computer addresses, the most salient feature of which was a device to separate the name of the user from the machine the user was on. Tomlinson needed a character that could never be found in any user's name, and looking at the keyboard, he happened upon the "@" sign. It was perfect: it meant "at" and had no other use. Tomlinson, a big, cheerful bear of a man, has been pestered ever since to say what this first email consisted of, but he confesses it was so inconsequential he has forgotten it.[7] He died in 2016.

By 1975 the Net community had grown to more than a thousand, but the next real breakthrough was Vint Cerf's idea, which came as he sat in the lobby of a San Francisco hotel, waiting for a conference to begin. Born in New Haven, Connecticut (the site of Yale University), he grew up with an intense interest in music and played the cello. When he was fifteen he was taken to a master class with Pablo Casals, but in the same year his father also took him to see a system called SAGE, which stood for Semi-Automatic Ground Environment, a computer that was used for radar tracking to detect Russian bombers coming from the North Pole. Cerf diverted off into computers and studied mathematics at Stanford before joining IBM.

By then, ARPANET was no longer the only computer network. Other countries had their own Nets, and other scientific-commercial groups in America had initiated theirs. Cerf began to consider joining them all together, via a series of what he referred to as gateways, to create what some people called the Catenet, for Concatenated Network, and others called the Internet. This required not more machinery but the design of TCPs, or transmission control protocols, a universal language. In October 1977 Cerf and his colleagues demonstrated the first system to give access to more than one network. The Internet as we know it was born.[8]

Growth soon accelerated. It was no longer purely a defense exercise, but, in 1979, it was still largely confined to (about 120) universities and other academic/scientific institutions. The main initiatives, therefore, were now taken over from ARPA by the National Science Foundation, which set up the Computer Science Research Network or CSNET, and in 1985 created a "backbone" of five supercomputer centers scattered around the United States and a dozen or so regional networks. These super-computers were both the brains and the batteries of the network, a massive reservoir of memory designed to soak up all the information users could throw at it and prevent gridlock. Universities paid $20,000 to $50,000 a year in connection charges.

In January 1986, a grand summit was held on the West Coast and order was put into email, to create seven domains or "Frodos."

These were universities (edu), government (gov), companies (com), military (mil), nonprofit organizations (org), network service providers (net), and international treaty entities (int). It was this new order which, as much as anything, helped the phenomenal growth of the Internet between 1988 and 1989. The final twist came in 1990 when the World Wide Web was created by researchers at CERN, the European Laboratory for Particle Physics near Geneva. This used a special protocol, HTTP, devised by Tim Berners-Lee, and made the Internet much easier to browse, or navigate.[9]

Tim Berners-Lee (knighted in 2004) read physics at Oxford, worked for Plessey, the telecommunications company headquartered in Poole, Dorset, then went to CERN in Geneva. And it was there, with CERN being the largest Internet node in Europe, that he saw an opportunity to join hypertext with the Internet. We take hypertext for granted now, with secondary levels of information being embedded in what we read, but it was a major software breakthrough all by itself. It was Berners-Lee who put all this together with "http," standing for hypertext transfer protocol, the first-ever website being built at CERN (in the part of it in France) and put online on August 6, 1991. The phrase the "World Wide Web" was thought up in the CERN cafeteria and "info.cern.ch" was the address of the world's first-ever website. Mosaic, the first truly popular browser, devised at the University of Illinois, followed in 1993. It is only since then that the Internet has been commercially available and easy to use, and we can all be said to be inhabitants of a digital universe.[10]

The Link Between Information and Thermodynamics

"Information is as real and concrete as mass, energy, or temperature. You cannot see any of these properties directly, but you accept them as real. Information is just as real." This is Charles Seife in his book *Decoding the Universe*, where he goes on to highlight how, when Claude Shannon published his seminal paper, in which he realized that information could be quantified and measured, he also realized that it was intimately linked to thermodynamics. "There is something about information that transcends the medium it is stored in. It is a physical property of objects akin to energy or work or mass. . . . Nature seems to speak in the language of information."

More prosaically, Shannon began by showing engineers how they could use Boolean logic. This was the mathematics of manipulating *1s* and *0s*—named after George Boole, the British nineteenth-century mathematician who conceived this approach—to design better switches for electrical equipment, including computers. When Shannon transferred to Bell Labs, the research arm of AT&T, the first binary computer and the transistor were being developed there.[11] This was the birth of "information technology," and what he realized was that written language is a stream of symbols and that symbols can be written as a stream of bits, so that any question that has an answer that can be written in a language can be written as a stream of *0s* and *1s*.

For many the real excitement came when Shannon and his colleagues tried to tackle very practical problems with communications technology—the mistakes made (e.g., noise on a telephone line) and the elimination of redundancy (not always so advisable because redundancy in a message allows it to be depleted without loss of meaning). In making calculations, it was found that the equations were similar to those Ludwig Boltzmann had derived when he was analyzing the entropy of gases. Thus it emerged that information displays entropy as a measure of disorder. And that it costs more energy to keep order than disorder, which is why disorder, entropy, is the natural state, both in thermodynamics and information.*

Also, in effect, *0s* and *1s* are quantum states. You have to be one or the other—intermediate states are not allowed. And in passing

---

* Some physicists regard this similarity between thermodynamics and information as an analogy, nothing more.

between the sender of information and the receiver, physical change takes place: information is thus physical. Further, on this account, the act of living (and that phrase, "the act," is crucial) can be seen as replicating and preserving the information that a living body is comprised of. Living organisms, as Schrödinger noted (chapter 9), are continuously fighting off decay, maintaining an internal order despite the fact that the rest of the universe is always increasing its entropy. By eating food, consuming energy that for the time being comes from the sun, an organism keeps itself in equilibrium. Life, as Schrödinger said, is a delicate dance of energy, entropy, and information. Life exists to propagate and reproduce the information about itself. This, on reflection, is close to what a gene is and does.[12]

Among other things, this has led to a new speciality—DNA computing. "DNA computing," says Lila Kari, computer scientist and language expert, "is an area of natural computing based on the idea that molecular biology processes can be used to perform arithmetic and logic operations on information encoded as DNA strands." This is a genuinely interdisciplinary approach in which computer scientists and molecular biologists collaborate to investigate computations taking place naturally in the living cell.

And this puts a new light on reproduction. It means that the aim of reproduction is not the reproduction of the self but the reproduction of the *information* within the self. This is easier to see with ant colonies where, typically, only one organism is fertile— the queen. All the other thousands of ants are sterile. Yet these sterile ants nurture the queen's eggs and rear them to adulthood. It is the *information* inscribed in their genes that instructs them to obey the queen and forfeit their own reproduction.[13] The information is the entity that is being preserved.

It also seems that the information in our brains is not dissimilar to the information in our genes. William Bialek, a biologist at Princeton, has been investigating the brains of flies—because they are relatively simple. He has shown that basic brain activities comprise a complicated electrochemical process: sodium and potassium atoms on opposite sides of a cell membrane switch places, and the neuron goes from *0* to *1* and then, after a fraction of a second, switches back. In other words, basic neuron activity is pretty much a binary activity. And by showing the fly very primitive, bold, and simple images (a white bar, a dark bar, a moving bar), Bialek has investigated what you might call the "alphabet" of neural signals. He finds that the fly brain is able to transmit something in the region of five bits of information per millisecond. "Even though it is an extraordinarily complex bit-processing machine, the fly's brain is a bit-processing machine nonetheless."[14]

So far so good. However, information theory becomes much more complex, and far weirder, when applied to the latest developments in physics and cosmology.

Physics Becomes Mathematics

We now enter the realm of "ironic science," one of the latest and more intriguing phases of thinking in physics. Some critics argue that "ironic science" is speculation as much as experimentation, where there is no real evidence for the (often) outlandish ideas being put forward. But that is not quite fair. Much of the speculation is prompted—and supported—by mathematical calculations that point toward solutions where words, visual images, and analogies all break down. And that is why this development belongs in this chapter: in some ways physics has *become* mathematics. Throughout the twentieth century, physics came up with ideas— of new particles, for instance—that only found experimental support much later, so perhaps there is nothing very new here. At the moment we are living in an in-between time, and have no way of knowing whether many of the ideas current in physics will endure and be supported by experiment.

A good place to start is with black holes since, by definition, hardly anything can be known about these entities. The general public was introduced to the idea of black holes in 1988, in *A Brief History of Time: From the Big Bang to Black Holes*, by the Cambridge cosmologist Stephen Hawking. This idea, as mentioned earlier, was first broached (in modern times) in the 1960s. Black holes are envisaged as super-dense objects, the result of a certain type of stellar evolution in which a large body collapses in on itself under the force of gravity to the point where nothing, not even light— no information at all—can escape. The discovery of pulsars, neutron stars, and background radiation in the 1960s considerably broadened our understanding of this process, besides making it real, rather than theoretical.

Working with Roger Penrose, then at Birkbeck College in London, Hawking first argued that at the center of every black hole, as at the beginning of the universe, there must be a "singularity," a moment when matter is infinitely dense, infinitely small, and when the laws of physics as we know them break down. Hawking added to this the revolutionary idea that black holes could emit radiation (this became known as Hawking radiation) and, under certain circumstances, explode. He also believed that, just as radio stars were discovered in the 1960s thanks to new radio telescopes, so X-rays should also be detectable from space via satellites above the atmosphere, which otherwise screen out such rays. Hawking's reasoning was based on calculations that showed that as matter was sucked into a black hole, it would get hot enough to emit X-rays. Sure enough, four X-ray sources were subsequently identified in a survey of the heavens and so became the first candidates for observable black holes. Hawking's later calculations showed that, contrary to his first ideas, black holes did not remain stable but lost energy, in the form of gravity, shrank, and eventually, after billions of years, exploded, possibly accounting for occasional and otherwise unexplained bursts of energy in the universe.[15]

Links to Other Universes?

In the 1970s Hawking was invited to Caltech, where he met and conferred with Richard Feynman. Feynman was an authority on quantum theory, and Hawking used this encounter to develop an explanation of how the universe began. It was a theory he unveiled in 1981 in, of all places, the Vatican. The source of Hawking's theory was an attempt to conceive what would happen when a black hole shrank to the point where it disappeared, the troublesome fact being that, according to quantum theory, the smallest theoretical length was the Planck length, derived from the Planck constant, and equal to 10^{-35} meters. Once something reaches this size (and though it is small, it is not zero), it cannot shrink further but can only disappear. Similarly, the Planck time is, on the same basis, 10^{-43} of a second, so that when the universe came into existence, it could not have done so in less time than this.

Hawking resolved this anomaly by a process that can best be explained by an analogy. Hawking asks us to accept, as Einstein said, that spacetime is curved, like the skin of a balloon, say, or the surface of the earth. (Remember that these are analogies only.) Using another, Hawking said that the size of the universe at its birth was like a small circle drawn around, say, the North Pole. As the universe—the circle—expanded, it was as if the lines of latitude were expanding around the earth, until they reached the equator, and then they began to shrink, until they reached the South Pole in the "Big Crunch." But, and this is where the analogy still holds in a useful way, at the South Pole wherever you go you must travel north: the geometry dictates that it cannot be otherwise. Hawking asks us to accept that at the birth of the universe an analogous process occurred. Just as there is no meaning for *south* at the South Pole, so there is no meaning for *before* at the singularity of the universe. Time can only go forward.[16]

Hawking's theory was an attempt to explain what happened "before" the Big Bang. Among the other things that trouble physicists about the Big Bang theory is that the universe as we know it appears much the same in all directions. Why this exquisite symmetry? Most explosions do not show such perfect balance—what made the "singularity" different? An answer came jointly from Alan Guth (rhymes with "truth") of MIT and Andrei Linde, a Russian physicist who emigrated to the United States in 1990. Both men argued that at the beginning of time—i.e., $T = 10^{-43}$ seconds, when the cosmos was smaller even than a proton—gravity was briefly a *repulsive* force, rather than an attractive one. Because of this, they said, the universe passed through a very rapid inflationary period, increasing by a factor of 10^{-30} in a period of 10^{-30} seconds, until it was about the size of a grapefruit, when it settled down to the expansion rate we see (and can measure) today. The point of this theory (some critics call it an "invention") is that it is the most parsimonious explanation required to show why the universe is so uniform. The rapid inflation would have blown out any wrinkles. It also explains why the universe is not completely homogenous. There are chunks of matter, which develop into galaxies and stars and planets, and other forms of radiation, which form gases.[17]

Linde went on to theorize that our universe is not the only one spawned by inflation. There is, he contends, a "megaverse," with many universes of different sizes, and this was something that Hawking also explored. Baby universes are, in effect, black holes—bubbles in spacetime. Going back to the analogy of the balloon: imagine a blister on the skin of the balloon, marked off by a narrow isthmus, equivalent to a singularity. No information, and none of us, certainly, can pass through the isthmus, and none of us is aware of the blister, which can be as big as the balloon, or bigger. In fact, any number may exist. They are a function of the curvature of spacetime and of the physics of black holes. By definition we can never experience them directly. In this sense they have no meaning.

Another theory of scientists like Hawking is that "in principle" the original black hole and all subsequent universes are actually linked by what are variously known as "wormholes" or "cosmic string." Wormholes, as conceived, are minuscule tubes that link different parts of the universe, including black holes, and therefore in theory can act as links to other universes. They are so narrow, however (a single Planck length in diameter), that nothing could ever pass through them, without the help of cosmic string—which, it should be stressed, is an *entirely theoretical* form of matter, regarded as a relic of the original Big Bang. Cosmic string also stretches across the universe in very thin (but very dense) strips and operates "exotically." What this means is that when it is squeezed, it expands, and when it is stretched it contracts. In theory at least, therefore, cosmic string could hold wormholes open. This, again in theory, makes time travel possible, for some future civilization.[18]

Martin Rees's "anthropic principle" of the universe is somewhat easier to grasp. Rees, a onetime astronomer royal, offers indirect evidence for "parallel universes." His argument is that for mankind to exist, a very great number of coincidences must have occurred, if there is only one universe. In an early paper he showed that if just one aspect of our laws of physics were to be changed—say, gravity was increased—the universe as we know it would be very different. Celestial bodies would be smaller, cooler; they would have shorter lifetimes, a very different surface geography, and much else. One consequence is that life as we know it can in all probability only form in universes with the physical laws that we enjoy. This means, first, that other forms of life are likely elsewhere in the universe (because the same physical laws apply). But it also means that many other universes "probably" exist, with other physical laws, in which *very* different forms of life, or no forms of life, exist. Rees argues that *we* can observe our universe, and conjecture others, because the physical laws exist to allow it. He insists that this is too much of a coincidence. Other universes, very different from ours, must almost certainly exist.[19]

The Philosophy of Cosmology

Joseph Silk agrees. He directs the international Philosophy of Cosmology project, from where he adds the idea that "everything that *can* happen *will* happen, infinitely many times." This builds on another original idea of Ludwig Boltzmann, who thought "every conceivable particle position and momentum would exist somewhere."

This view takes some getting used to, but, Silk says, it would be the climax of the great shifts that dislodged the earth, then the sun, and then our own galaxy from a special position at the center of physical reality. And it would help explain one of the great mysteries of cosmology—that the predicted acceleration of the expansion of the universe is greater than the observed value by a factor of 10^{120}. This is an enormous disparity, the most comfortable explanation being "eternal inflation"—that the brief phase of ultra-rapid expansion occurs again and again, budding off an infinity of new expanding universes.[20]

Outlandish as this may seem, it is not the end of the matter. David Bohm's idea of the cosmos, difficult as it is, seems to entail an extension of the "field" idea to encapsulate what Ernest Nagel described as a "mathematically continuous space." Bohm explains his view via an analogy—a thought experiment that proposes a device where there are two concentric glass cylinders, the outer one fixed, the inner one made to rotate about its axis. Between them is a viscous fluid, such as glycerine, and into this fluid is inserted a droplet of insoluble colored ink. As the inner radius moves faster than the outer, the ink is drawn out into a finer and finer thread. Eventually the thread becomes so fine as to be invisible. However, should the inner cylinder be turned in the reverse direction, the parts of this thread retrace their steps, and eventually the thread comes together to reform the ink droplet "and the latter suddenly emerges into view."[21] This is what Bohm means by "the implicate order" of the cosmos—that, in the continuous mathematical space, "there is no real division between mind and matter, psyche and soma." The field, the mathematical space, contains an invisible order that *emerges* under suitable circumstances. There is a lot more to the theory than this, of course, but Bohm's idea is that we are all related to a "whole" and that eventually quantum entanglement and quantum tunneling, particles, and organic entities, will be related to the whole in this sort of way.*

In his 1997 book, *The Fabric of Reality*, David Deutsch, an Oxford physicist, goes even further and proposes physics as a form of theology, incorporating the work of other scientists such as Frank J. Tipler, Roger Penrose, Alan Turing, and Kurt Gödel.

Deutsch also starts from the point that we all inhabit "parallel universes," that, as Silk says, there is a "multiverse," made up of many universes and that we—or copies of us—inhabit many of these universes, of which we are only intermittently and dimly aware. He bases his argument, in the book anyway, on a series of patterns thrown by light onto a screen after it has passed through a number of slits. He shows that, depending on the number of slits the light has passed through, some areas on the final screen are white and some are dark. This pattern can be explained, he says, only if we assume that, besides the photons that we *can* see—that are "tangible"—there are also "shadow" photons that are dark and intangible and "interfere" on occasions. This leads him to the idea of parallel universes, and it is profound, he says, because it explains so much that is otherwise incomprehensible.[22]

His other main concern is that computation, mathematics, also accords with physics (this is why the world is comprehensible). Moreover, and most importantly, it is the *only* form of knowledge. The increase of such computational knowledge, he says, is the "purpose" of life. He entertains the idea that, in a universe made up according to the laws of physics and patterned on mathematics, at some point in the future all of this mathematical,

---

* Quantum tunneling is another aspect of quantum weirdness. The most popular analogy used by physicists is of a ball bearing being rolled up a hill. In classical physics, unless the bearing is pushed with sufficient energy, it won't crest the hill and reach the other side. In the quantum world, however, where the bearing is both a particle and a wave, the wave element can tunnel through the hill to the other side by "borrowing" energy from the material it passes through.

computational knowledge will be known, and life "will have conquered."

From this starting point, he, along with Frank Tipler, of Tulane University in New Orleans, looks forward to a very distant future (billions of years ahead), when they both think that (computational) knowledge will have expanded immeasurably from where it is now, to the point, not only where space travel will be familiar, but possibly where time travel will be too and where we may be in a position to avert the final phase of our universe, which, according to the level of current knowledge, will end in a cataclysmic "Big Crunch." This is, roughly speaking, what Frank Tipler explores in his book *The Physics of Immortality*, in which his concept of the "omega" point is fleshed out.[23]

As the Big Crunch approaches, and the universe contracts, more and more energy will be concentrated in smaller spacetime, which will mean that "people's minds [by then] will be running as computer programs in computers whose physical speed is increasing without limit." By this point, billions of years ahead, with the computing power then available, experience will be determined not by elapsed time "but by the computations that are performed *in* that time." (Italics added.) "In an infinite number of computational steps there is time for an infinite number of thoughts—plenty of time for the thinkers to place themselves into any virtual-reality environment they like. . . . They will be in no hurry for subjectively they will live forever. Within one second, or one microsecond, they will have 'all the time in the world' to do more, experience more, create more—infinitely more—than anyone in the multiverse will ever have done before then."

There are some preparations that will have to be made, Deutsch says. But again—and the point is crucial to this theory—the physical knowledge we have today means that all this reasoning is just that, reasoning based on current knowledge, and is therefore *not* speculation. We shall need to "steer" the universe to the omega point, and, along the way, several deadlines will need to be negotiated. One is about 5 billion years from now, when the sun will—if left to its own devices—become a red giant star and wipe us out. As Deutsch says, insouciantly you might think, "We must learn to control or abandon the Sun before then." He goes so far as to say that the omega point theory "deserves to become the prevailing theory of the future."[24]

At this omega point, Tipler says, everything about the universe will be known. Whatever exists then will, therefore, be omniscient, from which it follows that it will be omnipotent and omnipresent. "And so [Tipler] claims that at the omega point limit there is an omniscient, omnipotent, omnipresent society of people. This society, Tipler identifies as God."

Deutsch emphasizes that there are very great differences between Tipler's idea of God and what most religious people believe in today. The people near the omega point would be so different from us that they couldn't communicate with us. And they couldn't work miracles—they did not create the universe or the laws of physics, so they could never violate those laws. They would be opposed to religious faith and have no wish to be worshipped (who would do the worshipping?). But he does think that technology would be so advanced at that point that they could resurrect the dead. This would be so because, by then, computers would be so infinitely powerful that they could create any virtual world that ever existed, including our world in which humans have evolved, which would, in an infinite system, enable them to improve our world materially, making it one in which people will not die. This, Tipler says, is a form of heaven.

What people would actually *do* at the omega point (people very different from us, beyond what we can imagine) is a matter of informed speculation, say both Tipler and Deutsch. This is because the omega point is a singularity, in which by definition the laws of physics break down. But they insist that present-day physics and mathematics support the narrative up to the omega point.

It is a breathtaking vision and, needless to say, both Deutsch and Tipler have been criticized, and criticized heavily, for yet another "ironic" conception, "unwarranted speculation" about events so far in the future as to be meaningless for most people. But they insist their theories are based on real knowledge of physics and computation—mathematics—that exist today. Moreover, life has proliferated on earth for roughly 3.5 billion years and it has taken that time for us to become aware of, for example, the future demise of the sun. We must learn to think in such ways and in such time frames.[25]

The importance of computation is also underlined by Stephen Wolfram's "new kind of science." Wolfram, a British-born physicist, and the creator of Mathematica, a technical software system, goes so far as to say that "all processes, whether they are produced by human effort or occur spontaneously in nature, can be viewed as computations." Computation provides "a uniform framework" with which to explore the different processes that occur in nature. There is, he insists, a "fundamental equivalence between many different kinds of processes."[26]

Wolfram arrives at this view of convergence and emerging order by way of what he calls "a new kind of science," but which is in fact a clever analogy. In an extraordinary publishing venture, rather than outline his evolving views gradually through professional journals, which could be subjected to peer review, Wolfram sequestered himself for ten years until he could produce one masterwork—rather like Darwin produced *On the Origin of Species* after years of gestation—where the argument is, allegedly, fully worked out. Wolfram's masterwork, *A New Kind of Science*, privately published by him in 2002, runs to 1,274 pages, including a paragraph on why it was necessary to write the book in an immodest style. At the time *Convergence* was being finalized, the book had sold near to 150,000 copies, no small achievement.

Wolfram's analogy involves something called cellular automata (first conceived by Stanley Ulam at the IAS in Princeton). To grasp this, imagine a grid made up of many small squares, all the same color and laid out rather like a computer screen consisting of very many minuscule pixels. The major discovery he made, Wolfram says, started from his idea to have one square in the top row of the array which is a different color (black on white, for example), and this is linked with a computer program—in effect a set of simple rules factored into the system, which determine what color the squares are in the rows lower down the grid and then on, beyond that. At its simplest, the rules might stipulate that a cell is black if either of its neighbors was black on the step before, and makes the cell white if both its neighbors were white. A more complex rule would be that a cell be black if one or the other, but not both, of its neighbors was black on the step before.

Starting from this simple beginning, Wolfram found that there are 256 possible sets of rules of this kind but, in addition, and what he found most surprising (and to him most important), is that over time, as these programs are run through the computer, very often in *millions* of iterations, two basic results emerge. One, the patterns that are produced are, a lot of the time, unbelievably complex, random, and unrepeating, while at other times distinctive patterns emerge and are repeated at regular intervals.

For Wolfram, this was a revelation, a revelation he wants us all to share in. The idea that a few simple rules can lead *both* to great complexity and to order down the line, that order and complexity are different sides of the same coin, is for him the most important thing about the universe, because the universe itself is both ordered and random, containing both simple features and complex ones. For Wolfram, these simple rules explain everything.

Throughout his book he describes the behavior of cellular automata only, patterns in black and white, and occasionally other colors, like gray. But Wolfram finds in the patterns many analogies to real life. For example, some of the patterns resemble the order we see in nature, the stripes on tigers, for example, or the shape of sand dunes, the whorls of snails and shells. Elsewhere, he argues that space itself may well be made up of discrete units, much like the cells of cellular automata, and that this helps us explain fundamental particles, which are essentially "tangles" in this network of units. It is the tangle that moves and interacts, and this is what mass is (not a million miles from the Higgs boson—see later in this chapter).

Looking at space in this way, he says, helps us understand quantum mechanics and such phenomena as "superposition," the idea that entities can be in two places at once, because they are essentially connected via the space network. He also shows that if the basic shape of the cells or "pixels" (for want of a better word) is, say, a hexagon rather than a square, then space will naturally be curved, just as Einstein said. Another analogy is with human memory: memories are patterns in the pixels of the brain that embody, conform to, or derive from the simple rules that give rise to the patterns. This is why we can retrieve memories, because similar memories conform to similar mathematical configurations.

Finally, and possibly most fundamentally, he says that his approach helps us understand the mystery of free will. The simple rules with which life began led to great complexity. Moreover, with many of the rules he describes, it is not possible to know, from the state of complexity at any one time, how that current state was arrived at, how the rules brought the chaotic or ordered pixels to that point. So the rules cannot be traced backward. This, he says, is analogous to the situation we find ourselves in with regard to free will. We know that certain rules have brought us to this point, but we don't know what these rules are, going forward, and can never find out. So, to all intents and purposes, we feel as if we have free will, for we will never know the underlying rules of our behavior. The mathematics of the history of life are beyond us.

Wolfram's ideas have had a mixed reception.[27] Partly this is because cellular automata are analogies of life, not life itself, and he claims just too much for them. Partly it is because his book abounds in phrases like "I strongly suspect," "it seems possible," "it seems likely," "if I am correct." Speculation piled on analogy is not everyone's idea of science.

But few dismiss Wolfram's huge work out of hand, and he fits in here because of his conclusions, the most relevant of which are:

- "It seems . . . plausible that both space and its contents should somehow be made of the same stuff—so that in a sense space becomes the only thing in the universe."[28]
- "A fundamental unity exists across a vast range of processes in nature and elsewhere."[29]
- "At some rather abstract level one can immediately recognize one basic similarity between nature and mathematics. . . . This suggests that the overall similarity between mathematics and nature must have a deeper origin."[30]

Mathematics, Psychology, and Evolution

Lee Smolin and Roberto Mangabeira Unger beg to differ. Smolin is a well-known physicist, whose work involves quantum gravity, string theory, and refinements of relativity theory, while Unger is a philosopher. In their book, *The Singular Universe and the Reality of Time* (2014), they argue—against the views we have just been considering—that the link between mathematics and nature is deceptive, and that there is only one universe at a time, though there may have been successive universes (with succession either linear or cyclic), and that, as universes grow, other universes can branch off them. This is a greater distinction than might appear at first. Behind the idea of a multiverse, or parallel universes, is the theory that they would have different laws, whereas in the Unger-Smolin view all branching universes would have the same laws, so presumably communication—information exchange between them—would be possible. Parallel universes are in principle beyond the reach of empirical inquiry. Our singular universe is "extraordinarily homogenous and isotropic and is therefore a setting hospitable to repeated phenomena and regular connections."[31] If the universe developed out of something, however, something preceded the Big Bang, when "causal connections may not yet have assumed law-like form and the division of nature into enduring natural kinds may not yet have taken shape."[32]

Their other arguments are that time is real, the most important consequence of this being that everything in the structure and regularities of nature changes sooner or later, and that "becoming" takes precedence over "being," process takes precedence over structure. They point out that, as things stand now, although the state of the cooled-down universe is remarkably stable, the laws of nature have changed at least once in the history of the universe— during the era of inflation when, as we have seen, gravity may have been, temporarily, a repulsive force. Smolin and Unger go on to consider whether other laws of nature may have changed over time (which, it will be remembered, was Schrödinger's idea about DNA). Unger and Smolin specifically mention—although almost in passing—that these changes may have occurred with the invention of sex and the invention of mind. (They are not alone in wondering whether mind is a special case when it comes to the laws of physics.) Unger and Smolin ask whether there may be deeper "meta-laws" of nature, which underlie how change happens. (There is a link here to Wolfram's cellular automata.)[33] And space, they say, may be emergent from time.

But they also argue, against Deutsch and Wolfram, that mathematics is only selectively real. ("The infinite is a mathematical contrivance . . . illegitimately applied in cosmology.") Mathematics, they say, has two concerns: the laws of nature, and itself. "It begins in an exploration of the most general relations in the world abstracted from time and of phenomenal particularity, but it soon escapes the confines of our perceptual experience. It invents new concepts and new ways of connecting them, inspired by its previous ideas as well as by the riddles of natural science."[34] But, they insist, mathematics offers us no shortcuts to timeless truth about incorruptible objects, either about nature or about some "special realm" of mathematical objects outside nature.

"[Mathematics] never replaces the work of scientific discovery and of imagination. . . . Its ulterior subject matter is the eviscerated natural world—eviscerated of time-bound particulars—that mathematics addresses."[35] Its being outside time, its *aim* of being outside time, limits its usefulness and, indeed, its reality.

The enigma of mathematics, "the unreasonable effectiveness of mathematics," as Eugene Wigner originally put it, is explained, they say, by psychology and by evolution. Mathematics is effectively "blind" to the phenomenal distinctions and the temporal variations of nature, and we must exercise self-critical vigilance to avoid the pseudo-Platonism that has, they say, been rife among mathematicians. This is the unwarranted supposition that abstractions are objects of a special type, "inhabiting a distinct part of reality."[36] Prime numbers are as factual as the feature of the world where water freezes at thirty-two degrees Fahrenheit. But they are not true or real in the same way. The links between the reality of the world, its plurality and its connections, can be represented mathematically (or some of it can) but this is no more than "a game of our invention," a "second-order" reality ("vulgar Platonisms"). The phenomenal particularity of the world ("of time and fuzzy distinction") can never be expunged and is just as interesting and every bit as important as are abstractions. "There is no preestablished harmony between physical intuition, or experimental discovery, and mathematical representation."[37] Reality is impermanent, and the emergence of the new "is a repeated event in the history of the universe."

From the standpoint of this book, Unger and Smolin's thesis is important in two senses. One, obviously enough, it calls into question the emerging consensus on parallel universes. And two, no less obviously, it casts a shadow over the all-important role of mathematics in the understanding of order.

Theories of Everything: "The Virtually Complete Amalgamation of Physics and Mathematics"

The so-called theory of everything is also an ironic phrase, but in a slightly different way. It refers to the attempt to describe all fundamental physics by one set of equations: nothing more. Physicists have been saying this "final solution" is just around the corner for decades (and remember that Einstein failed in his quest for a unified field theory half a century and more ago). In fact the theory of everything is still elusive.

Until the 1960s there were four forces that needed to be reconciled: gravity, electromagnetism, the strong nuclear force, and the weak radioactive force. In the 1960s a set of equations was devised by Sheldon Glashow, at Harvard, and built on by Abdus Salam of Imperial College, London, and Steven Weinberg, at Texas, which described both the weak force and electromagnetism and posited three new particles: $W+$, $W-$, and Zo. These were experimentally observed at CERN in Geneva in 1983. Together with the new theory for accounting how quarks interact (chapter 11), therefore, electromagnetism, the weak force, and the strong force have all been joined together in one set of equations.[38] This is a remarkable achievement, but it still leaves out gravity, and it is the incorporation of gravity into this overall scheme that would mark, for physicists, the theory of everything.

At first they moved toward a quantum theory of gravity. That is to say, physicists theorized the existence of one or more particles that account for the force and gave the name "graviton" to the gravity particle, though the new theories presuppose that many more than one such particle must exist. (Some physicists predict 8, others 154, which gives an idea of the task that still lies ahead.) But then, in the mid-1980s, physics was overtaken by the "string revolution" and, in 1995, by a second "superstring revolution." In an uncanny replay of the excitement that gripped physics at the turn of the twentieth century, a whole new area of inquiry blossomed into view as the twenty-first century approached.

The string revolution came about because of a fundamental paradox. Although each was successful on its own account, the theory of general relativity—explaining the large-scale structure of the universe—and quantum mechanics—explaining the minuscule subatomic scale—were, as reported earlier, mutually incompatible. Physicists could not believe that nature would allow such a state of affairs. As Newton said, "Nature is pleased with simplicity."

There were other fundamental questions too, which the string theorists faced up to: Why are there four fundamental forces? Why are there the number of particles that there are, and why do they have the properties that they do? (Questions that would, no doubt, have fascinated Mary Somerville.) The answer that string theorists propose is that the basic constituent of matter is not, in fact, a series of particles—point-shaped entities—but very tiny, one-dimensional strings, as often as not formed into loops. These strings are very small—about 10^{-33} of a centimeter—which means they are beyond the scope of direct observation by current measuring instruments, and are, therefore, for the moment anyway, entirely theoretical—mathematical—constructs. Notwithstanding that, according to string theory, an electron is a string vibrating one way, an up quark is a string vibrating another way, and a tau particle is a string vibrating in a third way, and so on—just as the strings on a violin vibrate in different ways so as to produce different notes. As the figures show, we are dealing here with exceedingly small entities indeed—about 100 billion (10^{20}) times smaller than an atomic nucleus. But, say the string theorists, at this level it is possible to reconcile—unite—gravity and quantum theory. As a by-product and a bonus, they also say that a gravity particle—the graviton—emerges naturally from the calculations.[39]

In dealing with such tiny entities as strings, possibilities arise that physicists did not earlier entertain, one being that there may be "hidden dimensions," and to explain this another analogy is needed. Start with the idea that particles are seen as particles only because our instruments are too blunt to see that small. To use an example from Brian Greene's *The Elegant Universe: Superstrings, Hidden Dimensions and the Quest for the Ultimate Theory*, think of a hosepipe seen from a distance. It *looks* like a filament in one dimension, like a line drawn on a page. In fact, of course, when you are close up it has two dimensions—and always did have; you just weren't close enough to see it. Physicists say it is (or may be) the same at string level. There are hidden dimensions curled up of which we are not at present aware. In fact, they say there may be *eleven* dimensions in all, ten of space and one of time. This is a difficult if not impossible idea to imagine or visualize, but the scientists make their arguments for mathematical reasons (math that even mathematicians find difficult). When they do make this allowance, however, many things about the universe fall into place. For example, black holes are explained—as perhaps similar to fundamental particles, and as gateways to other universes. The extra dimensions are also needed because the way they curl and bend, string theorists say, may determine the size and frequency of the vibrations of the strings; in other words they explain why the familiar "particles" have the mass and energy and number that they do. In its latest configuration, string theory involves more than strings: two-, three-, and more dimensional membranes, or "branes"—small packets, the understanding of which will be the main work of the twenty-first century.[40]

String theory stretches everyone's comprehension to its limits. As yet it is 99 percent theory (i.e., essentially mathematics). Physicists are beginning to find ways to test the new theories experimentally, but as of now there is no shortage of skeptics as to whether strings even exist (see the Conclusion). At these very small levels we may also enter a "spaceless and timeless realm." That stretches understanding too.

Greene believes this may be a major step forward philosophically as well as scientifically, a breakthrough "that is capable of giving us an answer to the question of how the universe began and why there are such things as space and time—a formalism that will take us a step closer to answering Leibniz's question of why there is something rather than nothing." Finally, in super-string theory we have the virtually complete amalgamation of physics and mathematics. The two have always been close, but never more so than now, as we approach the possibility that, in a sense, and despite what Unger and Smolin say, the very basis for reality is mathematical.

Quantum Mechanics and Information Theory Converge via Mathematics

If that is so, it doesn't necessarily make for a neat resolution. In fact, it may make things a lot more complicated, and it would be wrong to duck that question. Quantum mechanics and information theory have come together, in a mathematical way, that—whether ironic or not—is downright confusing for most of us. In 1982 the physicist Alain Aspect demonstrated "quantum entanglement" for the first time, but his experiment has now been repeated often.[41] In fact, it is regularly done in Geneva, where Nicolas Gisin and his colleagues "zap" a crystal made of potassium, niobium, and oxygen with a laser. As Charles Seife tells the story, when a crystal absorbs a photon from the laser, it splits into two entangled particles that shoot away in opposite directions, which are then piped into glass cables, cables which run through several towns around Lake Geneva. In the year 2000, Gisin's team shot entangled photons to the villages of Bernex and Bellevue, more than six miles from each other. Using an incredibly accurate clock, they showed that the particles "behaved in the manner Einstein had predicted: the two were in superposition and always seemed to conspire to have equal and opposite properties upon measurement."

In quantum mechanics, quantum weirdness reaches its peak. The superposition principle means that a particle can be in two quantum states at once. It can have up and down spin, in information terms it can be 1 and 0 at the same time, and it can be in two places at once. As if this is not mysterious enough, the

Geneva team determined that if some sort of "message" was sent between the two particles, it had to travel at 10 million times the speed of light.[42]

Physicists can explain this mathematically, in what they call a theory of quantum information, which, they also say, "is more closely tied to the fundamental laws of physics." Quantum bits, or qubits, occupy superposition states, and may even be split more than once, to be, in effect, in, say, four places at once. And, if their entanglement can be captured (not at all certain), they will be able to make calculations much faster. A quantum computer would outstrip regular computers by an order of magnitude.[43]

Quantum theory at this level, like superstrings, stretches everybody's imagination (and especially those without the mathematical skills to follow the equations). But there are problems with it. One is the fundamental question as to why microscopic entities should behave differently from macroscopic ones (and, after all, the instruments with which we measure microscopic entities are themselves macroscopic entities). It is also the case that information theory does not agree with relativity. A third problem is that entanglement implies an exchange of information, but that is forbidden by the theory. And a fourth is that if black holes consume information (cosmic rays, for example, or photons) this would violate the law of conservation of information, which is equivalent to the law of the conservation of energy (which is where we came in, and see chapter 1). So we cannot say that we are there yet in regard to the total unification of physics by way of mathematics.[44]*

"Convergence—the Deepest Thing about the Universe"

Despite the opposition to the unity of the sciences that developed in the late twentieth century on the part of many postmodern

---

* In October 2015, a conference at Oxford heard that several teams are working on quantum computers, aiming to have one up and running by 2020.

critics—usually historians of science, or philosophers, or sociologists—most scientists themselves took a very different view. Freeman Dyson, professor of physics at the IAS in Princeton, is not himself a reductionist. But here he is in his book *Infinite in All Directions*, published in 1988: "Now it is generally true that the very greatest scientists in each discipline are unifiers. This is especially true in physics. Newton and Einstein were supreme as unifiers. The great triumphs of physics have been triumphs of unification. We almost take it for granted that the road of progress in physics will be a wider and wider unification bringing more and more phenomena within the scope of a few fundamental principles." [45]

And here is Abdus Salam who, it will be remembered, shared the 1979 Nobel Prize for work on the unification of the electromagnetic and weak interactions, in his 1988 Paul Dirac Lecture, delivered at Cambridge University, and entitled "The Unification of Fundamental Forces":

> Another area where particle physics has provided us with important inputs is the subject of early cosmology, so much so that early cosmology has become synonymous with particle physics. This is because phase transitions, which separate one era of cosmology from another, are also the mechanism through which the one ultimate unified force is converted into two (gravity plus electronuclear), into three (electroweak plus strong nuclear plus gravity), and finally into four forces (electromagnetic plus weak plus strong nuclear force plus gravity as the overall temperature of the universe goes down). [46]*

These weren't the only figures stressing unification (see, for

---

* Incidentally, Salam also took it upon himself to unify physicists as well as physics. He held that Wolfgang Pauli venerated Paul Dirac, that Dirac venerated Heisenberg, that Heisenberg venerated Bohr, and that Bohr venerated Einstein.

example, John C. Taylor's *Hidden Unity in Nature's Laws*, published in 2001, albeit very technical). But probably the most prominent advocate pushing a theory of everything, from a strictly reductionist point of view, was and is Steven Weinberg, professor of physics at the University of Texas at Austin, and who, it will again be recalled, shared the Nobel Prize in physics in 1979 with Salam and Glashow. In 1993 Weinberg published *Dreams of a Final Theory*, in which he argued for a more or less traditional reductionist view of science, that fundamental particles are fundamental in a way that nothing else is, that physics seeks not only to describe the world at this basic level but also to explain it (*why* is there gravity, *why* is there quantum mechanics?).

He acknowledged that, by then, reductionism had become "a standard Bad Thing" in science, but he insisted that "by tracing [the] arrows of explanation back toward their source we have discovered a striking convergent pattern—*perhaps the deepest thing we have yet learned about the universe*." (Italics added.) We now know enough about the "sheer connectedness" of knowledge, he said, for us to expect that convergence will grow and become more precise.[47]

There was a specific backdrop to the publication of his book—it was the time when funding for the Superconducting Super Collider was being considered by the United States Congress. The SSC was a piece of equipment with no practical purpose that was expected to cost the US taxpayer in the region of $8 billion. Other physicists, such as Philip Warren Anderson, emeritus professor at both Cambridge and Princeton, argued that other aspects of physics—condensed matter physics, for example—were just as fundamental and yet potentially more useful in the long run, so that they deserved funding as much as, and maybe more than, particle physics.

While agreeing that condensed matter physics was important, and underfunded, Weinberg insisted that particle physics was *more* fundamental, more fundamental than anything else in fact. At the same time he accepted that many aspects of science—turbulence theories, for example, consciousness, memory, high-temperature conductivity—would be affected not at all by the discovery of a theory of everything.

In 1929, he said, physics had turned to a "more unified world view" when Werner Heisenberg and Wolfgang Pauli described both particles and forces as manifestations of a deeper reality, "the levels of quantum fields." [48] This was a "stunning synthesis," he said, synthesis being the crucial word. No one then could see where it would lead, and by implication, he argued, we were in a similar situation at the turn of the twenty-first century. No one could tell where a theory of everything would lead, should it ever be discovered. But he thought that the string revolution would prove to be another huge step forward.

Although he accepted there was no guarantee that progress in other fields of science would be helped directly by the discovery of a theory of everything, he was convinced that there is a "logical order built into nature itself," and that reductionism is, in effect, an attitude toward nature. "It is nothing more or less than the perception that scientific principles are the way they are because of deeper scientific principles (and, in some cases, historical accidents) and that all these principles can be traced to one simple connected set of laws." [49]

Despite the doubts of others (some of whom thought that the laws we study are imposed on nature by the way we make observations), Weinberg said that his own guess was that there *is* a final theory out there, waiting to be discovered, and that we are capable of discovering it. Perhaps his most fundamental feeling, in this most fundamental of books, was that the reductionist attitude provides "a useful filter" that saves scientists in all fields from wasting their time on ideas that are not worth pursuing. "In that sense," he insisted, "we are all reductionists now." [50]

At the time he delivered his Dirac Lecture, in 1988, on "The Unification of Fundamental Forces," Salam provided two tables that set out clearly the particles and forces that made up the Standard Model of Physics. He added that all the particles were known to exist directly except for the top quarks and the Higgs particle or boson. The top quarks were subsequently identified in 1995, and then, much later, the final piece of the jigsaw came into view.[51]

"The End of an Era of Unification, the Threshold of a New One"

Melbourne, Australia, July 4, 2012. The opening day of the International Conference on High Energy Physics. It is a biannual jamboree, routinely shifting from one city to another over the years. Today, however, news has leaked out that something "big" is being announced, and the main auditorium is filled long before the lectures start. Halfway round the world, at CERN in Geneva, from where the two main lectures will be live-streamed, the same rumor has aired and hundreds of physicists have lined up hours before the start.

The two main lectures are delivered by Joe Incandela, an American physicist, and Fabiola Gianotti, an Italian colleague. They are joint leaders of the two rival teams at CERN, named ATLAS and CMS, who—in a deft arrangement that links collaboration and rivalry—have been pitched against each other ever since CERN set up the two groups in 2009.

The substance of the talks comprises a small number of graphs that would mean very little to any but the most informed high-energy physicists. But, as Sean Carroll tells the story, the graphs show more events than expected, with a certain particular energy (collections of particles streaming from a single collision). "All the physicists in the audience know immediately what it means: a new particle. The LHC [Large Hadron Collider] has glimpsed a part of nature that had heretofore never been seen."[52]

In Melbourne there is wild applause. In Geneva, even journalists who pride themselves on having seen it all applaud no less enthusiastically. All are overwhelmed by the "exquisite agreement" between the two teams.

What had been found was the "Higgs boson." Or, at least, something that had the right mass, the right energy level, and decayed in the expected way, according to theory. This was named after the Scottish physicist Peter Higgs, who was in the auditorium that day, aged eighty-three, and had never imagined the particle would be found in his lifetime. He had first proposed the existence of the particle back in 1964, along with several other colleagues.*

The importance of the Higgs isn't just that it is a fundamental particle, but that it is a very special entity. As we have seen, there are three kinds of particle—those that make up matter (electrons and quarks, for example), the force particles (photons, gluons), and then there is the Higgs. This arises from a field known as the Higgs field, which pervades all of space, and provides, as it were, a background that gives other particles their mass. Without the Higgs, matter could not exist.

The discovery was a triumph for CERN and the Large Hadron Collider, and the organization of the scientists at Geneva and elsewhere. The LHC is an admittedly prosaic name for an incredible piece of human activity, incredible in the sense that it cost $9 billion to build and was not intended to produce anything useful in, say, a medical or transport sense, though it did, almost inadvertently, give rise to the World Wide Web.**

The collider is about seventeen miles round, with powerful magnets maintained in such a way that inside the machine it is actually cooler than outer space, with a temperature lower than that of the cosmic background radiation left over after the Big Bang. CERN carries out many experiments, but there were two involved in the search for the Higgs, known as ATLAS (for

---

* Peter Higgs and François Englert received the Nobel Prize in Physics in 2013, "for their theoretical discovery of a mechanism that contributes to our understanding of the origin of mass on subatomic particles."

** Of course the ability to manipulate particles has had applications in medicine, food technology, and elsewhere in biology.

A Toroidal LHC ApparatuS) and CMS (for Compact Muon Solenoid), basically two ways of observing proton collisions. The idea of two experiments doing the same thing, albeit in slightly different ways, was part of an approach designed to achieve two complementary aims. One, it would inject some rivalry into the system, spurring the rate of activity, and two, if both teams came up with the same answer, it would greatly confirm the veracity of the results. Each team consisted of about three thousand personnel: these were large experiments by any standard.[53]

The fact that both experiments came up with the same answer, a particle with a mass of 100 GeV, equal to a 100 billion electron volts, and at almost exactly the same time, was one of CERN's greatest achievements, and part of the romance of the discovery of the Higgs. Another measure of the achievement arises from the fact that a Higgs lifetime is estimated to be "somewhat less" than a zeptosecond (10^{-21} seconds), meaning that it travels less than a billionth of an inch before it decays.[54]

For most of us it is difficult to get our heads around these dimensions, but the discovery of the Higgs is important in a raft of ways. In the first place, it completes our understanding of everyday reality. "This is a towering achievement in human intellectual history," in Sean Carroll's words. At the same time, the fact that the discovery has been made means that, perhaps, the era of Big Science is not over. With the Higgs now a fully verified part of the picture, and with the Standard Model in physics now more or less complete, we have reached the end of an era of unification, but stand at the threshold of a new one.

This is because the Higgs may be the gateway toward an understanding of dark matter. And dark matter and dark energy are evidence that the Standard Model is not enough.

When Max Planck began to study physics in 1875, he was warned that the subject was virtually complete and nothing dramatic could be expected in the way of discovery. Within a few years, he had himself identified the quantum, the electron had been isolated, radioactivity discovered, and relativity conceived.

Although none of these intellectual breakthroughs was made in order to exploit its practical or commercial possibilities, they soon transformed our lives as much as—or more than—earlier discoveries.

There are those who doubt the nature of dark matter, who say that it may be like phlogiston or the luminiferous ether of earlier times, more imaginative than real. Rather more think that, since not even the LHC has come up with anything to support such an idea, it is too expensive to continue the search. And that is the measure of the importance and the dilemma of the Higgs. Since we now think there is more dark matter in the universe than any other kind, and that the Higgs may be the vehicle by which we can explore what it is, perhaps Weinberg's dream of a final unification has come a lot closer. But at what price?

Even more problematic, in the long run perhaps, is the argument of two leading researchers, Joseph Silk and George Ellis. In *Nature*, at the end of 2014, they argued to the effect that "some scientists" appear to have "explicitly set aside" the need for experimental confirmation of our most ambitious theories, "so long as those theories are sufficiently elegant and explanatory." They further complain that we are at the end of an era, "breaking with centuries of philosophical tradition," of defining scientific knowledge as empirical.

On this account, part of the problem is in fact the so-called Standard Model itself, which despite the "glories of its vindication" (in particular the discovery of the Higgs) is also a dead end. It is a dead end because it offers no path forward to unite the quantum world and Einstein's relativity. Physicists, as Silk, Ellis, and others point out, have struggled to go beyond the Standard Model, notably with the idea of Supersymmetry, which posits partner particles for every particle we currently know. This theory is elegant mathematically and could account for dark matter. The problem lies in the fact that no Supersymmetric particles—not one—have been found.

This is a direct challenge to the Large Hadron Collider, but, the argument goes, if the failure extends much into the future, what will the response be? Many physicists will conclude that Supersymmetry is a nice idea but just that. Yet the risk, as Silk, Ellis, and those who think like them argue, is that others will simply "retune" their models "to predict supersymmetric particles at masses beyond the reach of the LHC's power of detection." For such theorists, elegance comes before empirical confirmation. And they too ask: how long should we wait for such experimental confirmation? The Higgs took half a century to be detected. As physics gets more expensive, how long can we afford to wait?[55] There is a discernible sense of crisis.

The Great Frustration

Perhaps dilemma is a better word. The problem with such a process as convergence, which is a powerful idea, is that it presupposes a final end point. Convergence is happening all over the sciences, as this book is designed to show, but now the problem in physics is that it may be just too expensive to build the equipment that might—but only might—bring that final convergence about. We might call this "The Great Frustration."

18

SPONTANEOUS ORDER: THE ARCHITECTURE OF MOLECULES, NEW PATTERNS IN EVOLUTION, AND THE EMERGENCE OF QUANTUM BIOLOGY

I n chapter 10 we saw that George Gaylord Simpson, in his book *This View of Life*, remarked that Bacon's idea of the unity of nature and Einstein's attempts to seek the unification of scientific concepts "in the form of principles of increasing generality" had been by and large "worthy and fruitful." But he then went on to insist that biology offered the opportunity to add an extra and very different principle to our understanding. On top of reduction—explanation in terms of physical, chemical, or mechanical principles—biology, "which is more complicated than physics," invites us to introduce a second kind of explanation, what he called *compositionist*. We need to understand structures not just in reductionist terms but also in terms of the adaptive usefulness of those structures and the processes in the whole organism and the species of which it is a part. And, still further, in terms of the ecological function in the communities in which the species occurs.[1]

His further, allied, point, it will be remembered, was that on the scale running from subatomic particles to multispecific communities, a "sharp dichotomy somewhere along that scale" occurs at the level of molecular biology. This, and the organization of the organism, was the crucial—central—fact in science from then on, in his view. As an evolutionist, Gaylord Simpson was concerned to emphasize the adaptionist view, not just of science but of history and even of philosophy. A reductionist approach, he said, made its own kind of sense—food chains could be explained in one way by the enzymes mediating underlying chemical and physical properties, and DNA could specify enzymes, and of course DNA molecules were made of atoms. But this could never be a complete explanation, because no predictions about outcomes could be made on this basis, and furthermore it neglected the adaptive context in which an animal lived and reproduced.[2] It will be recalled that he thought it just as essential for understanding to insist that "lion enzymes" digest zebra meat *because* that enables the lion to survive. He accepted that the discoveries of genetics were deepening reductionist understanding but insisted that paleontological discoveries were deepening compositionist understanding.[3] The two approaches were needed equally.

The Doctrine of Emergence

Gaylord Simpson's point was well made, but he wasn't quite the first to point this out. In his book *The Nature of the Physical World*, published a little earlier, in 1928, Arthur Eddington introduced a distinction between primary and secondary laws of nature. "Primary laws," he argued, control the behavior of single particles, while "secondary laws" are applicable to collections of atoms or molecules. "To insist on secondary laws is to emphasize that the description of elementary behaviours is not sufficient for understanding a system as a whole."[4]

This was picked up on by Ernest Nagel, one of a number of Austro-Hungarian émigrés to the United States, whose masterpiece, published in 1961, was *The Structure of Science*.[5] In this he explored Eddington's ideas of primary and secondary science, arguing that some areas of science are reducible to others, but not all. That reductionism at its most successful led to new discoveries ("intimate and frequently surprising relations of dependence"); that reduction may be a "significant advance" in the organization of knowledge or merely a formal exercise; and that the very point of reductionism is to provide "increased power" for significant research.[6]

But his main concern was the *doctrine of emergence*—that there are higher levels of organization in nature "that are not pre-dictable from properties found at 'lower' levels" and that "the future will continue to bring forth unpredictable novelties."[7] "Emergent Evolution," he felt, was the most important discovery of science—the fact that, for example, "water is translucent" cannot be deduced from any set of statements about hydrogen and oxygen. "It is of the essence of Emergent Evolution that nothing new is added from without, that 'emergence' is the consequence of new kinds of relatedness between existents."[8] No one, for example, in predicting that nitrogen and hydrogen would combine to produce ammonia (if they could predict such a thing) could forecast its distinctive smell. The laws of nature, he said, are endlessly fascinating, not least for the fact that, so far as we can tell, they have evolved simultaneously. How then, historically speaking, is one law related to another? What is the law, if any, underlying a melody?

His basic idea was that we inhabit what he called a "mathe-matically continuous space" and that, one day, this would explain laws, particles, and macroscopic entities.[9] (And compare the ideas of David Bohm, discussed in chapter 17.)

This approach, focusing on emergence and spontaneous organization, grew in popularity and stature throughout the latter decades of the twentieth century, as mathematics was applied more and more to biological processes, and was associated, at least to begin with, with two well-known Nobel Prize–winning physicists.

The first was Philip Warren Anderson. A Nobel Prize winner in physics and professor at both Cambridge and Princeton, Anderson published a highly influential article in *Science* in 1972, entitled "More is Different." He began by saying that "we must all start with reductionism, which I fully accept. . . . The reductionist hypothesis may still be a topic of controversy among philosophers, but among the great majority of active scientists I think it is accepted without question." But he then went on to introduce a concept that essentially built on Simpson's "compositionist" viewpoint and has become increasingly popular. This was his idea of "emergence," the process whereby, in complex systems, the simple symmetry of elementary particles breaks down, and reductionism has to be replaced by what he preferred to call "constructivism," that there are in nature rules—laws—of complexity that, while they never break the laws of particle physics, *add* to them new laws of construction *which are every bit as fundamental* as the simpler laws governing electrons and photons, for example. As we go on up the hierarchy of sciences, "we expect to encounter . . . very fundamental questions at each stage in fitting together less complicated pieces into the more complicated system and understanding the basically new types of behavior that can result." [10] Both superconductivity and molecular biology were topics he cited as examples.

In this new context, superconductivity took on a fresh significance. It had been known for some time that, for example, there is a qualitative difference between water and ice. In a crystal, the molecules are arranged in a regular array, held in place by the forces between them. There is a long-range order not present in a liquid. In the same way, in a grain of magnetized nickel, to use John Taylor's example, the individual atoms are like tiny magnets, and at low temperatures the atomic magnets are all lined up in the same way, so their effects add up, making the whole ensemble a macroscopic magnet. Warm up the nickel, however, and the magnetization starts to decrease, until at 631°K it drops to zero. In a third case, when the temperature of metals is reduced to a critical level, superconductivity appears, whereby the electrical resistance totally vanishes and electrical currents, once started, persist indefinitely.[11] Whole technologies have been built on these phenomena, but their significance here lies in the fact that they show spontaneous order.

"The Evolutionary Aspects of Nature Have to Be Expressed in the Laws of Physics"

The second Nobel Prize winner who espoused this view, and built on it, was Ilya Prigogine, aided by Isabelle Stengers. In a series of books, including *Order Out of Chaos* (1984) and *The End of Certainty* (1997), Prigogine, from the aptly named International Solvay Institutes for Physics and Chemistry, in Brussels, who won the 1977 Nobel for work on the thermodynamics of non-equilibrium systems, put forward his view that we are living in a new scientific era. It is an era of complexity, in which the laws of nature within the range of low energies (that is to say, the domain of macroscopic physics, chemistry, and biology, the domain in which human existence actually takes place) need to be reconfigured, or understood in a new way. "We need a new formulation of the fundamental laws of physics."

In particular, he accepted that "the evolutionary aspects of nature have to be expressed in the basic laws of physics." But the central element in all this, he said, was that nature involves both time-reversible and time-irreversible processes, "but it is fair to say that irreversible processes are the rule and reversible processes the exception."[12] The new unification we need, therefore, in his view, is between dynamics (essentially Newtonian mechanics) and thermodynamics, which, as we saw in chapter 1, encompass the all-important phenomenon of entropy, which is time-related and irreversible. The man who first appreciated this, in Prigogine's view, was Ludwig Boltzmann, whose identification of the *statistical* nature of nature embodied the new attitude, that "the arrow of time" enters into everything.

Both Boltzmann and Darwin, Prigogine noted, replaced the study of individuals with the study of populations "and showed that slight variations . . . taking place over a long period of time can generate evolution at a collective level." Biological evolution cannot be defined at an individual level, he said, and this too is all-important. "We are in a world of multiple fluctuations and the fluctuations are the macroscopic manifestations of fundamental properties of fluctuations arising on the microscopic level of unstable dynamical systems." In this way, he said, "biology and physics begin to converge."[13]

Non-equilibrium thermodynamics proceed in a probabilistic way, as Boltzmann (and Maxwell) were the first to realize, and this ensures they are irreversible. What was new, what Prigogine was pointing out perhaps more clearly than Eddington, Simpson, or Anderson, was that these processes lead to *spontaneous self-organization*, which accounts for, and reflects, both the unity and the diversity of nature. The maintenance of organization in nature, he said, is not—and cannot be—achieved by central management: "order can only be maintained by self-organisation."[14] This is why nature is both ordered and yet diverse.

Despite the fact that they were in some ways coming together, the essential difference between physics and biology is that biological organisms and processes express much more complexity. This realization had led, Prigogine said, to a new chapter in mathematics and theoretical physics—the study of chaos and complexity, the important point being that "although the problem of chaos cannot be solved at the level of individual trajectories, it can be solved at the level of ensembles."[15] There are, he insisted, new properties of matter in non-equilibrium situations where "dissipative structures" and self-organization appear. (Living organisms are dissipative structures, which can only maintain equilibrium by taking on new sources of energy, like food.)

In order to appreciate this fully, we need a brief account of "chaoplexity," which is itself a synthesis—this time of chaos and complexity. In 1987, in *Chaos: Making a New Science*, James Gleick introduced this new area of intellectual activity to a general audience. Chaos research starts from the concept that there are many phenomena in the world that are, as the mathematicians say, nonlinear, meaning they are *in principle* unpredictable. The most famous of these is the so-called butterfly effect, whereby a butterfly fluttering its wings in, say, the Midwest of America can trigger a whole cascade of events that might culminate in a monsoon in the Far East.

A second aspect of the theory is that of the "emergent" property, which refers to the fact that, as Prigogine also said, there are on earth phenomena that "cannot be predicted, or understood, simply by examining the system's parts." Consciousness is a good example here, since even if it can be understood (a very moot point), it cannot—critics say—be understood from an inspection of neurons and chemicals within the brain. However, this only goes halfway to what the chaos scientists are saying. They also argue that the advent of the computer enables us to conduct much more powerful mathematics than ever before, with the result that we shall eventually be able to model—and therefore simulate—complex systems, such as large molecules, neural networks, population growth, and weather patterns. In other words, the deep order underlying the apparent chaos will be revealed.[16]

The basic idea of chaoplexity comes from Benoit Mandelbrot, a Polish Jew who escaped to France with his parents when he was eleven, in 1936. After the war Benoit became an applied mathematician at IBM, where he identified what he called the "fractal," taken from the Latin *fractus*, meaning "shattered" or "broken glass." (His book, *The Fractal Geometry of Nature*, appeared in 1982.) The perfect fractal, or "Mandelbrot set" (one of the most astonishing discoveries in mathematics, according to Arthur C. Clarke, the science-fiction writer), is a coastline, but others include snowflakes and trees. However close you go, the more intricate the outline, with, often, the patterns repeated at different scales. The familiar leg and boot–like outline of Italy, say, viewed from an aircraft eight miles up, is both different and yet the same as the outline viewed from an aircraft three miles up. Because these outlines never resolve themselves into smooth lines—in other words never conform to some simple mathematical function—Mandelbrot called them the "most complex objects in mathematics." At the same time, however, it also turns out that simple mathematical rules can be fed into computer programs that, after many generations, give rise to complicated patterns, patterns that never quite repeat themselves. From this, and from their observations of real-life fractals, mathematicians now infer that there are in nature some very powerful rules governing apparently chaotic and complex systems that have yet to be unearthed. (And compare the work of Stephen Wolfram, chapter 17.)

But—and this is what especially interested Prigogine, and then many others—when we move from equilibrium to far-from-equilibrium conditions, we move from the repetitive and universal to the specific and the unique. "Matter near equilibrium behaves in a 'repetitive' way. On the other hand, far from equilibrium there appears a variety of mechanisms. . . . We may witness the appearance of chemical clocks, chemical reactions which behave in a coherent, rhythmical fashion. We may also have processes of self-organisation leading to nonhomogenous structures and nonequilibrium crystals."[17]

Two simple examples will highlight the phenomenon. The first is the so-called Rayleigh-Bénard instability, when a liquid is heated. To begin with the lower area is heated, and a permanent heat flux is established, moving from the bottom to the top. When this gradient reaches a threshold value, the fluid's state of rest—the stationary state in which heat is conveyed by conduction alone, without convection—becomes unstable, and a convection corresponding to the coherent motion of *ensembles* of molecules is produced, increasing the rate of heat transfer. The convection motion produced actually consists of the complex spatial organization of the system. Millions of molecules move coherently, forming hexagonal convection cells of a characteristic size.[18] A second example is the Belousov-Zhabotinsky (or B-Z) reaction, named after Boris Belousov and Anatol Zhabotinsky, and refined by Art Winfree and Jack Cohen. If four particular chemicals are mixed together in fairly precise proportions, in a shallow dish, at first they form a uniform blue layer. Suddenly, however, it turns reddish-brown. After a while, and for no apparent reason, a few small blue spots appear. They grow, and their centers turn red. As what are now blue circles expand, the red spot inside also expands, and more blue spots appear in the red. Soon the dish is filled with concentric rings of red and blue, all slowly expanding and colliding with each other. Under certain other circumstances, the B-Z reaction can also produce spirals, but the important point—and widely influential—is the spontaneous order that is generated automatically, and by *inorganic* matter.[19]* The designs of the B-Z reaction are in fact mirrored in the patterns that appear in colonies of amoebae, *Dictyostelium discoideum*, a slime mold that when it is starved of heat or moisture, aggregates into a multicellular body which is then able to move in search of a more favorable environment. Aggregation is triggered by the release of a special substance by certain cells, known as "pioneer" cells. Prigogine thought that life must have started in this way, by the self-organization of inorganic substances, one indirect reason being that the approximately 15 billion years separating us from the Big Bang is "surprisingly short. To express it in years, we use the notion of Earth as a clock. Fifteen billion revolutions is indeed a small number if we remember that in the hydrogen atom, the electron rotates some 10,000 billion times per second!"[20]

Two people have taken this reasoning forward, in a further amalgamation of mathematics and biology.

---

* The starting materials are malonic acid and salts containing bromate and bromide ions. During the reaction, malonic acid is converted to bromo-malonic acid.

"The Modern Renaissance of Mathematical Biology": Mathematics as "Natural Language"

"Biology," insists Stuart Kauffman, "is surely harder than physics." He has done varied work at institutions in Santa Fe, Seattle, Vermont, and Oxford, but in his main book, *The Origins of Order: Self-Organization and Selection in Evolution*, he sets out to answer the question: "What are the sources of the overwhelming and beautiful order which graces the living world?"[21] He does this from the perspective of mathematical biology, Kauffman being, in the words of one reviewer of his book, "one of the pioneers in the modern renaissance of mathematical biology." (How Mary Somerville would have loved to have been alive at such a time.)[22]

The underlying biological order, the book concludes, obeys the first and second laws of thermodynamics and it is molecular in character (remember that Oppenheim and Putnam, and Gaylord Simpson, had identified the molecular level as the critical juncture in the reductionist hierarchy). "Its conceptualization rests firmly in physics and chemistry." Kauffman's ideas fall into two main areas: the emergence of spontaneous order within cells, in particular among proteins, and the "ordered morphology" of organisms, which also have certain inherent architectural (and therefore mathematical) rules that natural selection has to operate within. While this may limit the diversity of forms, it also helps speed up the evolutionary process.

Kauffman's overall thesis is that Darwinian natural selection only goes so far in explaining the diversity of the organic forms we see about us. Natural selection was aided by the inherent physical and mathematical properties of molecules, complexity theory producing over time a "profoundly immanent" complex order that we do not yet understand.[23]

Molecular biology, Kauffman says, is driving us to the "innermost reaches" of the cell's ultimate mechanisms, complexity, and capacity to evolve. At the same time "mathematics, physics, chemistry and biology are revealing how far-reaching the powers of self-organization can be. In phylogenetic comparisons of the amino acid sequences of a particular protein, we see that the evolution rate is nearly constant over extended periods of time, so much so that the amino acid substitution process has been called a 'molecular evolutionary clock.'"[24] This is self-organization of a remarkable kind.

Kauffman's achievement is not just his argument, but his relating of that argument to specific biophysical, or physiological, processes *in detail.* For example, he shows how the genomic system of any higher metazoan (differentiated) cell encodes the order of 10,000 to 100,000 structural and regulatory genes "whose joined orchestrated activity constitutes the development program underlying ontogeny from the fertilized egg." The human immune system has perhaps 100 million distinct antibody molecules in "harmonized patterns," and neural systems have perhaps billions of neurons to assess and categorize the exterior and interior milieu. "How, we must ask, can such wonderful systems emerge merely through random mutation and selection?" His point is that, "contrary to our deepest intuitions, massively disordered systems can spontaneously 'crystallize' a very high degree of order."[25] He applies the mathematical theory of dynamical systems, which he says is the "natural language" of integrated behavior.

This is a development of Newton differential equations, which lead to the emergence of what mathematicians call *attractors,* where systems "freeze," and in doing so form large connected clusters that are either stable or oscillate in a regular way. These attractors are entirely natural features of spontaneous organization and are the building blocks of evolution (see below for specific examples).

But Kauffman's further achievement is to show, again in great detail, how, using Boolean mathematics, or the concept of bits, *0* and *1*, molecules are first built up; then genes (how recognizable patterns of amino acids *recur* in the genes of different animals: sea urchin, *Drosophila*, mouse, human); and how cells are formed (yeast has 3 different types of cell, jellyfish 20 to 30, adult humans 254). He gives "fate maps" of larvae, for instance, which show which larval cells will form which organs. And finally he shows how the cells in the fate map order themselves mathematically to form organs. All along, the processes can be shown to *conform* to mathematical principles.

The Basic Patterns of Evolution

Much the same problem—the sheer complexity of evolution—was taken up by Robert Wesson in *Beyond Natural Selection* (1991).[26] A senior research fellow at the Hoover Institution in Stanford, he had two aims in his book. One was to show, in convincing detail, that natural selection cannot by itself possibly account for the diversity we see around us (for one thing, so many features contradict others, or are clearly maladaptive). And, second, to suggest other dynamics of evolution that might reinforce natural selection.

Wesson, like D'Arcy Wentworth Thompson, thought evolution had, in effect, had too easy a ride from biologists. "We can always invent a plausible adaptive advantage for an observed or supposed trait."[27] But in fact the evidence against certain aspects of Darwinism, he thought, was everywhere. Some plant families—horsetails, clubmosses, ginkgoes—have been unmodified for tens or hundreds of millions of years. Abundant fossils give little evidence of gradual change. He found the rise of birds remarkable because the air was already occupied by numerous and "apparently efficient" flying lizards and pterosaurs. The stages by which fish gave rise to amphibians are unknown, and the earliest land animals appear "with four good limbs, shoulders and pelvic girdles, ribs and distinct heads. . . . In a few million years, over 320 million years ago, a dozen orders of amphibians suddenly appear in the record, none apparently ancestral to any other."[28]

Other examples: birds fly, without previous experience, on immensely long journeys, often without landmarks. So, for instance, the young of the bronze cuckoo, a month after being left behind by their parents in New Zealand, fly twelve hundred miles over water to Australia and then one thousand miles north to the Solomon and Bismarck islands. "The journey seems unnecessary, since there is no great need to escape the mild winter . . . and it is unclear why the parent birds require their offspring to navigate by themselves."[29] Thousands of mutations of *Drosophila* have been observed, but they are generally trivial or pathological—none has been suggestive of a new organ. The water ouzel or dipper (*Cinclus*), a small bird that finds insects on the bottom of brooks, looks like an ordinary land bird. Although it uses its wings to swim, it, like the marine iguana, has failed to acquire webbed feet. The many legs of centipedes are a poor idea for locomotion.[30] Snakes have largely given up their hearing, despite its obvious utility. And why are birds toothless? Why do stags grow new antlers every year? Sheep, oxen, and antelope do not.

Many adaptations seem pointless (the male lion's mane, for example, which makes it harder to stalk game). And others are wasteful (avocados have a hundred flowers for every one that matures). There are many polygynous species where the males do little beyond insemination. Why must the bamboo die upon flowering?

With such pointless and wasteful aspects, are evolution and natural selection really the efficient powerhouses we suppose?

It is not that Wesson doubts evolution. He does not. He just thinks that it often does things the hard way, and that is because he doesn't think natural selection—the classical Darwinian device—is anywhere near enough to account for the diversity we see. Instead, his answer too is mathematical. He felt strongly that the idea of the "attractor," as used by Prigogine and Kauffman, was helpful, that information is transmitted by the genes in such a way as to naturally form patterns, self-organization derived from the physical and mathematical properties of matter, "not a grand single blueprint but a set of many overlapping partial blueprints."

There is clearly, for instance, a self-directed capacity for restoration of pattern following injury, and some patterns—some attractors—are, equally clearly, stronger than others. And it is the pattern that comes before the parts. For example, the "six-leg attractor" is stronger than the "four-leg attractor." Six legs feature among many millions of insect species in a very similar way, whereas the body plan for four limbs has been modified in very different ways by dinosaurs, birds, fish, and mammals.[31]

In other words, evolution proceeded by the development of a series of "basic plans," spontaneously organized regularities among the proteins, and this is how the higher-order taxa became established. To underline this, he points out that no new phyla have appeared since the early Cambrian (540 to 480 million years ago), that no new classes have arisen for at least 125 million years, and that there have been no new orders since the post-dinosaurian radiations, 65 million years ago. "Evolution has left behind the stages when it was possible to bring forth new phyla and new classes."[32] The oceans, where life began, have long been barren of important innovations. And, as structures became more evolved, the more complex innovations were less profound. Is this a biological manifestation of entropy?

In the end, even we humans are an amalgamation of biology and mathematics. As he puts it, neither a rat-sized nor an elephant-sized creature could have mastered fire, smelted metals, and produced the rest of the requisites of a technological civilization. Natural selection was aided by the self-organization of matter based on mathematical and physical principles to an extent that we are only just beginning to realize.

Chlorophyll's Quantum Walk

It will be recalled from chapter 7 that Pascual Jordan first had the idea of a possible marriage of quantum physics and biology, which he discussed with Niels Bohr in Copenhagen just as the 1930s were getting under way. It will also be remembered that

Jordan got caught up in the insidious involvement of Nazi ideology in science, with the result that he was sidelined after World War II.

It took a while for quantum biology to be taken up again, but its achievements received some recognition at a small conference on the subject held at the University of Surrey in 2012, and by the establishment of the Models and Mathematics in Life and Social Sciences project, which began, also at the University of Surrey, in 2010.[33]

Jim Al-Khalili and Johnjoe McFadden, in their book on quantum biology, which came out of the Surrey conference, urge us to consider that there are three "strata" of reality. In the top layer, the visible world is filled with objects such as falling apples, cannonballs, steam trains, and airplanes, "whose motions are described by Newtonian mechanics." Below that is the thermodynamic layer of "billiard-ball-like particles," whose motion is almost entirely random and is responsible for generating "order from disorder" laws that govern the behavior of objects such as steam engines. The bottom layer comprises the layer of fundamental particles "ruled by orderly quantum laws." The visible features of most of the objects we encounter appear to be rooted either in Newtonian or thermodynamic layers "but living organisms have roots that penetrate right down to the quantum bedrock of reality."[34] They also say that the three greatest mysteries in science are the origin of the universe, the origin of life, and the origin of consciousness. Quantum mechanics, they argue, is involved in all three.

The fields of inquiry that quantum biologists have pioneered include the navigational behaviors of certain migratory birds, insects, and fish, the explanation of enzyme catalytic action, photosynthesis, genes, and consciousness. This is a not unambitious list for such a young science. We shall concentrate on photosynthesis and genetics, since the work on consciousness is mainly speculative for the moment.

The main arguments of the quantum biologists are that, at the molecular level, many important biological processes are very fast (of the order of trillionths of a second) and are confined to short atomic distances—just the dimensions where the properties of superposition, entanglement, and quantum tunneling enable certain processes to occur that are denied in classical mechanics.

Take chlorophyll, for example. This substance, they point out, is probably the second most important molecule on our planet (after DNA). It is best thought of as a two-dimensional structure made up of "pentagonal arrays" of mostly carbon and nitrogen atoms enclosing a central magnesium atom with a long tail of carbon, oxygen, and hydrogen atoms. The outermost electron of the magnesium atom is only loosely bound to the rest (as per Bohr's orbital laws and Pauli's exclusion principle) and can be knocked into the surrounding carbon "cage" by absorption of a photon of solar energy. Simplifying somewhat, when this happens the excited (and now homeless) electron, the "exciton," has to be transferred to a molecular manufacturing unit known as a reaction center, which helps the electron drive the all-important photosynthesis process.

Traditionally, the problem in understanding this scenario was that the reaction centers are quite distant (in nanometer terms) from the excited chlorophyll molecules and no one knew what route the electron would take, or how it "knew" which route to take. One theory was that it just went on a "random walk," though this was hardly efficient and in that sense unlike the rest of the biological world, where photosynthesis is known to be very efficient indeed.

The solution to the problem was traced by a team at the University of California in 2007. Using a special kind of electron spectroscopy (see chapters 5 and 9), which can probe the inner structure and dynamics of the smallest molecular systems, they fired three successive impulses of laser light into a protein made by photosynthetic microbes called green sulphur bacteria. The laser pulses load their energy in very rapid and precisely timed bursts, and, say Khalili and McFadden, whose account this is, they generate a light signal that is picked up by detectors. Pulsed lasers can emit, within a single second, about 2,000 million million pulses, the crucial point being that this is many thousands of times shorter than the time it takes for a molecule to execute a single rotation of vibration. The team examined a special set of results covering fifty to sixty femtoseconds (a femtosecond is one-millionth of one-billionth of a second, or 10^{-15} seconds). What they found was a rising and falling signal "that oscillated for at least six hundred femtoseconds."[35] The significance of this very technical result is that these oscillations are analogous to the interference of the light and dark patches in the two-slit experiment that are such a feature of quantum activity. In other words, the exciton wasn't taking a single route through the "chlorophyll maze" but was instead "following multiple routes simultaneously." By extension of the concept of random walk, this process was given the name quantum walk. Since then this phenomenon has been observed also in algae and in spinach.[36]

In the case of genes, quantum activity concerns a process called tautomerization, which was initially observed by Watson and Crick in their second paper on the structure of DNA, published in 1953. This process involves the movement of protons within a molecule—essentially between alternative proton positions—that can generate mutagenic forms of nucleotides. The protons that connect the hydrogen bonds are quantum entities—both particles and waves—and they can move in opposite directions. If they happen to be in the tautomeric form when reproduction occurs, then, via the process of quantum tunneling, the energy differentials that normally exist naturally could be overcome. This would mean that, for example, a thymine molecule could pair with a guanine—which in classical terms is impossible—and a mutation occur. This has been observed indirectly, if not yet directly. In a study of single yeast genes, mouse genes, and human genes, it was found that when the genetic information is transcribed from DNA to RNA, ahead of making proteins, the genes that are "read" most often (and "reading" is a kind of measurement which, we know from quantum theory, "interferes" with the quanta being measured) mutate more often, thirty times more often in fact. This research is all very new but appears to be pointing to the fact that genetic mutation may itself be a quantum process.

The emergence of quantum biology—new as it is—is in some ways the climax of the route we have been traveling throughout this book. If quantum processes *are* shaping mutation, then this is one way in which energy directly drives evolution and the two great strands with which we began have come together.

One thing is certainly clear. Whether biology is more complex than physics, it is quite complex enough. More than that, though, these recent advances in biology promise further convergence to enrich evolutionary theory. If the quantum nature of mutation is confirmed, this will help explain how diversity occurs in the first place—making evolution possible at all. Evolution in a real sense will be a quantum process. And the mathematical work on spontaneous order, the example of the attractor in particular, shows how evolution was facilitated—and accelerated—by entirely natural organized processes to produce levels of diversity that some have doubted could be achieved in the time available.

Taken together, what this means is that we may be on the verge of the most important revision of Darwinism since his original idea of natural selection appeared in 1859. This is a good example of how reductionism—convergence—leads to new areas of inquiry, and to new levels of understanding. A link between quantum processes and evolutionary theory is as great a convergence as one could hope for.

19

THE BIOLOGICAL ORIGIN OF THE ARTS, PHYSICS AND PHILOSOPHY, THE PHYSICS OF SOCIETY, NEUROLOGY AND NATURE

Here is a fascinating and yet provocative list. Languages around the world vary in the extent to which they make use of color terms, but where they do so there is an order. Languages with only two color terms use them to distinguish black from white. Languages with only three terms have words for black, white, and red. Languages with only four terms have words for black, white, red, and either green or yellow. Languages with only five terms have words for black, white, red, green, and yellow. Languages with only six terms have words for black, white, red, green, yellow, and blue.

Edward O. Wilson, the sociobiologist we met in chapter 13, produces this list in his 1998 book, *Consilience: The Unity of Knowledge.*[1] His point is that if basic color terms were combined at random, human vocabularies would be drawn "helter-skelter" from among a mathematically possible 2,036 instances, whereas in fact studies show that vocabularies for the most part draw on only twenty-two.

This is a minor—but vivid—part of Wilson's wider argument, that "genes keep culture on a leash."[2] It is a contentious argument in some quarters, but Wilson attempts to draw on the widest context possible, and it introduces this last chapter. If the sciences have converged over the past two hundred years—and given the successes of science more generally, and of the reductionist approach—is it any surprise that they should seek to bring the rest of human activity into their ambit?

"The greatest enterprise of the mind," Wilson says in the same book, "has always been and will always be the attempted linkage of the sciences and the humanities."[3] Moreover, philosophy, "the contemplation of the unknown, is a shrinking dominion. We have the common goal of turning as much philosophy as possible into science." This is not simply arrogance. During the past two generations, he says, "the ideal of the unity of learning, which the Renaissance and the Enlightenment bequeathed us, has been largely abandoned," one result of which has been that the social sciences have in many places replaced the humanities.[4]

Rather than decrying this, Wilson sets out to show how the order revealed by the sciences can help explain and enrich culture and cultural appreciation. Societies, he says, are ordered around six sociobiological principles (alluded to in chapter 13): *kin selection*; *parental investment*; *mating strategy*; *status*; *territorial expansion and defense*; and *contractual agreement*. On top of which, there are four "bridges" between natural science and social science: *cognitive neuroscience*; *human behavioral genetics*; *evolutionary biology*; and *environmental science*. Via these intermediaries, he says, there has been gene-culture co-evolution, and it is in this respect above all that humans differ from other animals.[5]

Over time, these intermediaries have brought about a situation where universals or near-universals emerged in the evolution of culture. The arts, Wilson argues, are "innately focused" toward certain forms and themes but are otherwise "freely constructed."[6] He goes much further, asserting that "the arts are not solely shaped by errant genius out of historical circumstances and idiosyncratic personal experience. The roots of the inspiration date back in deep history to the genetic origins of the human brain, and are permanent." More even than that, "metaphors, the consequences of spreading activation in the brain during development, are the building blocks of creative thought." Early temples were built as metaphors of mountains, rivers, animals. Studies of successful paintings reveal they are the images most likely to be most arousing to the brain's EEG activity. Art imitates, intensifies, and "makes geometrical," for clarity's sake. Art "stays true to the ancient hereditary ground rules that define the human aesthetic."[7]

Even in fiction, he says, we see a small number of basic "archetypes" that have evolved under epigenetic rules, recognizable by their repeated occurrence in narratives:

In the beginning . . .
The tribe emigrates . . .
The tribe meets the forces of evil . . .
The hero descends into hell or is exiled . . .
The world ends in apocalypse . . .
A source of great power is found (tree of life, river of life) . . .
The nurturing woman is apotheosized as the Great Goddess
    or Mother . . .
The seer has special knowledge . . .
The Virgin has the power of purity . . .
Female sexual awakening is bestowed . . .
The trickster disturbs established order . . .
A monster threatens humanity . . .

What spawned the arts was the need to "impose order on the confusion caused by intelligence."[8] The ancients sought to express and control—sometimes by magic—the sheer abundance of the environment. And the arts still perform this "primal function."

The biological origin of the arts, although only a working

hypothesis (he says), is supported by the fact that, for example, Hollywood plays well in Singapore, or that Africans and Asians as well as Europeans win Nobel Prizes in literature—near-universal themes are, well, near-universal. By the same token, brain arousal activity is at its greatest when shown pictographs taken from Asian languages, the glyphs of ancient Egyptians and Mayans, and photographs of what most people consider to be beautiful women. On this account, there does appear to be a biology of beauty.

Many people will and do say that this is crass, that it takes science too far, into areas where it doesn't belong. But Wilson insists we are still in a Paleolithic world, emotionally, and the ever-closer link between the natural sciences and the social sciences, he says, offers the best—indeed the only—way forward to revitalize the arts. Theology no longer suffices, he concludes, and philosophy offers "no promising substitute. Its involuted exercises and professional timidity have left modern culture bankrupt of meaning."[9]

"Morality, an Undeveloped Branch of Science"

The largest-ever survey of gay brothers, published in February 2014, discovered that two regions of the genome—an area on the X chromosome called Xq28, and the second in a twist in the center of chromosome 8, known as 8q12—provided the strongest indication yet that sexual orientation in men is biologically determined. In a quite separate experiment, it was found that apes (capuchin monkeys and chimpanzees, but not owl monkeys and rhesus macaques), as well as humans, will give up part of a reward and cede it to a partner if it equalizes the total reward and preserves the partnership.[10]

Given that there are still areas of the world where homosexuality is a crime, the first result may go some way to changing attitudes. The second experiment would appear to be evidence for the evolution of a sense of fairness. Both are important for our understanding of morality and further evidence of science extending its reach in important ways and bringing different aspects of human activity closer together.

Sam Harris, the American philosopher and neuroscientist, gave his 2010 book, *The Moral Landscape*, the subtitle "How Science Can Determine Human Values." His argument was that meaning, morality, and "life's larger purpose" are really questions about the well-being of conscious creatures. Values, therefore, "translate into facts that can be scientifically understood. . . . Just as there is no such thing as Christian physics, or Muslim algebra, we will see that there is no such thing as Christian or Muslim morality. Indeed, I will argue that morality should be considered an undeveloped branch of science." He claimed that science will "gradually encompass life's deepest questions"—what he termed "the moral landscape" of human flourishing—whose peaks correspond to the heights of potential well-being and whose valleys represent the deepest possible suffering.[11]

He conceded that it is too early to say that we have a full understanding of how people flourish but insisted that a piece-meal account is emerging. We know, for instance, that there is a connection between early childhood experience and a person's ability to form healthy relationships in later life (chapter 13). And we know that nurturing is related to certain substances in the brain (considered later in this chapter).

These arguments enable us to say, he concludes, that goodness may be defined not as what any particular religion or philosophy claims, but as that set of attitudes, choices, and behaviors that potentially affect our well-being as well as that of other conscious minds. Look around us, he says, and we will see that some people lead better lives than others and that these differences relate, in some ordered "and not entirely arbitrary way, to states of the human brain, and to states of the world." This is all we need as a starting point, he says, which leads to his claim that there are right and wrong answers to moral questions just as there are right and wrong answers to questions of physics, "and such answers may one day fall within reach of the maturing sciences of mind."

Human consciousness is the only intelligible domain of value, and "well-being" captures all that we can intelligibly value. This has to be where we concentrate our activities and concerns.[12]

In this approach, he thinks that there may be nothing more important than human cooperation, and that a full understanding of our moral evolution—in which kin selection, reciprocal altruism, and sexual selection are the chief ingredients (all sociobiological areas)—helps explain how we have evolved to be not "merely atomised selves in thrall to our self-interest but social selves disposed to serve a common interest with others."

This means that our own happiness requires that we extend the circle of our self-interest to others—to family and friends and even perfect strangers whose pleasures and pains matter to us. We know this, he says, partly because it is easy to observe a moral hierarchy in human societies. Many "primitive" tribes really are savages, where cruelty and suspicion, for example, are much more in evidence than in more evolved societies. He shows that fairer, more egalitarian societies (Sweden, Norway, Holland) are characterized by less violence (rape and robbery), less crime, better life expectancy, and reduced child mortality, lack of corruption, and greater charity to poorer nations. These trends and correlations are not arbitrary.[13]

He lists the regions of the brain involved in moral cognition (the prefrontal cortex and the temporal lobes) and argues that neuroimaging studies show that cooperation is associated with heightened activity in the brain's reward regions. He says that fMRI studies of the brain show that the same areas are involved, in the same way, when believing a mathematical equation (versus disbelieving another) and believing an ethical proposition (versus disbelieving another). This shows, he argues, that there is no real difference between scientific dispassion and judgments of value. Our well-being does not depend on religion but on the interaction between "events in our brains and events in the world," and the more we understand this interaction, the more we will tend to produce lives that are more worth living than others.[14]

The Chemistry of Caring

Whereas Harris took a broad-brush approach, Patricia Church-land, a Canadian-American neurophilosopher, is more focused. In her book *Braintrust: What Neuroscience Tells Us About Morality* one of her starting points is the "nightmare" (her word) when assessing rival moral questions so as to maximize happiness for ourselves and the greatest number. "No one has the slightest idea how to compare the mild headaches of five million against the broken legs of three . . . the needs of one's own two children against the needs of a hundred unrelated brain-damaged children in Serbia."[15] Instead, she says that morality begins in evolution and in attachment theory, which is itself, of course, not unrelated to evolution.

The family is the "motherboard" of human sociability, she says, and caring for and trust in the family is the foundation of that sociability. Morality, on this account, is essentially about maintaining sociability in the widest possible context. The family teaches us how and whom to trust, and experience teaches us how to extend that trust—to kin and then beyond kin, to fellow members of the various in-groups we belong to, and then on to, say, business partners, potential rivals, and competitors.

Psychologically, morality is a four-dimensional scheme for social behavior that is shaped by interlocking processes: caring, recognition of others' psychological states, problem-solving in a social context, and learning social practices. According to this hypothesis, she says, values are more important than rules.[16]

But how can neurons care? she asks. Her answer is largely given through a description of the action of two neuropeptides in the brain, oxytocin and arginine vasopressin. These, along with other hormones, affect our social behavior, in particular our "maternalizing" behavior, making mothers care for their offspring, and the rest of us trust others. Along the way she points out how this behavior evolved, by highlighting that an "earlier version" of oxytocin and vasopressin—vasotocin—plays a role in amphibian mating behavior. By a similar process, another chemical, naloxone, this time one which blocks the receptors for opioids, has the effect in monkeys of making them indifferent to their babies.

Receptors for these substances are found in very specific areas of the brain and play a part in social behavior and, where impaired, affect behavior in deleterious ways.[17] Churchland's underlying point is that "social behaviour and moral behaviour appear to be part of the same spectrum of actions." In laboratory games of economic rivalry and cooperation, subjects given oxytocin by nasal spray before the games showed more trusting behavior—greater cooperation—than rivals who had no such treatment. Administering oxytocin in other cases reduced levels of fear and anxiety in social situations and, in certain circumstances (and mainly among men), helped people assess what was in other people's minds.[18] In still other experiments, neuroscientists have shown that decreases in serotonin correlate with increased rejection of "unfair" offers in a game transaction and that again these episodes occur in specific areas of the brain.

Other experiments described by Churchland show that we unconsciously mimic people we admire, that babies are innately drawn to people like them, and that when people are with others whom they trust, their oxytocin levels rise. Morality—our values—concludes Churchland, grows out of family experience and is mediated by a series of identifiable substances operating in specific areas of the brain. This highly reductionist account is not a complete explanation, as she is the first to admit. But it does vitiate ideas that morals come from God or any non-evolutionary philosophy.

The Physics of Society

Could there be such a thing as a "calculus of society"? The science writer Philip Ball considers this in his 2004 book, *Critical Mass: How One Thing Leads to Another*, in which he notes that physics has come to have the confidence, "perhaps even the arrogance," to venture into social science. No one has deliberately set out to construct a physics of society, he says. "It just so happens that physicists have realized that they have at their disposal tools which can be applied to this new task."[19] He went on to consider a variety of phenomena: people's behavior in crowds, traffic movements, the stock markets, crime, neighborhood interactions, the mathematical rules governing the growth of cities, the spread of disease, and even marriage.

He found a good deal of spontaneous organization in human behavior. For example, people walking down a corridor in opposite directions tend to organize themselves spontaneously into counter-flowing streams, which reduces the need for last-minute collision-avoiding maneuvers. When trees or other objects are in the middle of walkways, people spontaneously use them as separation markers, even though nothing is ever specified. In a dangerous situation, a fire in a nightclub for example, when people panic and all rush for the exits, this actually increases the amount of time needed to empty the room. Computer simulations of this emergency show that if people move at a calm speed—less than a meter and a half per second—they are able to evacuate in an orderly manner. At speeds greater than this, the people press against one another and friction takes hold; they become locked shoulder to shoulder, "unable to pass through the door even though it stands open in front of them."[20] Interestingly, exactly the same thing can happen in a salt shaker even though none of the grains is bigger than the hole.

So far as traffic is concerned, statistical physics can now help us make sense of how vehicles move and how they clog. The physics of liquids and gases are also helpful, and it has been shown that there are some similarities between how traffic movements develop and bacterial growth.[21] Although this may not solve traffic problems at the drop of a hat, it does help safety measures and more accurate forecasting of tailbacks.

Drivers are assumed to share certain general characteristics, which turn out to be all-important. On the open road they accelerate to a preferred speed, they brake to maintain a velocity-dependent distance from the vehicle in front, and they have imperfect responses that make overreaction likely. This then has to be combined with the identification of three states of flow of traffic, which is very similar to fluid dynamics: free flow (gas), congested (liquid), jammed (solid). The crucial factors in the flow of traffic are speed and density. As traffic density increases, the flow can be maintained if drivers collectively decide to risk maintaining their speed as traffic gets heavier. It is as if all the drivers are on the same cruise control. But this state is fragile and is known to physicists as "metastable"—gases, liquids, and solids can all be metastable. For example, a liquid can be cooled below its freezing point without it suddenly turning rigid. It passes through the usual phase transition, becomes metastable, and is then known as supercooled. But if there is any sudden perturbation, it freezes immediately. Another similarity with physics is that metastability is a one-way affair. A free-flow state can be achieved by increasing traffic density from low values (slower speeds), but not by decreasing it from high values. The same is true of liquids: one can make supercooled water by cooling it, but not by warming up ice.

In traffic in a metastable condition, where there is relatively high speed and high density, it behaves like a metastable liquid—any perturbation can provoke congestion. This is most often caused by one driver suddenly slowing down. That may be because the driver suddenly thinks he or she is too close to the vehicle in front, is making a call on a mobile phone, or gives way to someone joining the highway from a side road. The briefest of disturbances can destroy the flow and precipitate congestion.

A couple of practical possibilities arise from this analysis of the physics of flow. Since the single biggest cause of people suddenly slowing down is other drivers changing lanes, confining vehicles to one lane only in rush hours should prevent perturbations and ease flow. And it suggests that the American system, where people can do any speed in any lane, is better than the European system, where slower vehicles keep in the "slow" lanes. Since most lorries keep to the slow lanes, cars coming up behind them will often be changing lanes. If, however, lorries are allowed in other lanes, then provided they keep to a constant (and relatively high) speed with which car drivers are comfortable, and provided everyone keeps his or her lane discipline, high-speed flow should result.[22]

Much the same has been found with crime and the extent to which it pervades neighborhoods. It is as if neighborhoods can be in metastable conditions too, where sudden changes—"pertur-bations"—in, say, social conditions, or the punishment regimes of offenders, can produce rapid change in which crime suddenly increases or decreases. Ball quotes the changes that took place in various neighborhoods in New York City in the 1990s where, over a five-year period, total crime fell by a half and murders declined by 64 percent. If the metastable analogy with physics holds here, a low-crime area with a small locality harboring a high crime rate poses a threat to the wider society. Metastability in society is a useful concept, providing it can be identified.[23]

One of Ball's conclusions converges on Patricia Churchland's. He refers to William Newmarch, a British statistician addressing the Statistical Society of London in 1860:

The rain and the sun have long passed from under the administration of magicians and fortune-tellers; religion has mostly reduced its pontiffs and priests to simple ministers with very circumscribed functions . . . and now, men are gradually finding out that all attempts at making or admin-istering laws which do not rest upon an accurate view of the social circumstances of the case, are neither more nor less than imposture in one of its most gigantic and perilous forms.[24]

Mathematics and Art

In a series of huge cultural histories, *Exploring the Invisible: Art, Science and the Spiritual* (2002) and *Mathematics + Art* (2015), Lynn Gamwell, at the School of Visual Arts in New York, has explored in particular the influence of mathematical ideas on artistic practices. She traces how modern mathematics and abstract art emerged together, in predominantly German spheres of influence, arguing that German idealism, which led to abstract art, is itself a form of abstraction, which dominated German thought for two centuries at least, from 1750 on. She argues that the "meaning-free colours and forms" of early abstraction and the formality of constructivism parallel the ideas of mathematician David Hilbert, that mathematics is itself a system of self-contained meaning-free signs. She argues that the symbolic logic worked out by Gottlob Frege and Bertrand Russell led to analytic philosophy, "which was expressed by the sculptors Henry Moore and Barbara Hepworth, and authors T. S. Eliot and James Joyce." The search for symmetry—in, say, Einstein's work with mass and energy—inspired Swiss concrete artists, led by Max Brill, "to create art with striking symmetry." And she argues that the mathematical findings of Kurt Gödel, for example that no bedrock of axioms can exist, because there are limits to artificial, symbolic languages, shared common ground with the paradoxical images of René Magritte and M. C. Escher.

Lavishly illustrated, and thoroughly researched, her arguments grow on you.[25]

Neurophilosophy and Outdated Metaphysics: The Mathematics of Muscles

We no longer believe in demons, the four humors, phlogiston, the vital force, or the luminiferous ether. These are outdated prescientific concepts that science has effectively eliminated. Could the same reasoning be applied to our understanding of human nature? Many people believe so, and this is an area that has seen the coming together of two very different disciplines: philosophy and neurology.

The basic idea is this: that our traditional, common-sense understanding of the mind, of human nature, our "folk psychology" as it is called, is prescientific, and has been stagnating for twenty-five hundred years. On this account, even the categories we think in—mental states like beliefs and desires—are misconceived, or simply do not exist. Instead, they will eventually be explained in terms of lower-level neurophysiological processes. After all, ancient, prescientific theories of folk physics, folk biology, and folk cosmology have all been supplanted, and there is no reason to believe in folk psychology just because it has lasted longer. We know that our intuitions are often mistaken.

The unswallowable term given to this approach is "eliminative materialism." At its strongest it contends that such words and concepts as "intend," "love," and "consciousness" do not refer to anything real and will eventually be replaced as neuroscience progresses. In part, this new interdisciplinary approach has seen philosophers taking a long look at neurology, to explore whether their training in various approaches to thinking can combine with the latest findings in brain science, to take us forward.

In some senses, part of this process could have been included in chapter 16 because the philosophers have given some of the clearest accounts of recent brain research, showing how several aspects of psychology have been hardened into neurology. For example, in an earlier book than *Braintrust*, called *Neurophilosophy: Toward a Unified Science of the Mind-Brain*, the same Patricia Churchland gives a clear account of the work by Andras Pellionisz and Rodolfo Llinás and their tensor network theory of how sensory input in an organism is converted into a muscular response.[26] This work shows that if the pattern of neurological input is regarded as one type of (mathematical) vector, and the muscular movement as another type, then there is a mathematical relationship between them that can be calculated (this function is what mathematicians call the tensor). In a second example, she shows that in vision research on amphibia, the input network of neurons forms a more or less two-dimensional grid or map that overlies, literally, a broadly coexistent grid or map that stimulates muscular movement. The links between them allow the organism to respond in space.[27]

All this has to be put alongside copious research that shows how the brain's synapses change—grow or disappear—in response to alterations in the environment. The fact that neurons are known to fire in binary fashion, and that geometry (albeit a very complex geometry) now comprises a feature of brain structure, these are mathematical entities and offer confirmation, for eliminative materialists, that we shall one day be able to dispense with "folk psychology" and move forward.

Many people flatly disagree with this premise and think that, in fact, folk psychology is very useful and, moreover, a tolerably efficient system of explanation. That is one reason why it has survived.

The most intense debate is about "qualia," how the *quality* of yellowness is represented in the neurons, say, or how we distinguish the taste of a plum from an apple with only a limited number of taste buds. Some eliminative materialists say these are semantic problems that will disappear the more we learn. Others believe they are part of a complex whole that we don't even know how to talk about yet.

Advance, if there *is* advance, appears to be piecemeal and haphazard. But perhaps that's what you would expect. Some of the results—such as those of Pellionisz and Llinás—are fascinating, and that must count for something.

Philosophy in an Age of Science: The Nature of Human Nature

The rapid and seismic advances in science, in neurology as much as anything, have brought with them many new political, social, and moral dilemmas: global warming, reliable contraception, overpopulation, obesity, not least the implications of evolution itself. But the dilemma that caps the story that has unfolded in this book—as Patricia Churchland realizes as much as anyone—is the very nature of human nature itself. Now that we are approaching—may already have reached—a time when we can interfere in, amend, even direct the biological order of ourselves, where do we go, what order do we seek? How, even, does that affect (if at all) the science we pursue now? There are four philosophers who may be highlighted as epitomizing those who have thought hardest about the wider effects of recent science on our understanding of who we are and the way that we think: Daniel Dennett, Hilary Putnam, Jonathan Glover, and Jürgen Habermas.

Dennett has prominently considered the issue of free will, a matter that has fascinated—even obsessed—theologians and philosophers for centuries, with names as distinguished as Descartes, Kant, and Schopenhauer all trying to get to grips with the problem. The issue has actually grown more intractable since the successes of science because, in a wholly material world, in which the basic building blocks—particle physics and its interactions—have been shown to behave according to invariant laws (the four forces), "determinism" seems established, and therefore, in theory, it would appear that everything in the universe is, and always has been, set out since the word go. Extremely complicated, of course, but nonetheless inevitable.

Dennett's aim, in *Freedom Evolves*, 2003, is to show that determinism and inevitability or fatalism are not the same thing at all and to do this he emphasizes how freedom itself has evolved. A bird, in flying around in the skies, is freer than a jellyfish, swimming in the narrower confines of the ocean. We humans have more freedom than a "chemically switched bacterium" or a clam that closes up by reflex when something strikes its shell. Adults have more freedom than a small child.[28]

Because we have evolved into self-conscious, self-monitoring agents, because we are language users, we humans are able to reflect consciously and deliberately on alternative courses of action before choosing between them. We live in complex societies, among others who are equally intelligent and reflective, and our knowledge of this equality brings with it a moral sense, a conscience, which is another aspect of freedom not possessed—so far as we know, and so far as we observe—in other creatures.

In other words, the freedom that (we think) we have has been arrived at by a natural process. There is no question that we are freer than the lower animals, so determinism (and evolution) have produced varying degrees of (relative) freedom, and therefore need there be any question about our own? We may not be completely free (whatever that means), but we are freer than other species. Fatalism teaches that human effort makes no difference to what happens, yet clearly we can see that human effort often does make all the difference. We have no need to "interrupt" determinism, Dennett says, by imagining "patches of quantum indeterminancy," which could only produce "spasms of randomness, not freedom." Once you have human thoughts and feelings as evolved parts of nature, you realize that nature has evolved freedom. That that is one evolutionary purpose of thoughts and feelings, to help people decide what their best course of action is.[29]

The philosopher (but also mathematician and computer scientist) Hilary Putnam has ranged more widely than Dennett. We met him in chapter 10, considering the unity of science. In *Philosophy in an Age of Science* (2012), he considered, among many other topics, the implications of quantum mechanics, the Gödel theorem and its lessons for human nature, the status of cognitive science, and the problem of cloning.

Putnam was notorious for changing his mind. This can make it hard to give a short, accurate account of his position, but at the same time it is refreshing for a noted thinker to admit so often in public that he has been wrong in his earlier views. He probably reflects the human condition more accurately than most of us are prepared to admit.[30]

The philosophical study of science, he says, combines the study of nature "and some of the limitations of human reason."[31] He is exercised by whether our "mental states" are "just our *computational* states (as implicitly defined by a 'program' that our brains are hard-wired to 'run')," whether the "software" analogy fails to capture either our intentions or our intuitions, and whether truth can ever be identified under "epistemically ideal conditions," since that may involve "infinitely prolonged scientific inquiry," making such an ideal unattainable.[32] "The truth may outrun what we can as a matter of fact verify." He is likewise exercised by the claims of the Oxford philosopher John Lucas and Roger Penrose, the Oxford-based mathematical physicist, that, essentially, Gödel's theorem, which shows that "non-computational processes" (processes that in principle cannot be carried out by a digital computer, even if its memory space is unlimited) mean that our brains cannot be identical with our minds, "and hence that our minds are immaterial."[33]

Putnam was also very involved in discussions over quantum mechanics, in particular the extent to which our measuring instruments limit what we can know and the extent to which our intellects limit what we can know. Some scientists, he felt, since the quantum revolution, were too ready to accept that, perhaps, the human mind has reached its limit with quantum weirdness, that it is essentially beyond our understanding. Many physicists take the view that, so long as we can use quantum theory to make predictions that are borne out by experiment (as they have been borne out to a very accurate extent, as described in chapter 11), then this is enough. But Putnam was not satisfied by this. He agreed with Einstein that physical theories are not merely formal systems that produce predictions, but that we have to *understand* why the predictions work and what the underlying reality is. In Putnam's view, so long as what he calls the micro-observable world operates on different principles from the macro-observable world, something is missing in our understanding. The wave function in the wave-particle duality, for example, is essentially a mathematical entity, so in what sense may it be said to exist? What can it mean to have statements about unobservables that don't apply to observables? There was, to his mind, a problem with understanding quantum mechanics in a way that is incompatible with scientific realism.[34]

Likewise, what does it mean to say, as some physicists argue, that micro-observables don't exist until they are measured, whereas macro-observables retain sharp values at all times? (He quotes Einstein as saying, on the one occasion they met, that he didn't believe his bed disappeared, or changed its shape or measurements, the minute he left the room.) After all, measuring micro-observables is a certain kind of interaction between a micro-observable and a macro-observable. Could it be, Putnam asks, not entirely rhetorically, that in the quantum world the principle of the conservation of energy is violated but at too small a level to be observed?[35] He argues that there is no absolute distinction between fact and value, even in science. We have to know what facts and observations we can trust, and that depends on our experience and on our reasonable judgments.[36]

Like many philosophers interested in science, Putnam was intrigued by the announcement, in February 1997, by Ian Wilmut and colleagues at the Roslin Institute in Edinburgh, that a sheep, "Dolly," had been successfully cloned. In fact, what Putnam was chiefly worried about, of course, was the prospect of humans cloning children, because there seemed to be no traditional moral principle available to guide people as to what to do. It was, of course, an opportunity for infertile couples to explore a new way of conceiving a child, but it was the opportunities for normally fertile people to "design" their children that, he thought, most filled people with concern—even dread. He thought that the dignity of being human might be jeopardized by cloning and that, as Patricia Churchland also said, natural families are moral images on which society is based, and that that image should be diverse and "unexpected" if it is to be fully autonomous and dignified. He thought that there was one novel human right introduced by the possibility of cloning: "the 'right' of each newborn child to be a complete surprise to its parents."[37]

What Sort of People Should There Be?
The Grown and the Made

The possibility that—at least in theory—we could create people "to order," that we could in effect "play God" in respect of future generations, was also addressed by the Oxford philosopher Jonathan Glover, in two books, *What Sort of People Should There Be?* and *Choosing Children: Genes, Disability and Design.*[38]

Put bluntly, the main questions were: should we try to alter the genetic composition of future generations, how far should we go, who would make the decisions? But Glover showed that the questions don't stop there. Although many people found—and continue to find—the idea of intervention in this way wrong, he reminded us that we have been intervening in human lives for many years, in the form of medical research, anti-poverty programs, tax regimes, and various forms of eugenic breeding patterns. And, while many people would approve of using DNA research to remove disability from the population, this in itself raises the question as to what extent disability is a shortcoming. He quotes the case of a lesbian couple who were both deaf and decided to use the sperm of a man with congenital deafness so that their child would be born deaf. What attitude are we to have to this? Does it contravene "natural order"?

The German philosopher Jürgen Habermas also considered these matters. In his book, *The Future of Human Nature* (2003), he anticipated what he called new forms of "damaged life" that we may be about to inflict on ourselves. He cautions us that here a line may be being crossed, with profound implications for our understanding of freedom, and that it requires a philosophical resolution, not a technical-scientific-psychiatric one.[39]

In the future, children of one generation will be given characteristics by another generation (their parents') that are irrevocable.

What, he asks, will this do to an individual's understanding of him- or herself, his or her sense of being? For Habermas, this new technology blurs the line between the "grown" and the "made," between chance and choice, all of which are essential ingredients in who we are, who we feel ourselves to be. For Habermas, if these processes are allowed to continue then future generations may become *things* rather than *beings*. As he puts it, the ethics of "successfully being oneself" will have been compromised.[40]

For Habermas, not only does this pose a threat to our essential "sense of being," it poses a threat to our capacity to see ourselves as equally free and autonomous as the next individual, to the idea of "anthropological universality," that man is everywhere the same. For Habermas the evolution of the species is a matter for nature. To intervene in this process at the very least marks a new epoch in the history of humanity and perhaps something much worse. Evolution, he insists, should not be a matter of *bricolage*, however well intentioned parents may be.[41]

His worry is that such intervention amounts to nothing less than a third "decentration" of our worldview, after Copernicus and Darwin, so that a person's sense of "I" and his or her understanding of "we" would be changed irrevocably, with incalculable consequences for our shared moral life.

The Greatest Innovation, the Greatest Dilemma

These are big questions, enough to make anyone and everyone breathless. And they won't go away. Our power to intervene in our own evolution is, arguably, the greatest innovation of all time, the greatest dilemma of all, because it is also, in theory at least, the greatest of opportunities. As the convergence between molecular biology, quantum biology, genetics, and behavior becomes ever clearer, as we become more and more informed about the nature of disease, the opportunities for improving the human condition will grow.

Some decisions will be easier than others, or will appear to be.

What reasonable person would not want to see a world in which HIV/AIDS had been eradicated? Yet, at the same time, the world is already overpopulated. . . .

Other decisions will be much more difficult, and, between them, Jonathan Glover and Jürgen Habermas have highlighted several of the most poignant. Tampering with a sense of self, for example, which worries Habermas, probably worries most of us, if we stop to think about it. But this is the point to which the converging sciences have brought us. In one way they clarify or help us recalibrate some of our traditional moral dilemmas. In another way they introduce a whole raft of *new* dilemmas. What sort of people should there be in the future? Who would have thought such a question could even have been posed a quarter of a century ago? Could there *be* a more difficult dilemma?

Conclusion

OVERLAPS, PATTERNS, HIERARCHIES: A PREEXISTING ORDER?

G od is love. Love is blind. Ray Charles is blind. Ray Charles is God. We owe this—shall we say somewhat mischievous?— syllogism to Robert Laughlin. Professor of physics at Stanford University in California, winner of the Nobel Prize in physics for 1998, Laughlin is one of the wisest and wittiest scientists writing at the moment, whose theories about the nature and meaning of science are as profound as they are frequently hilarious. In 2005 he published an important book, whose unwieldy and meaningless title, *A Different Universe: Reinventing Physics from the Bottom Down,* hardly did it justice. Apart from its collection of wise jokes and meaty anecdotes, it offered up a groundbreaking argument that is directly relevant to the story this book has been telling, and especially to where we might go from here. At the same time, Laughlin dumped a hefty shower of cold water and common sense on some cherished ideas of twenty-first-century scientists. (String theory is the study of "imaginary" matter, built out of extended objects, rather than point particles, "as all known kinds of matter have been shown experimentally to be." Theories of a Big Bang origin of the universe are "inherently unfalsifi- able," notwithstanding widely cited supporting "evidence"—his punctuation—such as the cosmic microwave background radiation.)[1] But before we get to Laughlin properly, so to speak, we need to set the scene.

The theme of this book has been the great convergence, the growing interconnections between the various sciences, and the interwoven, interlocking, coherent narrative they tell together. Along the way, we have mentioned several of the main figures who have stressed the intriguing and instructive overlaps of the various disciplines, who have discovered patterns and hierarchies in surprising places. Some of them—Mary Somerville, Hermann von Helmholtz, Albert Einstein, Erwin Schrödinger, Otto Neurath, Rudolf Carnap, Hilary Putnam, Philip Warren Anderson, Edward O. Wilson, Ilya Prigogine, Abdus Salam, Stuart Kauffman, Steven Weinberg—have explored the implications of these interconnections, and what shape any final unity might have and what it could mean. These individuals are called reductionists, as we have seen, and the approach they represent, to remind ourselves, is known as reductionism.

It is important at this point to say that, in recent years, beginning in the 1970s and 1980s, but taking wing in the 1990s, a counter-movement proliferated. It consisted mainly of philosophers, historians, sociologists, and even a few computer specialists, rather than mainstream scientists themselves. This camp argues that reductionism is little short of a sin, that the sciences can never constitute a single unified project—that, in fact, the sciences are a *dis*unity and that it is "imperialistic" and "patriarchal" to say otherwise.[2] These individuals are skeptical that there is a *preexisting* order to the world we see around us. They argue that the "apparent order" is in many ways an artifact of the methods we use to study the world, and they insist that the sciences offer us no privileged view as to how the world "really" works. Instead, there is, as one philosopher puts it, a "promiscuous realism" in the world, an "antecedent awareness of diversity." There are many different ways of understanding the observable world, where no one method or account takes precedence.[3]

In a famous example, the difference between rabbits and hares is highlighted, comparing a hunter's understanding of rabbits, which retire to their holes when sensing danger, and hares, which rely on their speed for safety—to outrun foxes. (No jokes about splitting hares, please.) This, it is claimed, is just as "essential" a difference between the two species as is an evolutionary understanding of their relationship. Moreover, it is more important to the survival of rabbits and hares than is their theoretical position on any evolutionary tree. There is no one God-given way to classify "the innumerable and diverse products of the evolutionary process." Instead, "There are many plausible and defensible ways of doing so," depending on circumstances.

On this view too there are grades of essentialism, depending on purpose and use, and it is pointed out that there are in any case not many sharp divisions in biology, which put normal species, say, into doubt, with many similarities between individuals being due to historical circumstances as much as evolutionary genetic detail.[4] The argument moves on to say that, for example, cats and mountains, dogs and molehills, exist in "just as metaphysically robust a sense as do electrons and quarks (assuming of course that the latter exist at all)."[5]

All of which, it is held, is a reflection of the "exaggerated deference" that is paid to the physical sciences and fits with a "gross exaggeration of their achievements." Everything is made of its own "unique kind of stuff," and there is no reason for us to attach "any kind of ontological primacy to those things that are made of anything." Reduction helps us explain *how* certain kinds of things happen but does not help us to understand *what* a complex thing will do.[6]

Another argument of the disunifiers is that, for example, molecular genetics is too complex to permit reduction. "Reductionism is a local condition of scientific research, not an irresistible tide sweeping the whole of science into an increasingly orderly pattern." "We cannot [build] a theory of complex systems by the addition or aggregation of simple ones."[7] The claim that any sequence of gene frequencies can be fitted to some function is "mathematically trivial" and in any case it is not possible. "Mental states instantiate different levels of abstraction from physical descriptions of neural states so there cannot possibly be a law of nature equating one with the other."[8] "Mental entities—if, indeed, the mental can even be said to constitute 'entities'— would not seem to be made of anything." "The relationship of science and metaphysics in contemporary philosophy is often an unhealthy one. Much contemporary metaphysics has an almost fetishistic reverence for science."[9]

More analytically, we can divide anti-reductionism into the "soft" and the "hard" kind. In the soft kind (Ian Hacking, Richard Creath), it is pointed out that there is, for example, no theoretical link between, say, electronics and cultural anthropology, that they are just too far apart to be evidence for the unity of anything. That though the sciences are unified from time to time in one respect or another, the unity of science is really a "preference for coherence," not a fact about current science. And that James Clerk Maxwell, the greatest of unifiers, really meant "harmony" not singleness.[10] Reference is often made to the "Poincaré-Duhem-Quine" thesis, that it is very difficult to argue that any theory has been falsified based on one experiment alone, because so many alternative explanations are usually available. The soft anti-reductionists argue that there is no consensus on the scientific method—methodologically speaking, it is not unified—and that there is no single "disciplinary matrix" that determines specifically scientific practice. They point out that while unification can work up to a point, so can diversification. In biology, for example, in several universities, micro- and macrobiology are treated differently for administrative purposes. Finally, they say that the unity that mathematics *appears* to confer on science is spurious, because, as Wittgenstein insisted, mathematics is itself a "motley," not one thing.[11]

The "hard" anti-reductionists (John Dupré, Richard Rorty, Peter Galison) use much stronger language. Dupré finds all attempts to privilege science over all other forms of knowledge a species of scientism, a term he intends, he says, to be deliberately "abusive." He rages against the "imperialism" of science, against the "tyranny" and "causal imperialism" of the microphysical. Rorty makes the same point and finds no essential difference between the activities of literary criticism and chemistry, or between biology and morality. Peter Galison argues that particle physics changed fundamentally in the 1960s, transformed by computer science, which created a new kind of nuclear scientist, "neither a theoretician, nor a data-taker, but a data-processor specialized in using computers." In the process, he says, the computer ceased to be a tool and became a substitute for nature itself, artificially bringing together scientists who had never had dealings before. In this context, according to one view, mathematics became the analog of mining diamonds—"finding extraordinary theorems among the dross of uninteresting observations."[12]

As this shows, hard anti-reductionists pass up no opportunity to be rude about science and scientists. Dupré argues that mathematics can be a "mystifying veneer" and that without mathematics, some disciplines might even disappear.[13] As if that is not enough, he adds: "Rejecting all forms of reductionism, and rejecting the assumption of a complete causal nexus, leaves it entirely open how much order there may prove to be in the world. There may be many kinds of phenomena in which no interesting kinds of patterns can be found; and even when there are patterns, it remains an open question how pervasive they may be."[14] He takes the view that laws might be local to populations and their contingent factors—for example, there are some people who do not smoke who nonetheless get lung cancer. On this view, we can never *know*, in any one case, what causes something.[15] And this leads him to conclude one part of his argument: "Thus insistence on contextual unanimity [that cause and effect should apply everywhere and at all times, without exception, if a reductionist account of a cause is to be insisted upon] really does drive us to agree with the tobacco companies that there is no evidence that smoking causes disease." And finally: "Perhaps overarching theories of sociology, human psychology, or meteorology will eventually be accepted. But I see no a priori reason to anticipate or even welcome such developments."[16]

There are various observations to be made about this debate. The first concerns its temperature. It is hot, intemperate, abusive. Why should that be? One can understand certain scientific debates being sharply divided and divisive. The debate over climate change, for example: whether or not the world is warming and whether or not human actions are causing any such warming. Irrespective of the merits of the argument, one can easily see that it potentially affects the lives of millions of people—people who work in fossil fuels, travel and transport businesses, tourism, not to mention people living on islands or in low-lying coastal regions where a rise in sea levels would threaten their very homes. Sea levels have risen and fallen before. It is a potentially existential debate. The same is true concerning the seemingly endless wrangles over race and IQ. People's earning power, their ability to feed their families, their self-esteem, the sheer unpleasantness of the experience of discrimination may all depend on the outcome of this debate, where the very integrity of the measures is itself an important bone of contention.

But the unity or the disunity of the sciences? Does it reverberate on the livelihoods of most of us? The converging story that the sciences tell is amazingly coherent, as I hope to have shown. It is, as was said at the beginning, the greatest story there ever could be and embodies as many pleasures as it does insights. It tells us who we are, and, in some very general ways, there is nothing more important than that. But does it impinge on our lives quite as immediately as climate change, or as unpleasantly as racial differences (if there are any)? Not really.

A clue to the stridency of this debate lies in the language used by some of the hard anti-reductionists, when they speak of the "imperialism" of physics, of their *fear* of this imperialism, the *tyranny* of the microphysical, that it is *hegemonic* and *patriarchal*. These are all familiar terms of abuse used by postmodern critics of essentially modern Western thought processes, deriving mainly from the French author Michel Foucault, whose central message was that "power is omnipresent" in life in general and in Western life in particular. As he himself put it, "Knowledge is always intertwined and mutually reinforcing with relations of power."[17] And, most famously, "Truth is merely a mask for power."

Some of the anti-reductionists do admit that there is in place a "current rage" against reason, of which postmodernism, when it was a force at the end of the twentieth century, was a part, and that there is a "current mood for skeptically undermining the sciences."[18] John Dupré goes so far as to say, repeatedly, that the sciences are "too immature" to be regarded as unified, and that the sheer difficulty of mathematics for many people acts as a barrier to entry to the discipline, which, he insists, "serves to increase the financial and other rewards that accrue to membership of scientific professions."[19]

Given the free-flowing venom sluicing from the postmodern critics of unified science, it therefore comes as something of a shock to find that what they end up concluding is so—and there is no other word for it—lame. The sciences are not unified, they insist, instead they "show a family resemblance," they approximate a rope made up of smaller threads. Others say the sciences are "orchestrated" rather than unified. "Integrated harmony" is allowable, if not singleness. One might wonder what all the fuss and vituperation have been about.

It is perfectly true that we do not have, as yet—and may never have—a unified causal story from the Big Bang 13.8 billion years ago to the Taj Mahal or Carnegie Hall or Sydney Opera House or the Large Hadron Collider. Nor should we allow the search for unity to mesmerize us, as it appears to have mesmerized Einstein near the end of his life, which can lead to a situation where we identify links that aren't really there. But the narrative outlined in this book, as pieced together by overlapping, hierarchical, and clearly patterned and interlocking scientific discoveries, could not exist in the way that it does without reductionism being true at some fundamental—and widespread—level.

A further problem with the anti-reductionist standpoint is that new findings are occurring all the time in science that pointedly counter their argument. We have known for some time, for example, that there are photocells in the eye, known usually as rods, which consist of layers of membranes, which capture photons. We know that bats and dolphins use a form of sonar, involving ultrasound, to find their way around. We know that cows and peas have an almost identical gene, called the histone H4 gene. Its DNA text is 306 characters long, and there is a length in this text that is almost identical in cows and peas—cows and peas differ from each other in only 2 characters out of the 306. The fossil evidence suggests that the common ancestor of cows and peas lived about 1.5 billion years ago.[20] We know that an evolutionary earlier version of the behavioral neurotransmitters oxytocin and vasopressin—vasotocin—played a role in the mating behavior of amphibians. How are any of these processes—at varying levels of complexity—to be explained other than in a reductionist way? Eight species of bacteria have been discovered that exist only on electricity. They generally live in the sediment of seabeds and show that some forms of life can do away with sugary middlemen and handle energy in its purest form—electrons harvested from the surface of minerals.

In the chapter on the hardening of psychology, we showed that magnetic resonance imaging works by showing the different behavior of healthy and diseased tissue under magnetization. SQUIDS, superconducting quantum interference devices, can now pinpoint the sites of brain tumors and the anomalous electrical pathways associated with cardiac arrhythmias and epileptic foci (the localized sources of some types of seizures). And in a direct response to John Dupré's point about smoking, we can repeat the study mentioned in chapter 16, that smokers developed coronary heart disease, *or did not*, depending on their lipoprotein lipase genotype and their apolipoprotein E4

(APOE4) genotype. His objection has already been answered by reductionist science.

Then there is the recently identified phenomenon of reductive evolution, where several parasites that cause diseases have lost their former complexity while specializing to live in simpler niches. All of them retain structures that are now accepted as deriving from mitochondria—these structures are known to biologists as hydrogenosomes and mitosomes.[21] In other words, these organisms have gone from being simple to being complex to being simple again, traveling both ways on the reductive hierarchy.

Finally, we may return again to results reported in May 2015, introduced in the Preface, in which researchers inserted two small silicon chips into the posterior parietal cortex of a tetraplegic individual, ninety-six microscopic electrodes that could record the activity of about one hundred nerve cells at the same time.[22] As noted, the researchers found that they could reliably read out where the patient *intended* to move his arm by analyzing the differing activity patterns of these one hundred cells. They could even predict how fast he wanted to move, and whether he wanted to move his left or right arm. One nerve cell would increase its activity when he imagined rotating his shoulder, and decrease its activity when he imagined touching his nose. The fact that this experiment could throw light on the man's *intentions* is a real challenge for the anti-reductionists, who have simply been overtaken by developments in science. John Dupré, one of the toughest anti-reductionists, says that mental entities "would not seem to be made of anything." Can he mean this? Where would the energy come from to support such entities?

The experiment with the tetraplegic man would appear to show (if replicated) that his intentions—mental entities—were reflected in the patterned activity of just one hundred cells in the brain (the brain contains 86 billion neurons). We can, then, after due consideration, dismiss the anti-reductionist arguments, and move on.

The Emergence of Accuracy: The Most Important Emergence of All?

It is time now to head back to Robert Laughlin, and we do so, first, via Stuart Kauffman, who we met in chapter 18. Based at the Santa Fe Institute, which is devoted in its entirety to an examination of "complex adaptive systems," Kauffman has attempted to draw out Philip Warren Anderson's ideas so far as molecular biology is concerned. His basic contention is that there are in the world around us many examples of "spontaneous order" (chapter 17), that the physics and mathematics of matter provide, to a degree, "order for free." This is close to—but not quite the same as—saying that there is indeed a preexisting order in the world, at some level. Kauffman's main contribution, as was introduced earlier, has been the development of a mathematics of biological order, stemming from his view, already noted, that natural selection, by itself, could never have accounted in the time available for the vast amount of organized diversity we see around us. As was mentioned in an earlier chapter, Kauffman argues that life originated by a process of the collective autocatalysis of primary molecular structures similar to enzymes. On this basis, he argued, life was not just a chance occurrence, but was to be *expected* from the principles of spontaneous order.

In a later book, *Reinventing the Sacred: A New View of Science, Reason, and Religion*, Kauffman argues that *emergence* is the most important and radical new scientific view, moreover a heretical view which helps us beyond the "newly discovered" limits to reductionism, to explain "the furniture of the universe."[23] But he still entertains hope of unification, even if it is a new unification. The principles of emergence, he says, "may again be able to link matter, energy, and information into a unified framework."[24]

Aspects of this vision are exciting (spotting genuine overlaps and interconnections is pleasurable in itself, as Einstein noted). But there are two caveats. Kauffman presents himself as a heretic in the world of science, but he is in fact just a little too pleased with being a heretic to actually *be* one. Real heretics spend years in the wilderness, not at a well-funded institute on the edges of the New Mexico desert. He is not so much a heretic as an evangelist for what is itself emerging as a new orthodoxy. More important still, perhaps, his view that the discovery of how life originated was "just around the corner" is not as impressive-sounding now as it was in the early 1990s, when he first made this claim. On top of which, his mathematization of molecular biology remains just that—in effect, a form of theoretical biology with no practical or wider spin-offs yet. (For example, his observation that the number of types of cell in an organism is the square root of the number of genes is tantalizing but, so far, no more than that.)

Steven Strogatz has shown that many aspects of the natural world do indeed evince spontaneous order—from fireflies that wink in the night in concert, to spontaneous synchronized clapping in auditoria, to menstrual synchrony among nuns, female prisoners, girls' school roommates, and so on. And that there are spontaneous patterns in traffic, and that hearts sometimes lose their synchronicity, with catastrophic effects. Spontaneous order is therefore established and seems to confirm Gaylord Simpson and Anderson's original point that there are laws of constructivism, over and above the laws of elementary particles.[25] This is a crucial point.

The Age of Emergence: Another Watershed Moment

But what then?

Like Anderson, Robert Laughlin begins with reductionism. "All physicists are reductionists at heart, myself included. I do not wish to impugn reductionism so much as to establish its proper place in the grand scheme of things. . . . All of us secretly wish for an ultimate theory, a master set of rules from which all truth would flow. . . . [Our] concern for ultimate causes gives theoretical physics a special appeal even to nonscientists, even though it is by most standards technical and abstruse. . . . Physicists assume the world is precise and orderly."[26] And he goes on, making it clear that "the natural world is regulated both by the essentials and by powerful principles of organization that flow out of them. These principles are transcendent. . . . The laws of nature that we care about . . . emerge through collective self-organization and really do not require knowledge of their component parts to be comprehended and exploited."[27] This is the most important point of all, Laughlin thinks, and, because of this, he argues that we are at the end of an era, and that the "frontier" of reductionism has been "officially closed."[28]

The natural world can still be looked at as an "interdependent hierarchy" of descent, and this state of affairs is in fact "why the world is knowable." But there is a further all-important link between precision, organization, and order. It is part of a process where all areas of life—economics no less than psychology or quantum biology—are getting "harder," by which he means more accurate and predictive. The speed of light in a vacuum is now known to an accuracy of better than one part in 10 trillion, atomic clocks are accurate to one part in 100 trillion.[29]

Even the phases of matter—solid, liquid, vapor—are organizational phenomena. The property we value in any case is order. And this is where the statistical nature—the uncertainty—of elementary particles becomes more certain, as phenomena get larger and more complex, the important point being that microscopic rules can be perfectly true and yet quite irrelevant to macroscopic phenomena.[30] To repeat: *emergent exactness, growing out of the uncertain, probabilistic statistical nature of particles, may be the most important emergence of all.* The link between the statistical uncertainty of elementary particles and organized accuracy to a fantastic degree *is spontaneous*: a function of the way molecules *naturally* organize themselves.

Laughlin puts the watershed moment for this new view of nature at 1980, when the German physicist Klaus von Klitzing, working at the High Magnetic Field Laboratory in Grenoble, discovered a very high level of reproducibility in certain electronic measurements. Moreover, they were stable across a range of magnetic field strengths, and produced a resistance, he found, that was a combination of three fundamental constants: e, the electron charge, h, the Planck constant, and c, the speed of light. This convergence was an amazing feat in itself, but it was the extreme and astonishing accuracy and the reproducibility—the reliability—of the measurements that were important and which attracted global attention among physicists. The measurements of the resistance were accurate to one part in 10 billion—that was like counting everyone in the world and not making an error. No less astonishing was the fact that this remarkable reliability completely disappears if the sample is too small. This also led to the discovery of new phases of matter, one of which carried an exact fraction of e, the electron charge—$e/3$ as it happened. This fitted exactly with the charge carried by quarks.[31] More unlooked-for convergence, the mathematics of which would certainly have thrilled Mary Somerville.

How Nature Organizes Herself

The point of these experiments, and why they were a watershed, as Laughlin puts it, was because they showed that it was the *organization* of fundamental entities, their order, which produced the new phases, which had not been anticipated. This was a defining moment, Laughlin said, "in which physical science stepped firmly out of the age of reductionism into the age of emergence. This shift is usually described in the popular press as the transition from the age of physics to the age of biology, but that is not quite right. What we are seeing is a transformation of a worldview in which the objective of understanding nature by breaking it down into ever smaller parts is supplanted by the objective of understanding how nature organizes herself."[32] The principle of organization—emergent order—is at work in superconductors, as it is with sound, and with vacuum properties.[33] The same is true of the properties being seen in nano-engineering. Again, this is close to saying that there is a preexisting order to the world. The phenomenon of spontaneous order is built into nature.

And, after all this, it is reasonable to ask, says Laughlin, at what scale collective principles of organization begin to matter in life. His answer is that there is considerable circumstantial evidence that both stable and unstable emergence occurs already at the scale of individual protein molecules.[34] Proteins are big molecules, and in order to work effectively they need something analogous to mechanical rigidity, "an emergent property that occurs only in systems that are large." And the folding patterns of proteins, about which so much has been heard, shows that they cannot depend on interatomic forces, because otherwise they would have to conform to the equations of motion. Large-scale biological processes, like metabolism and gene expression, all depend instead on collective principles.

A person obsessed with ultimate truth, says Laughlin in another of his caustic asides, is a person asking to be relieved of money. Yet he did himself attend a daylong Interdisciplinary Workshop on Emergence held at Stanford University where the other members were equally distinguished and included Carl Djerassi, inventor of the birth control pill, the well-known cosmologist Andrei Linde, and the French-American anthropologist Denise Schmandt-Besserat, whose ideas about the birth of writing in the Middle East are close to the orthodoxy. After brainstorming, rambling over coastal hills, and drinking boxes of wine (boxes?), this group came up with, among other things: "Emergence means complex organizational structure growing out of simple rules. Emergence means stable inevitability in the way certain things are. Emergence means unpredictability, in the sense of small events causing great and qualitative changes in larger ones. Emergence is a law of nature to which humans are subservient."[35]

He didn't disagree with any of this, but put it himself this way: "The laws of quantum physics, the laws of chemistry, the laws of metabolism and the laws of bunnies running away from foxes in the courtyard of my university all descend from each other, but the last set are the laws that count, in the end, for the bunny."[36]

We have reluctantly to concede that Newton wasn't entirely right when he said that nature is pleased by simplicity. If Putnam and Laughlin and Kauffman and Prigogine and Anderson and Gaylord Simpson are right, there is a major hinge in nature that occurs between elementary particles and molecules, particularly organic molecules. The latter are spontaneously organized in a way that appears to have little to do with the properties of the particles themselves. Though they do not contravene those properties, the information in the particles does affect the behavior and form of the molecules, as the quantum nature of mutation suggests (chapter 18). And it is this spontaneous organization that has given rise to the diverse—and yet related, ordered—complex systems that comprise the world. It is in the nature of molecules, as Schrödinger was the first to point out, that the uncertainty of the quantum world becomes the certainty of the world as it appears to our senses. This is why organic molecules are so large and consist of so many atoms. And the spontaneous organization—which runs through nature—is why convergence exists and explains its importance. Convergence is evidence of spontaneous order.

Energy and Evolution, Again

One final step in the argument relates to Kauffman's wished-for unification of matter, energy, and information. All good stories should end, in some sense, back where they began, and this one aims to be no different.

The beginning of the end of this story—so far—took place in May 2010, when a team of Japanese biologists, trawling the Pacific waters near an underwater volcano named Myojin Knoll, collected some polychaete worms clinging to a hydrothermal vent. Such hydrothermal vents had first been discovered in the late 1970s, a seminal moment for geologists and oceanographers but also for biologists, on account of the discovery of strange forms of life at great depths, surviving enormous pressures in complete darkness, far away from the sun.

The polychaete discovered by the Japanese biologists, and which they named *Parakaryon myojinensis*, was large for a bacterium, and on closer inspection, it proved difficult—impossible—to classify. They themselves described it as a "parakaryote," signifying their view that it was intermediate in form between a prokaryote and a eukaryote (see chapter 12). For the British biologist Nick Lane, however, the new organism was much more interesting than that. He observed that it had a cell wall, a nucleus surrounded by a nuclear membrane, and several endosymbionts, internal features, and it struck him that it might be a prokaryote that had acquired endosymbionts and was "changing into a cell that resembles a eukaryote, *in some kind of evolutionary recapitulation*." (Italics added.)[37]

It is not hard to see why Lane was so excited. All biologists, not to say many of the rest of us, are interested in how and where and why life began, and whether it could do so again, either here on earth or on some other planet. More than this, if evolution were to begin again, would it take a different course? The idea that *Parakaryon myojinensis* was replaying one of the most important early evolutionary transitions was clearly of the first importance.[38]

Unfortunately, only one specimen of *P. myojinensis* has ever been found and so we know no more than this. Being so rare, it may already be extinct. (Although another microbe, found in the murky depths of the North Atlantic and published in May 2015, looked like another transitional form, and also an ancestor of the eukaryotes.)[39]

Nonetheless, *P. myojinensis* was also of particular interest to Lane since he is part of a group of biologists who have been exploring the origins of life in a research initiative that has been overshadowed by other areas of molecular biology but might, conceivably, be fundamental. The argument among this group is that the orthodox view that life originated when a "soup" of chemicals was struck by lightning, and turned into organic matter (amino acids), just doesn't stand up. Lane himself once calculated that the evolution of photosynthesis alone "would require four bolts of lightning per second, for every square kilometre of ocean. . . . There are just not all that many electrons in each bolt of lightning."[40]

Instead, his view, shared by numerous colleagues, and now thought through to the *atomic* level (anti-reductionists, please note), and backed up by a number of experiments and observations, is that life on earth began in these hydrothermal vents, deep in the ocean at the edges of the great tectonic plates, which are the areas where energy (radioactive decay) escapes from the center of the planet, in the form of heat. The vents come in two varieties, black smokers and alkaline vents, and can form great chimneys, some the size of twenty-story skyscrapers. (One, discovered in the mid-Atlantic Ridge in 2000, was dubbed the "Lost City" because of its huge calcium carbonate chimneys and towers.)[41] It is in the alkaline vents that Lane and his colleagues think life began.

This interests us because, as Lane puts it, "energy is central to evolution," and this of course leads us back to where we came in, to the two great unifying theories of the 1850s. Lane's idea, reflecting the views of many colleagues, is that the origin of life was driven by energy flux, that the phenomenon of proton gradients was central to the emergence of cells, and that their use constrained the forms of living things. The importance of proton gradients was first pointed out by the eccentric British biologist Peter Mitchell. They form most commonly during cell respiration, when the concentration of protons (H^+) is higher on one side of a lipid membrane than the other, and they act in a way analogous to water on one side of a hydraulic dam, driving turbines to produce electricity, a form of energy that Lane calls "proticity."

Why this matters is that the oceans 4 billion years ago were richer in CO_2 than they are now, making the water slightly acidic.

And so, when the alkaline solutions of the vents met the acidity of the oceans, a similar proton gradient would have automatically occurred. Moreover, this didn't happen only in the open seas but in "micropores," minuscule pockmarks that occur in the substance of the alkaline vents.[42] It is known that lipids form vesicles spontaneously, especially if confined in micropores, so this may tell us how and where the first mineral-rich solutions were concentrated and where and how breathing organic cells formed. Geochemistry gave rise to biochemistry, with protons (H^+) combining with CO_2, which is otherwise unlikely. It also explains why iron sulphide is found at the heart of respiratory enzymes. Iron and sulphur are common in the vents and these too would have helped create energy gradients.[43]

Lane also tells us that the early bacteria that would have emerged in this way didn't in fact evolve any further for 2 billion years—they stayed simple. The next change, as was discussed in chapter 12, was when one cell incorporated another inside it (an endosymbiosis), to create a eukaryotic cell, in which the endosymbionts became mitochondria and eventually gave rise to all other forms of complex life—plants, animals, fungi, and so on. Here again, the significant fact was that mitochondria (which are now emerging as far more important than anyone hitherto thought) enabled eukaryotic cells to metabolize far more energy than prokaryotes. On one calculation, eukaryotic cells can support a genome *five thousand times* larger than bacteria.[44] Complexity would not be possible without mitochondria to act as "batteries" for cells, providing copious energy.

There is much more to this "Copernican revolution," as it has been described. (This overused cliché may in this case be merited.) We now think we know why mitochondria are inherited only from the mother, why some animals live longer than others, why there are two sexes rather than more, all of which may well support the contention among many biologists that Peter Mitchell's original notion of a proton gradient, for which he received the Nobel Prize in 1978, is a biological idea that ranks with those of

Einstein, Heisenberg, and Schrödinger in physics and is the most "counterintuitive" idea in biology since Darwin.

If these ideas are confirmed, they too will add to our ideas of the converging order of the world. But the important basics to remember about the historical biological order are that it was a particular form of energy that kick-started life in the first place, and a second innovation in bioenergetics that enabled complex life to begin to form and proliferate, after 2 billion years of not very much going on. This two-stage process of bioenergy may well become the new orthodoxy.[45]

A Step Change in the Understanding of Order

It may need repeating just once more that these biological processes are now understood at the atomic level. The anti-reductionists are correct in saying that we don't know *why* the world takes the form that it does, but they can no longer hide from the fact that we understand increasingly well *how* the world works and that the links between the sciences are growing inexorably. Lane's work encompasses an intimate mix of geology, oceanography, physics, chemistry, and biology—far more so than Mary Somerville could have even begun to anticipate. And it supports Patricia Churchland's remark that "where one discipline ends and the other begins no longer matters."

The "why" question may in any case be a misleading chimera. The story told in these pages is not a straight line, as was noted at the very beginning, but it *is* a line, a narrative, which hangs together, and is not a mere artifact of the instruments with which the observations have been carried out. The coherence of the convergence is just too wide for that to be true. There *is* an order to our world, and how we got here. It is all the more amazing and interesting for being put together piecemeal, without any preconceived ideas.

And it is the convergence, the overlaps and interconnections— the unifying findings and their implications, the patterns and the hierarchies, the indications of a preexisting order—that hold our interest. That is why the arguments that run from Arthur Eddington, George Gaylord Simpson, Philip Warren Anderson, and Ilya Prigogine to Stuart Kauffman and Robert Laughlin are so welcome and so important. The very idea of amalgamating spontaneous order and self-organization together into the concept of emergence means nothing other than that those of us alive today are experiencing a step change in our understanding of order and its place in our lives.

We also know now that, if a preexisting order *is* to be glimpsed, recent developments—as Stuart Kauffman put it—mean that it will link matter, energy, and information. That would be the ultimate convergence.

Will More Sciences "Harden Up"?

We have already seen how one of the "softer" sciences, psychology, has hardened up in the last fifty years or so. Perhaps now, with this new perspective, some of the other, softer sciences—sociology, economics, anthropology, criminology—will become better organized, producing results that are more predictive. (Genetic fingerprinting has already revolutionized criminal detection, while Adam Benforado at Drexel University in Philadelphia proposes among other things using fMRI to screen jurors for inherent biases towards certain groups.) Will the ever-improving calculational capacity of computers, the proliferation of big data sets, some of them put together relatively cheaply using social-media outlets, combine to uncover ever-higher order in our teeming world?

For example, mathematicians, physicists, and psychologists have all examined various aspects of capitalism. One study (with an admittedly small sample) found that merchant bankers are more dishonest than the general population (no surprises there, perhaps). But if there is an overriding focus it has been what *Science* magazine, in a special issue, called "The Science of Inequality." This stems from the realization, belated maybe, that under capitalism, except for a few decades following the two world wars in the twentieth century, when many advanced industrial urbanized states were financially on their knees, the basic economic order of those countries has been for a growing wealth disparity within populations.[46]

This finding—which applies to many countries—appears solid and has emerged from a fresh wave of big data, examining tax returns for the past *two centuries*. In its potential usefulness, and with only slight exaggeration, it has been likened to other big data sets such as the Human Genome Project, which is itself producing a new order in medicine: treatment personalized according to someone's genetic makeup. Is this where science is going, or expected to go? The richness of the tax-return data means that, in the words of *Science*, the "stuff of science" can be applied to it—analysis, extracting causal inferences, formulating hypotheses.[47]

One explanation for the main finding—of evolving inequality—has been produced by mathematically sophisticated economists, who argue that, throughout history (except for the postwar period mentioned above), the rate of return to capital has always been higher than the overall rate of economic growth over the long run (currently, for example, 4 to 5 percent versus 2 percent). Therefore, holders of capital will inevitably extend their wealth more than people who must work for a living.[48]

Mathematicians have a different theory, though one that is no less gloomy. Victor Yakovenko, a theoretical physicist at the University of Maryland, likens income distribution to the entropy in gases (entropy as an analogy again). Wherever you have random activity, he says, you are bound to end up with an exponential (rather than an equal) distribution of whatever entity you are considering, as with the distribution of energy levels of different molecules in a gas. In other words, the very nature of a free market, which embodies elements of randomness, is bound to produce the inequalities we are seeing.[49]

The question now is: what attitude are we to have to such

results? Are people expected to accept such conclusions as having the imprimatur of scientific "laws," with the inevitability and inescapability implied? The recent furor over the scientific evidence for climate change, with accusations of deception and misrepresentation being thrown in all directions, suggests otherwise. Do the recent findings in fact take us any further than Adam Smith's metaphor of the "hidden hand" driving market capitalism?

Other sciences have examined the effects of great inequality on the physical and psychological well-being of those on the wrong side of the divide, and have shown those effects to be quite substantial. In time, policies based on these studies may well ameliorate some of the unfortunate effects, but they seem unlikely to overcome all of the deficits identified by the economists. Some of them have predicted that growing inequality will lead to social unrest, but in the poorer countries, mainly in Africa (where income inequality can be just as marked as in the more developed West), being on the wrong side of the divide is just as likely to lead to disease.

It makes uncomfortable reading, but the recent advances in the mathematics of order may suggest that the social problems we see around us are not so much the result of the *failure* of certain political policies, but are structural problems *built in* to the system. Or not only that. They are, if you like, a result of the preexisting order ingredient to the world. For many, perhaps for all of us, this is a difficult conclusion, and we can agree with Philip Ball when he says that one does not build an ideal world from scientifically based traffic planning, market analysis, criminology, network design, or game theory. But at the same time we must register what he also says: that science does show that the world of human social affairs is not open to all options. "The issue," he concludes, "is really whether we can trust ourselves to distinguish between moral and physical law," and that if scientific theory were to become a justification for moral choices "it has exceeded its brief."[50]

But is that true anymore? Science is advancing all the time, and so is the convergence. In his recent book *The Master Algorithm*, Pedro Domingos considers whether machines that learn might soon be able to discern patterns—order—in the masses of data that we now produce.[51] "Depending on your world view, the development of a master algorithm is either really thrilling or downright scary."

It is certainly true that understanding the laws of fundamental particles does not allow us to predict the course of evolution over millions of years, or the behavior of the rabbits in the courtyard of Professor Laughlin's university. At the same time, the great convergence of the sciences continues, as Nick Lane's research shows, as the advent of quantum biology shows, and the convergence is spreading to newer areas of our world, as the physics of traffic shows, and as the new exploration of morality shows. Moreover, if the likes of David Deutsch and Stephen Wolfram and Joseph Silk are right, and our powers of computation continue to grow, as seems likely if quantum computers really are just around the corner, then some of the more outlandish ideas we have encountered will not seem so far away as once they did. Can a master algorithm be ruled out for all time?

Thrilling or scary, to insist that science has nothing to say about morals or politics or social affairs is surely to beg the question, and it may be to avoid or delay facing difficult issues. If, to take just one example, wealth inequality *is* a function of free-market capitalism, if that view were to become convincing to a critical mass of people, what then? Would this newfound conviction spark radical thought processes designed to shake us out of what would in effect be a new reality?

What we can at least say is that new ideas about higher-order manifestations are urgently needed. Convergence implies order, order implies ever more accurate prediction, as Robert Laughlin says, and accurate prediction is the lifeblood of science. It has taken a while to arrive at the relatively new science of order, but it holds out great promise. Let us hope it is soon realized.

ACKNOWLEDGMENTS

Among the individuals whose brains I have picked to produce this book, and to whom I offer my sincere thanks, are the following: David Ambrose, Anne Baring, Peter Bellwood, Jonathan Cox, David Henn, Andrew John, Mike Jones, Thomas LeBien, Gerard LeRoux, George Loudon, Ingborg Müller, Brian Moynahan, Andrew Nurnberg, Nicholas Pearson, Navaratna Rajaram, Chris Scarre, Allan Scott, Robin Straus, Randall White, and David Wilkinson. Any errors or solecisms that remain are all my own work. I would also like to thank the authors of a number of books that have been produced recently and, between them and others, amount to nothing less than a golden age of scholarship in the history of science. First and foremost, my thanks go to John Gribbin, probably the most prolific and imaginative of science historians writing at the beginning of the twenty-first century, whose works range from general histories to specific studies of such entities as the Royal Society and the quantum. Helge Kragh, Paul Davies, and Philip Ball have between them explored thoroughly both the science *and* the institution of physics. Ian Inkster, Margaret Jacob, and Joel Mokyr have done much to reorient sensibly the relationship between science and technology in works that have inspired this one. As have Steven Shapin's ideas on the changing nature of science down the years. James Secord has enlivened the Victorian scientific scene more than anyone else.

Most of all perhaps, a debt is owed to those who have thought most deeply about order—order in the universe, among elementary particles, among molecules, in complex systems, in living things, in the brain, in mathematics: Philip Warren Anderson, David Bohm, Patricia Churchland, Brian Greene, George Johnson, Stuart Kauffman, David Knight, Robert Laughlin, Jacques Monod, Sebastian Seung, Edward O. Wilson. Thanks also to the many other authors whose works are referred to in the pages above and to the staffs of the London Library and the Haddon Library in Cambridge.

Having myself written a number of works in the history of ideas, I have covered some of the material in this book before in other titles (and in other ways). These overlaps are indicated at appropriate points in the Notes, but in all cases the material here has been updated and expanded.

NOTES AND REFERENCES

Preface: Convergence: "The Deepest Idea in the Universe"

1. Ruth Moore, *Niels Bohr: The Man and the Scientist*, London: Hodder & Stoughton, 1967, p. 51.
2. Helge Kragh, "The Theory of the Periodic System," in A. P. French and P. J. Kennedy (eds.), *Niels Bohr: A Centenary Volume*, Cambridge, MA, and London: Harvard University Press, 1985, p. 62.
3. C. W. Ceram, *The First Americans: A Study of North American Archaeology*, New York: Harcourt Brace Jovanovich, 1971, p. 126.
4. A. E. Douglass, *Climatic Cycles and Tree Growth*, Volumes 1–11, Washington D.C. Carnegie Institution, 1936, pp. 2 and 116–22; see also Ceram, *The First Americans*, p. 128.
5. John Bowlby, *Child Care and the Growth of Love*, London: Penguin Books, 1953.
6. Ibid., pp. 161ff.
7. Bowlby eventually wrote three large and important books on these studies: *Attachment and Loss: 1: Attachment* (1969); *Attachment and Loss: 2: Separation* (1973); and *Attachment and Loss: 3: Loss, Sadness and Depression* (1980). He maintained the ethological and evolutionary approach throughout, arguing that the behavior he and Ainsworth (and by now many others) were observing related to the evolutionary adaptedness of hunter-gatherers, the ultimate function of attachment being survival.
8. Bowlby, *Child Care*, pp. 181ff.

Introduction: "The Unity of the Observable World"

1. Allan Chapman: *Mary Somerville and the World of Science*, Bristol: Canopus, 2004, p. 23.
2. Chapman, *Mary Somerville*, p. 23. See also: Kathryn A. Neeley, *Mary Somerville: Science, Illumination and the Female Mind*, Cambridge, UK: Cambridge University Press, 2001.
3. Neeley, *Mary Somerville*, p. 23.

4. Chapman, *Mary Somerville*, p. 27.

5. Elizabeth Chambers Patterson, *Mary Somerville and the Cultivation of Science, 1815–1840*, Boston and The Hague: Martinus Nijhoff, 1983, p. 331.

6. Patterson, *Mary Somerville*, p. x.

7. Chapman, *Mary Somerville*, p. 38.

8. Mary Somerville, *Collected Works of Mary Somerville*, edited and with an introduction by James Secord, Bristol: Thoemmes Continuum (9 volumes), 2004, Volume 4, p. xxviii.

9. Patterson, *Mary Somerville*, p. 123.

10. James Clerk Maxwell, "Grove's 'Correlation of Physical Forces,'" *Nature* 10, no. 3024 (August 20, 1874): 303. Quoted in Secord (ed.), *Collected Works*, p. 266.

11. Joanna Baillie to Mary Somerville, February 1, 1832, in Martha Somerville (ed.), *Personal Recollections, from Early Life to Old Age, of Mary Somerville*, London: John Murray, 1873, p. 206; quoted in Secord (ed.), *Collected Works*, p. 267.

12. Secord (ed.), *Collected Works*, p. 413.

13. Neeley, *Mary Somerville*, pp. 35–37.

14. Ibid., p. 40.

15. Ibid., pp. 2–3.

16. Secord (ed.), *Collected Works*, p. 59.

17. Secord (ed.), *Collected Works*, Vol. 4, p. viii.

18. Ibid.

19. Neeley, *Mary Somerville*, pp. 121–23.

Chapter 1: "The Greatest of All Generalizations"

1. Iwan Rhys Morus, *When Physics Became King*, Chicago and London: University of Chicago Press, 2005, p. 63. Thomas Kuhn, *The Essential Tension, Selected Studies In Scientific Tradition and Change*, Chicago and London: University of Chicago Press, 1977, p. 68.

2. P. M. Harman, *Energy, Force and Matter: The Conceptual Development of Nineteenth-Century Physics*, Cambridge, UK: Cambridge University Press, 1982, p. 144. J. C. Poggendorff, *Annalen der Physik und Chemie*, Leipzig: J. A. Barth, 1824.

3. Harman, *Energy, Force and Matter*, p. 145.

4. Dan Charly Christensen, *Hans Christian Ørsted: Reading Nature's Mind*, Oxford: Oxford University Press, 2013, p. 4. Richard Holmes, *The Age of Wonder: How the Romantic Generation Discovered the Wonder and Terror of Science*, London: Harper Press, 2008, p. 208.

5. Thomas S. Kuhn, *The Essential Tension*, pp. 97–98.

6. Crosbie Smith, *The Science of Energy: A Cultural History of Energy Physics in Victorian Britain*, London: Athlone, 1998, p. 9. John Theodore Merz, *A History of European Thought in the Nineteenth Century*, New York: Dover, 1904 and 1965, Volume 1, pp. 23 and 204.

7. Crosbie Smith, *The Science of Energy*, p. 72.
8. Ibid.
9. John Gribbin, *Science: A History: 1543–2001*, London: Allen Lane, 2002, p. 586.
10. Crosbie Smith, *The Science of Energy*, p. 110.
11. Ibid., p. 8.
12. Ibid., p. 74.
13. Ibid.
14. *Dictionary of Scientific Biography*, III, pp. 303–310.
15. http://www-history.mcs.stand.ac.uk/Biographies/html.
16. Harman, *Energy, Force and Matter*, p. 149.
17. Morus, *When Physics Became King*, p. 65.
18. L. Campbell and W. Garnett, *The Life of James Clerk Maxwell*, London: Macmillan, 1882, p. 143.
19. Basil Mahon, *The Man Who Changed Everything: The Life of James Clerk Maxwell*, London and New York: Wiley, 2003, p. 2; Raymond Flood et al. (eds.), *James Clerk Maxwell, Perspectives on His Life and Work*, Oxford: Oxford University Press, 2014, p. 241.
20. Flood et al., *James Clark Maxwell*, p. 14.
21. Mahon, *The Man Who Changed Everything*, p. 36.
22. Flood et al., *James Clark Maxwell*, p. 14.
23. Mahon, *The Man Who Changed Everything*, p. 69.
24. Ibid., p. 61.
25. Gribbin, *Science*, p. 432; Flood et al., *James Clark Maxwell*, pp. 190–91.
26. Flood et al., *James Clark Maxwell*, p. 279.
27. C. Jungnickel and R. McCormmach, *The Intellectual Mastery of Nature*, Volume 1, p. 164. Quoted in Morus, *When Physics Became King*, p. 147. See also: Yehuda Elkana, *The Discovery of the Conservation of Energy*, London: Hutchinson, 1974.
28. Harman, *Energy, Force, and Matter*, pp. 148–50. Engelbert Broda, *Ludwig Boltzmann: Mensch, Physiker, Philosoph*, Vienna: Franz Deuticke, 1955, pp. 57–66 and 74ff for his views on heat death.
29. Carlo Cercignani, *Ludwig Boltzmann: The Man Who Trusted Atoms*, Oxford: Oxford University Press, 1998, especially pp. 120ff, for the statistical interpretation of entropy. This book also contains some amusing cartoons of Boltzmann by Karl Przibram.
30. Carl Boyer, *A History of Mathematics*, second edition, revised by Uta C. Merzbach, New York: Wiley, 1991, p. 497.

Chapter 2: "A Single Stroke Unifies Life, Meaning, Purpose, and Physical Law"

1. Richard Holmes, *The Age of Wonder: How the Romantic Generation Discovered the Beauty and Terror of Science*, London: Harper Press, 2008, p. 60.

2. Ibid., p. 65.

3. Ibid., p. 71.

4. Ibid., p. 81.

5. Ibid., p. 87. Sir William Herschel, *The Herschel Chronicle: The Life Story of Sir William Herschel and His Sister Caroline Herschel*, edited by Constance A. Lubbock, Cambridge, UK: Cambridge University Press, 1933.

6. Holmes, *The Age of Wonder*, p. 88.

7. Ibid., p. 91. For Flamsteed, see: *Flamsteed's Stars: New Perspectives on the Life and Work of the First Astronomer Royal, 1646–1719*, edited by Francis Willmoth, Woodbridge, Suffolk: Boydell Press in association with the National Maritime Museum, London, 1997.

8. Holmes, *The Age of Wonder*, p. 102.

9. Ibid.

10. Ibid., p. 112.

11. Ibid., p. 123.

12. Ibid., 192. For Laplace, see: Charles Coulston Gillispie, with the collaboration of Robert Fox and Ivor Grattan-Guinness, *Pierre-Simon Laplace, 1749–1827, a Life in Exact Science*, Princeton, NJ: Princeton University Press, 1997.

13. Holmes, *The Age of Wonder*, p. 203.

14. Maria Rosa Antognazza, *Leibniz: An Intellectual Biography*, Cambridge, UK: Cambridge University Press, 2009, p. 177.

15. Peter Bowler, *Evolution: The History of an Idea*, Berkeley: University of California Press, 1989, p. 40.

16. Mott T. Greene, *Geology in the Nineteenth Century: Changing Views of a Changing World*, Ithaca, NY: Cornell University Press, 1982, p. 36; Abraham Gottlob Werner, *Kurze Klassifikation und Beschreibung der verschiedenen Gebirgsarten*, 1789, republished: New York: Haber, 1971, with an introduction by Alexander Ospovat; Rachael Laudan, *From Mineralogy to Geology: The Foundations of a Science: 1650–1830*, Chicago and London: University of Chicago Press, 1987, pp. 48ff. For Werner's theory of color, see: Patrick Syme, *Werner's Nomenclature of Colours*, Edinburgh: W. Blackwood, 1821.

17. Charles Gillispie, *Genesis and Geology: A Study of the Relations of Scientific Thought, Natural Theology, and Social Opinion in Great Britain, 1790–1850*, Cambridge, MA: Harvard University Press, 1951, p. 48; Jack Repcheck, *The Man Who Found Time: James Hutton and the Discovery of the Earth's Antiquity*, London: Simon & Schuster, 2003, who says that Hutton's prose was "impenetrable" and that, at the time, people were not very interested in the antiquity of the earth: http://www-groups.dcs.stand.ac.uk/history/Printonly/Hutton James.html.

18. Gillispie, *Genesis and Geology*, pp. 41–42.

19. Ibid., p. 68.

20. Ibid., p. 84.

21. Ibid., p. 101.

22. Bowler, *Evolution*, p. 116.
23. Ibid., p. 110.
24. http://www.ucmp.berkeley.edu/.history/Sedgwick/html. For Buckland, see Nicolaas A. Rupke, *The Great Chain of History: William Buckland and the English School of Geology (1814–1849)*, Oxford: Clarendon Press, 1983.
25. Gillispie, *Genesis and Geology*, p. 133.
26. Bowler, *Evolution*, p. 138.
27. Ibid., p. 130.
28. Peter J. Bowler, *The Non-Darwinian Revolution*, Baltimore and London: Johns Hopkins University Press, 1988, p. 13.
29. James A. Secord, *Victorian Sensation: The Extraordinary Publication, Reception, and Secret Authorship of "Vestiges of the Natural History of Creation,"* Chicago and London: University of Chicago Press, 2000, p. 388; p. 526 for the publishing histories of *Vestiges* and the *Origin* compared.
30. Edward Lurie, *Louis Agassiz: A Life in Science*, Chicago: University of Chicago Press, 1960, pp. 97ff for Agassiz's development of the concept of the ice age. But there was something else too. Among the moraines were found considerable quantities of diamonds. Diamonds are formed deep in the earth and are brought to the surface in the molten magma produced by volcanoes. Thus, here was further evidence of the continuous action of volcanoes, reinforcing the fact that the discovery of the great ice age(s) confirmed both the antiquity of the earth and the uniformitarian approach to geology. J. D. Macdougall, *A Short History of Planet Earth*, New York and London: John Wiley & Sons, 1996, pp. 206–10.
31. Ernst Mayr, *The Growth of Biological Thought*, Cambridge, MA: The Belknap Press of Harvard University Press, 1982, p. 590.
32. Ibid., p. 321.
33. For Wallace's trip to the Far East, see: James T. Costa, *Wallace, Darwin and the Origin of Species*, Cambridge, MA: Harvard University Press, 2014, pp. 223–31.
34. For his interest in land reform, see Martin Fichman, *An Elusive Victorian: The Evolution of Alfred Russel Wallace*, Chicago and London: University of Chicago Press, 2004, pp. 145–46.
35. Peter Bowler, *Charles Darwin, the Man and His Influence*, Cambridge, UK: Cambridge University Press, 1990, p. 36.
36. Ibid., p. 39.
37. Ibid., p. 42. See also: Jonathan Conlin, *Evolution and the Victorians: Science, Culture and Politics in Darwin's Britain*, London: Bloomsbury Academic, 2014.
38. Bowler, *Charles Darwin*, p. 47.
39. Ibid.
40. Ibid., p. 60.
41. Ibid., p. 64. Peter Godfrey-Smith, *Darwinian Populations and Natural Selection*, Oxford: Oxford University Press, 2009.
42. Secord, *Victorian Sensation*, pp. 224 and 230.

43. Mayr, *The Growth of Biological Thought*, p. 501.
44. Bowler, *The Non-Darwinian Revolution*, p. 132.
45. Ibid., p. 145.
46. Ibid., p. 175. For a recent and remarkable turnaround in his view of evolution, see: Thomas Nagel, *Mind and Cosmos: Why the Materialist, Neo-Darwinian Conception of Nature Is Almost Certainly False*, Oxford: Oxford University Press, 2012.
47. David Dennett, *Darwin's Dangerous Idea*, New York and London: Simon & Schuster, 1995, p. 21.

Chapter 3: Beneath the Pattern of the Elements

1. Paul Strathern, *Mendeleyev's Dream: The Quest for the Elements*, London: Hamish Hamilton, 2000, p. 262.
2. Ibid., p. 265.
3. Ibid., p. 267.
4. Ibid., p. 275. Michael Gordin, *A Well-Ordered Thing: Dmitrii Mendeleyev and the Shadow of the Periodic Table*, New York: Basic Books, 2004, pp. 18ff.
5. Strathern, *Mendeleyev's Dream*, p. 282.
6. Ibid., p. 285.
7. Ibid., p. 292. Gordin, *A Well-Ordered Thing*, p. 182.
8. Helge Kragh, *Quantum Generations: A History of Physics in the Twentieth Century*, Princeton and Oxford: Princeton University Press, 2002, p. 8.
9. *New Dictionary of Scientific Biography*, III, pp. 291–94.
10. *Physicists' Biographies*, p. 2. http://physicist.info/.
11. Rollo Appleyard, *Pioneers of Electrical Communication*, London: Macmillan, 1930, p. 131.
12. Kragh, *Quantum Generations*, p. 30.
13. *Dictionary of Scientific Biography*, XI, pp. 519–21.
14. Kragh, *Quantum Generations*, p. 107.
15. Dennis Brian, *The Curies: A Biography of the Most Controversial Family in Science*, New York and London: Wiley, 2005, p. 52.
16. Ibid., p. 54.
17. Ibid., p. 55.
18. J. G. Crowther, *The Cavendish Laboratory: 1874–1974*, London: Macmillan, 1974, p. 107. See also: G. P. Thomas, *J. J. Thomson and the Cavendish Laboratory in His Day*, London: Nelson, 1964, and Richard P. Brennan, *Heisenberg Probably Slept Here: The Lives, Times and Ideas of the Great Physicists of the Twentieth Century*, New York and Chichester: Wiley, 1997, p. 115.
19. Steven Weinberg, *Facing Up: Science and Its Cultural Adversaries*, Cambridge, MA, and London: Harvard University Press, 2001, p. 71.
20. Brennan, *Heisenberg Probably Slept Here*, p. 94. See the Max Planck timeline in: Brandon R. Brown, *Planck: Driven by Vision, Broken by War*, Oxford: Oxford University Press, 2015, p. xvii.

21. Richard Rhodes, *The Making of the Atomic Bomb*, New York and London: Simon & Schuster, 1988, p. 50.
22. Brennan, *Heisenberg Probably Slept Here*, p. 109.
23. Ibid., p. 122.
24. Weinberg, *Facing Up*, p. 105. See also: Sir Mark Oliphant, *Rutherford: Recollections of the Cambridge Days*, Amsterdam and New York: Elsevier, 1972.
25. Kragh, *Quantum Generations*, p. 53.
26. Rhodes, *The Making of the Atomic Bomb*, p. 50.
27. C. P. Snow, *The Physicists*, London: Macmillan, 1981, p. 56.
28. Rhodes, *The Making of the Atomic Bomb*, p. 69; Snow, *The Physicists*, p. 58.
29. Moore, *Niels Bohr*, p. 71. See also Rhodes, *The Making of the Atomic Bomb*, pp. 69–70.

Chapter 4: The Unification of Space and Time, and of Mass and Energy

1. Walter Isaacson, *Einstein: His Life and Universe*, New York and London: Simon & Schuster, 2007, p. 92.
2. Ibid., p. 38.
3. Ibid., pp. 36–37.
4. Ronald W. Clark, *Einstein: The Life and Times*, London: Hodder & Stoughton, 1973, pp. 61–62.
5. Brennan, *Heisenberg Probably Slept Here*, p. 57.
6. Ibid., p. 89.
7. Isaacson, *Einstein*, p. 108.
8. Ibid., p. 111.
9. Ibid., p. 120. For general philosophical background, see: Jimena Canales, *The Physicist and the Philosopher: Einstein, Bergson and the Debate That Changed Our Understanding of Time*, Princeton, NJ: Princeton University Press, 2015, especially chapter 15 on "Full-Blooded Time."
10. Isaacson, *Einstein*, p. 126.
11. Brian Cox and Jeff Forshaw, *Why Does E=mc² (and Why Should We Care)?*, Cambridge, MA: Da Capo Press, 2009, p. 100.
12. Ibid., p. 130.
13. Ibid., p. 49.
14. Richard Wolfson, *Simply Einstein*, New York: Norton, 2003, p. 156; see also Isaacson, *Einstein*, p. 138.
15. Isaacson, *Einstein*, pp. 148 and 157.
16. Ibid., p. 193. Jeroen van Dongen, *Einstein's Unification*, Cambridge, UK: Cambridge University Press, 2010, especially chapters 3 and 6.
17. Albrecht Fölsing, *Albert Einstein: A Biography*, translated and abridged by Ewald Osers, New York: Viking, 1997, p. 374. See also Isaacson, *Einstein*, p. 224.
18. A. Vibert Douglas, *The Life of Arthur Stanley Eddington*, London: Thomas

Nelson & Sons, 1956, p. 38; L. P. Jacks, *Sir Arthur Eddington: Man of Science and Mystic*, Cambridge, UK: Cambridge University Press, 1949, pp. 2 and 17; John Gribbin, *Companion to the Cosmos*, London: Weidenfeld & Nicolson, 1996, pp. 92 and 571.

19. Douglas, *The Life of Arthur*, p. 39; Brennan, *Heisenberg Probably Slept Here*, p. 76.
20. Douglas, *The Life of Arthur*, p. 40; Brennan, *Heisenberg Probably Slept Here*, p. 77.
21. Kragh, *Quantum Generations*, p. 344.
22. Paul Halpern, *Einstein's Dice and Schrödinger's Cat: How Two Great Minds Battled Quantum Randomness to Create a Unified Theory of Physics*, New York: Basic Books, 2015.

Chapter 5: The "Consummated Marriage" of Physics and Chemistry

1. C. P. Snow, *The Search*, New York: Charles Scribner's Sons, 1958, p. 88.
2. Rhodes, *The Making of the Atomic Bomb*, p. 137.
3. David Wilson, *Rutherford: Simple Genius*, London: Hodder & Stoughton, 1983, p. 404.
4. Rhodes, *The Making of the Atomic Bomb*, p. 137; Moore, *Niels Bohr*, p. 21.
5. Emilio Segrè, *From X-Rays to Quarks*, London and New York: W. H. Freeman, 1980, p. 124.
6. Kragh, *Quantum Generations*, p. 145.
7. Arthur I. Miller, *Deciphering the Cosmic Number: The Strange Friendship of Wolfgang Pauli and Carl Jung*, New York and London: W. W. Norton, 2009, p. 55.
8. Moore, *Niels Bohr*, p. 137.
9. Strathern, *Mendeleyev's Dream*, p. 74.
10. Werner Heisenberg, *Physics and Beyond*, New York: Harper, 1971, p. 38.
11. Moore, *Niels Bohr*, p. 138.
12. Heisenberg, *Physics and Beyond*, p. 61.
13. Moore, *Niels Bohr*, p. 139.
14. Snow, *The Physicists*, p. 68.
15. Moore, *Niels Bohr*, p. 14.
16. John A. Wheeler and W. H. Zurek (eds.), *Quantum Theory and Measurement*, Princeton, NJ: Princeton University Press, 1983, p. 209. See also: Werner Heisenberg, "Theory, Criticism and a Philosophy," in Abdus Salam, *Unification of Fundamental Forces, The First of the 1988 Dirac Memorial Lectures*, Cambridge, UK: Cambridge University Press, 1990, pp. 85–124.
17. Moore, *Niels Bohr*, p. 138.
18. Gerald Holton, *Thematic Origins of Scientific Thought*, Cambridge, MA: Harvard University Press, 1973, p. 120.
19. Kragh, *Quantum Generations*, p. 167.
20. Wilson, *Rutherford*, p. 449.
21. Rhodes, *The Making of the Atomic Bomb*, p. 155.

22. Ibid., pp. 160–62.

23. Brian, *The Curies*, p. 209.

24. James Chadwick, "Some personal notes on the search for the neutron," *Proceedings of the Tenth Annual Congress of the History of Science*, 1964, p. 161.

25. Kragh, *Quantum Generations*, p. 185.

26. Ibid., p. 187.

27. Ibid. See also: "The First Cyclotrons," American Institute of Physics, www://aip.org.

28. Kragh, *Quantum Generations*, p. 189.

29. Thomas Hager, *Force of Nature: The Life of Linus Pauling*, New York: Simon & Schuster, 1995, p. 113.

30. Ibid., p. 138.

31. Ibid., p. 141.

32. Ibid., p. 148.

33. Ibid., p. 142. J. D. Dunitz, "Linus Carl Pauling," *Biographical Memoirs of Fellows of the Royal Society*, 1996, Volume 42, pp. 316–26.

34. Hager, *Force of Nature*, p. 143.

35. Ibid., p. 145.

36. Ibid., p. 159. Suman Seth, *Crafting the Quantum: Arnold Sommerfeld and the Practice of Theory: 1890–1926*, Cambridge, MA, and London: MIT Press, 2010, especially chapter 6.

37. Hager, *Force of Nature*, pp. 154–55. John Horgan, "Profile: Linus C. Pauling—Stubbornly Ahead of His Time," *Scientific American* 266, no. 3: 36–40.

38. Hager, *Force of Nature*, p. 159. For the impact of Weisskopf, see: Kurt Gottfried and J. David Jackson, "Mozart and Quantum Mechanics: An Appreciation of Victor Weisskopf," *Physics Today* 56, No. 2 (February 2003), pp. 43–47.

39. Hager, *Force of Nature*, p. 165.

40. Ibid., p. 166. A. Rich, "Linus Pauling (1901–1994)," *Nature* 371, Issue 6495, 1994, p. 285.

41. Hager, *Force of Nature*, pp. 182 and 282.

42. Ibid., p. 168. For an introduction to the work of George Wheland, see: Kostas Gavroglu and Ana Simoes, "From Physical Chemistry to Quantum Chemistry: How Chemists Dealt with Mathematics," *International Journal for Philosophy of Chemistry* 18, no. 1 (2012): 45–69.

Chapter 6: The Interplay of Chemistry and Biology: "The Intimate Connection Between Two Kingdoms"

1. Henry J. John, *Jan Evangelista Purkyně: Czech Scientist and Patriot, 1787–1869*, Philadelphia: The American Philosophical Society, 1959. Chapter VI is on Goethe and Purkyně and there is an appendix on Purkyně's contribution to physiology.

2. Henry Harris, *The Birth of the Cell*, New Haven, CT: Yale University Press, 1999, p. 88.
3. T. Schwann, *Mikroskopische Untersuchungen ber die Vebereinstimmung in der Struktur und dem Wachstum der Thiere und Planzen*, Berlin: Sanderschen Buchhandlung, 1839. Quoted in Harris, *The Birth of the Cell*, p. 100.
4. John Buckingham, *Chasing the Molecule*, Stroud, UK: Sutton, 2004, p. 109.
5. Diarmuid Jeffreys, *Aspirin: The Remarkable Story of a Wonder Drug*, London: Bloomsbury, 2004, pp. 56–57. See also: R. Benedikt, *The Chemistry of the Coal-Tar Colours*, translated by E. Knecht, London: George Bell, 1886, pp. 1–2.
6. John Joseph Beer, *The Emergence of the German Dye Industry*, Urbana: University of Illinois Press, 1959, p. 10.
7. Ibid., p. 33.
8. Ibid., p. 88.
9. *New Dictionary of Scientific Biography*, VII, pp. 157–61.
10. Ibid.
11. For the relationship between Virchow and Koch, see: Frank Ryan, *Tuberculosis: The Greatest Story Never Told*, Bromsgrove, UK: Swift Publishing, 1992, pp. 9f. Bernhard Möllers, *Robert Koch: Persönlichkeit und Lebenswerk, 1843–1910*, Hannover: Schmorl & von Nachf, 1950, chapter iv, pp. 512–17.
12. Ragnhild Münch, *Robert Koch und sein Nachlass in Berlin*, Berlin: Walter de Gruyter, 2003, pp. 41–46 for the cholera expedition. See also: Möllers, *Robert Koch*, pp. 139–47.
13. Thomas Dormandy, *The White Death: A History of Tuberculosis*, London: Hambledon, 1999, pp. 132, 199n, 265n.
14. Claude Quétel, *Le Mal de Naples: Histoire de la Syphilis*, Paris: Editions Seghers, 1986; translated as *History of Syphilis*, London: Polity Press in association with Basil Blackwell, 1990, pp. 2ff.
15. Martha Marquardt, *Paul Ehrlich*, London: Heinemann, 1949, p. 160.
16. Ibid., pp. 175–76.
17. Joseph Fruton, *Proteins, Enzymes, Genes: The Interplay of Chemistry and Biology*, Newhaven, CT: Yale University Press, 1999; Mayr, *The Growth of Biological Thought*, pp. 750–51.
18. Bruce Wallace, *The Search for the Gene*, Ithaca, NY: Cornell University Press, 1992, pp. 57–58.

Chapter 7: The Unity of Science Movement: "Integration Is the New Aim"

1. Bryan Magee, *Men of Ideas: Some Creators of Contemporary Philosophy*, Oxford: Oxford University Press, 1978, p. 96.
2. Ibid., pp. 102–3.
3. Rudolf Carnap, *The Unity of Science*, London: Kegan Paul, 1934, p. 10.

4. Otto Neurath, Rudolf Carnap, and Charles Morris (eds.), *International Encyclopaedia of Unified Science*, Chicago: Chicago University Press, 1938–1955, p. 2.
5. Ibid., p. 15.
6. Ibid., p. 20.
7. Ibid., p. 28. And see chapter 7 of: James Scott Johnston, *John Dewey's Earlier Logical Theory*, Albany, NY: State University of New York Press, 2014.
8. Neurath et al., *International Encyclopaedia*, p. 34.
9. Ibid., p. 46.
10. Ibid., p. 49. See also: chapter 3 of A. W. Carus, *Carnap and Twentieth-Century Thought: Explication and Enlightenment*, Cambridge, UK: Cambridge University Press, 2007.
11. Neurath et al., *International Encyclopaedia*, p. 59.
12. Ibid., p. 61.
13. Ibid., p. 28. And see chapter 11 of: Michael Friedman and Richard Creath, *The Cambridge Companion to Carnap*, Cambridge, UK: Cambridge University Press, 2007.
14. Neurath et al., *International Encyclopaedia*, p. 35.
15. Ibid., p. 64.
16. Ibid., p. 66. For general background, see: Hans Reichenbach, *Modern Philosophy of Science: Selected Essays*, translated and edited by Maria Reichenbach, London: Routledge and Paul; New York: Humanities Press, 1959.
17. Neurath et al., *International Encyclopaedia*, p. 97.
18. Jim Al-Khalili and Johnjoe McFadden, *Life on the Edge: The Coming of Age of Quantum Biology*, London: Bantam, 2014, p. 50.
19. Ibid., p. 51.
20. Ibid., p. 52.
21. D'Arcy Wentworth Thompson, *On Growth and Form*, abridged and edited by John Tyler Bonner, Cambridge, UK: Cambridge University Press, 1961, p. 178.
22. Ibid., p. 278.
23. Ibid., p. 2.

Chapter 8: Hubble, Hitler, Hiroshima: Einstein's Unifications Vindicated

1. Gale Christianson, *Edwin Hubble: Mariner of the Nebulae*, New York: Farrar, Straus & Giroux, 1995, p. 199. See also: John Gribbin, *Copernicus to the Cosmos*, London: Phoenix, 1997, pp. 2 and 186ff.
2. Clark, *Einstein*, p. 406.
3. Christianson, *Edwin Hubble*, p. 55.
4. Gribbin, *Companion to the Cosmos*, pp. 92–93.
5. Kragh, *Quantum Generations*, p. 230.

6. Philip Ball, *Serving the Reich: The Struggle for the Soul of Physics Under Hitler*, London: Bodley Head, 2013, p. 85.

7. Albert Einstein, *The Born-Einstein Letters: Friendship, Politics and Physics in Uncertain Times; Correspondence Between Albert Einstein and Max and Hedwig Born from 1916 to 1955 with Commentaries by Max Born*, translated by Irene Born, Basingstoke, UK: Macmillan, 2005, pp. 113ff.

8. John Cornwell, *Hitler's Scientists*, London: Viking, 2003, Penguin, 2004, p. 130. For an unusual view on Einstein, see Dennis P. Ryan (ed.), *Einstein and the Humanities*, New York and London: Greenwood Press, 1987, with chapters on: the moral implications of relativity, poetic responses to relativity, and relativity and psychology.

9. Ball, *Serving the Reich*, p. 133; Ruth Lewin Sime, *Lisa Meitner: A Life in Physics*, Los Angeles and Berkeley: University of California Press, 1996.

10. Cornwell, *Hitler's Scientists*, pp. 208–10.

11. Robert Jungk, *Brighter Than a Thousand Suns*, London: Victor Gollancz in association with Rupert Hart Davis, 1968, pp. 67–77.

12. Rhodes, *The Making of the Atomic Bomb*, p. 261.

13. Peter Watson, *A Terrible Beauty: The People and Ideas That Shaped the Modern Mind*, London: Phoenix, 2000, pp. 392–93.

14. David C. Cassidy, *Uncertainty: The Life and Science of Werner Heisenberg*, New York: W. H. Freeman, 1992, p. 420.

15. Ibid.

16. Ball, *Serving the Reich*, p. 188; Ruth Lewin Sime, "Lisa Meitner's Escape from Germany," *American Journal of Physics* 58, No. 3, 1990, pp. 263–67.

17. Rhodes, *The Making of the Atomic Bomb*, p. 119.

18. For more details about Peierls's calculations, see Ronald Clark, *The Birth of the Bomb*, London: Phoenix House, 1961, p. 323.

19. Rhodes, *The Making of the Atomic Bomb*, p. 212.

20. Laura Fermi, *Atoms in the Family*, Chicago: University of Chicago Press, 1954, p. 123.

21. Kragh, *Quantum Generations*, p. 265; and Rhodes, *The Making of the Atomic Bomb*, p. 379.

22. Rhodes, *The Making of the Atomic Bomb*, p. 389; also: Kragh, *Quantum Generations*, p. 266.

23. Rhodes, *The Making of the Atomic Bomb*, pp. 450–51; Leslie R. Groves, *Now It Can Be Told: The Story of the Manhattan Project*, New York: Harper Bros, 1962.

24. Jane Wilson (ed.), "All in Our Time," *Bulletin of the Atomic Scientists*, 1975; quoted in Rhodes, *The Making of the Atomic Bomb*, p. 440.

25. Rhodes, *The Making of the Atomic Bomb*, pp. 494 and 496–500.

26. Jungk, *Brighter*, chapters XI, XII, and XIV.

27. Paul Tibbets, "How to Drop an Atomic Bomb," *Saturday Evening Post*, June 8, 1946, p. 136.

Chapter 9: Caltech and the Cavendish: From Atomic Physics to Molecular Biology via Quantum Chemistry

1. Hager, *Force of Nature*, p. 184.
2. Ibid., p. 185.
3. Mayr, *The Growth of Biological Thought*, pp. 722–26.
4. Robin Marant Henig, *A Monk and Two Peas, The Story of Gregor Mendel and the Discovery of Genetics*, London: Weidenfeld & Nicolson, 2000, p. 16.
5. For the context of Mendel's discoveries, see: Peter J. Bowler, *The Mendelian Revolution: The Emergence of Hereditarian Concepts in Modern Science and Society*, London: Athlone Press, 1989, especially p. 100.
6. Ibid., p. 282.
7. See for example: Eileen Magnello, "The Reception of Mendelism by the Biometricians and the Early Mendelians (1899–1909)," in Milo Keynes et al. (eds.), *A Century of Mendelism in Human Genetics*, proceedings of a symposium organized by the Galton Institute and held at the Royal Society of Medicine, London/Boca Raton, FL: CRC Press, 2004, pp. 19–32.
8. Mayr, *The Growth of Biological Thought*, pp. 750–51.
9. Bruce Wallace, *The Search for the Gene*, Ithaca, NY: Cornell University Press, 1992, pp. 57–58; Mayr, *The Growth of Biological Thought*, p. 748.
10. Augustine Brannigan, *The Social Basis of Scientific Discoveries*, Cambridge, UK: Cambridge University Press, 1981, pp. 89–119.
11. Peter J. Bowler, *The Mendelian Revolution*, p. 132.
12. Raphael Falk, *Genetic Analysis: A History of Genetic Thinking*, Cambridge, UK: Cambridge University Press, 2009, pp. 81ff.
13. T. H. Morgan, A. H. Sturtevant, H. J. Muller, and C. B. Bridges, *The Mechanism of Mendelian Heredity*, New York: Henry Holt, 1915; see also Bowler, *The Mendelian Revolution*, p. 134.
14. David Kath (ed.), *The Darwinian Heritage*, Princeton: Princeton University Press, in association with Nova Pacifica, 1985, pp. 762–63.
15. Ian Tattersall, *The Fossil Trail*, Oxford and New York: Oxford University Press, 1995, pp. 89–94.
16. Ibid., p. 95.
17. Hager, *Force of Nature*, p. 189.
18. Ibid., p. 190.
19. Erwin Schrödinger, *What Is Life?*, Cambridge, UK: Cambridge University Press, 1944, p. 77.
20. Moore, *Niels Bohr*, p. 396.
21. Hager, *Force of Nature*, p. 285.
22. J. G. Crowther, *The Cavendish Laboratory*, p. 301; Soraya de Chadarevian, *Designs for Life: Molecular Biology After World War II*, Cambridge, UK: Cambridge University Press, 2002, p. 252.
23. De Chadarevian, *Designs for Life*, p. 7.
24. Ibid., p. 98.

25. Ibid., p. 109.
26. Crowther, *The Cavendish Laboratory*, p. 302.
27. Ibid., p. 61.
28. Ibid., p. 69.
29. Hager, *Force of Nature*, p. 373.
30. Ibid., p. 383.
31. Ibid., p. 399.
32. Paul Strathern, *Crick, Watson and DNA*, London: Arrow, 1997, pp. 37–38; James D. Watson, *The Double Helix*, London: Weidenfeld & Nicolson, 1968, passim.
33. Robert Oltby, *Francis Crick: Hunter of Life's Secrets*, Cold Spring Harbor, NY: Cold Spring Harbor Lab Press, 2009, p. 4.
34. Ibid., p. 128.
35. Strathern, *Crick, Watson*, p. 42.
36. Moore, *Niels Bohr*, p. 397. Matt Ridley, *Francis Crick: Discoverer of the Genetic Code*, London: Harper Press, 2006, pp. 32–33.
37. Watson, *The Double Helix*, p. 85.
38. Ibid., pp. 82–83; Strathern, *Crick, Watson*, pp. 57–58. For the clash of personalities, see: Anne Sayre, *Rosalind Franklin and DNA*, New York: W. W. Norton, 1975, especially chapters 5 and 8.
39. Oltby, *Francis Crick*, p. 59.
40. Ibid., p. 180.
41. Strathern, *Crick, Watson*, p. 82.
42. Oltby, *Francis Crick*, p. 185; De Chadarevian, *Designs for Life*, p. 184.

Chapter 10: Biology, the "Most Unifying" Science: The Switch from Reduction to Composition

1. Paul Oppenheim and Hilary Putnam, "Unity of Science as a Working Hypothesis," in Herbert Feigl, Michael Scriven, and Grover Maxwell (eds.), *Concepts, Theories and the Mind-Body Problem*, Minneapolis: University of Minnesota Press, 1958, p. 8.
2. Ibid., p. 16.
3. Ibid., p. 17.
4. Ibid., p. 21.
5. Ibid., p. 23. And see chapter 6, "The Diversity of the Sciences," in Hilary Putnam, *Words and Life*, edited by James Conant, Cambridge, MA: Harvard University Press, 1994.
6. Oppenheim and Putnam, "Unity of Science," p. 24.
7. Ibid., p. 28.
8. Ibid., p. 30.
9. George Gaylord Simpson, *This View of Life: The World of an Evolutionist*, New York: Harcourt Brace and World Inc., 1964.
10. Ibid., p. viii.
11. Ibid., p. 37.

12. Ibid., p. viii.

13. Ibid., pp. 172–73. See chapter 9 of: Leo F. Laporte, *George Gaylord Simpson: Paleontologist and Evolutionist*, New York: Columbia University Press, 2000.

14. Simpson, *The View of Life*, p. 93.

15. Ibid., p. 107.

16. Ibid., p. 105.

17. Ibid., p. 107. See for example, "Species as an Evolutionary Concept," pp. 109ff of Laporte, *George Gaylord Simpson*.

18. Simpson, *The View of Life*, pp. 110–11.

19. Ibid., p. 181.

Chapter 11: Physics + Astronomy = Chemistry + Cosmology: The Second Evolutionary Synthesis

1. John Gribbin, *Q is for Quantum: Particle Physics from A to Z*, London: Phoenix, 1998, pp. 381f.

2. George Gamow, *The Creation of the Universe*, New York: Viking, 1952. Page 42 for his discussion of the current temperature of space in the universe. See also: Gino Segrè, *Ordinary Geniuses: Max Delbrück, George Gamow, and the Origins of Genomics and Big Bang Cosmology*, New York: Viking, 2011, p. 13.

3. Gribbin, *Q is for Quantum*, p. 190.

4. Ibid., p. 547.

5. See under "quark," "baryon," and "lepton" in: Gribbin, *Q Is for Quantum*, and pp. 190–191 for the early work on quarks. See Murray Gell-Mann, *The Quark and the Jaguar*, New York: Little, Brown, 1994, p. 11, for why he chose "quark."

6. Gribbin, *Companion to the Cosmos*, p. 51; Kragh, *Quantum Generations*, p. 347.

7. Steven Weinberg, *The First Three Minutes: A Modern View of the Origin of the Universe*, New York: Basic Books, 1977, p. 47.

8. Ibid., p. 49.

9. Ibid., pp. 126–27.

10. Ibid., p. 52.

11. Kragh, *The Theory of the Periodic System*, p. 359.

12. Gribbin, *Q is for Quantum*, p. 371.

13. Gribbin, *Companion to the Cosmos*, p. 401.

14. Ibid.

15. Richard Mason (ed.), *Cambridge Minds*, Cambridge, UK: Cambridge University Press, 1994, p. 55.

16. Brian Cox and Jeff Forshaw, *Why Does $E=mc^2$*, p. 168.

17. Weinberg, *The First Three Minutes*, chapter 5 in essence, pp. 101ff. See also: John Gribbin, *The Birth of Time*, London: Weidenfeld & Nicolson, 1999, pp. 50–52, for another synthesis and more recent astronomical observations.

18. Gribbin, *Companion to the Cosmos*, p. 401.
19. Ibid., pp. 353–54.
20. Gribbin, *Companion to the Cosmos*, p. 5.
21. Ibid., p. 41.
22. Ibid., p. 108.
23. Ibid., p. 112.

Chapter 12: A Biography of Earth: The Unified Chronology of Geology, Botany, Linguistics, and Archaeology

1. W. Wertenbaker, *The Floor of the Sea: Maurice Ewing and the Search to Understand the Earth*, Boston: Little, Brown, 1974, p. 205.
2. Ibid., p. 205.
3. Roger M. McCoy, *Ending in Ice: The Revolutionary Idea and Tragic Expedition of Alfred Wegener*, Oxford: Oxford University Press, 2006, p. 7.
4. David. R. Oldroyd, *Thinking About the Earth*, London: The Athlone Press, 1996, p. 250.
5. McCoy, *Ending in Ice*, p. 22.
6. Ibid., p. 25.
7. Ibid., p. 29.
8. Oldroyd, *Thinking About*, p. 257.
9. R. Gheyselinck, *The Restless Earth*, London: The Scientific Book Club, 1939, p. 281. See map of geosynclines in Oldroyd, *Thinking About*, p. 275.
10. George Gamow, *Biography of the Earth*, London: Macmillan, 1941, p. 133.
11. Wertenbaker, *The Floor of the Sea*, p. 364.
12. Ibid., p. 371.
13. Ibid., p. 374.
14. Robert Muir Wood, *The Dark Side of the Earth*, London: Allen & Unwin, 1985, pp. 165–66.
15. For slimes, see: Richard Fortey, *The Earth: An Intimate History*, London: HarperCollins, 2004, pp. 81ff.
16. J. D. MacDougall, *A Short History of Planet Earth*, New York: Wiley, 1996, p. 52.
17. Fortey, *The Earth*, pp. 102ff. For Ediacara, see Ibid, pp. 86ff. The Ediacara are named after Ediacara Hill in South Australia, where they were first discovered.
18. John Noble Wilford, *The Riddle of the Dinosaurs*, London and Boston: Faber & Faber, 1986, pp. 221ff.
19. Walter Alvarez, *T. Rex and the Crater of Doom*, Princeton and London: Princeton University Press, 1997, p. 69.
20. MacDougall, *A Short History*, p. 160; and see the chart of marine extinctions on p. 162.
21. Donald R. Prothero, *After the Dinosaurs: The Age of Mammals*, Bloomington, IN: Indiana University Press, 2006, p. 130.

22. Ibid., p. 199.
23. Ibid., pp. 240–41.
24. Willard F. Libby, *Radiocarbon Dating*, 2nd edition, Chicago: Phoenix Science/Chicago University Press, 1965, p. 26.
25. Colin Renfrew, *Before Civilisation: The Radiocarbon Revolution and Prehistoric Europe*, London: Jonathan Cape, 1973.
26. Ibid., p. 93.
27. Ibid., pp. 160 and 170.
28. Virginia Morrell, *Ancestral Passions: The Leakey Family and the Quest for Humankind's Beginnings*, New York: Simon & Schuster, 1995, p. 57.
29. Mary Leakey, *Olduvai Gorge: My Search for Early Man*, London: Collins, 1979, pp. 52–53, for a detailed map of the gorge.
30. Morrell, *Ancestral Passions*, p. 181.
31. Leakey, *Olduvai Gorge*, p. 74.
32. L. S. B. Leakey, "Finding the World's Earliest Man," *National Geographic*, September 1960, pp. 421–35.
33. Walter Bodmer and Robin McKie, *The Book of Man: The Quest to Discover Our Genetic Heritage*, London: Little, Brown, 1994, p. 218.
34. Brian M. Fagan, *The Journey from Eden: The Peopling of Our World*, London: Thames & Hudson, 1990, pp. 27–28; Bodmer & McKie, *The Book of Man*, pp. 218–19.
35. Colin Renfrew, *Archaeology and Language*, London: Jonathan Cape, 1987, pp. 9–13.
36. Fagan, *The Journey*, p. 186.
37. Renfrew, *Archaeology and Language*, pp. 9–13.
38. Ibid., p. 205.
39. Dan O'Neill, *The Last Giant of Beringia: The Mystery of the Bering Land Bridge*, New York: Westview, 2004, p. 6.
40. Ibid., p. 8. For more recent research on the area, see: "Welcome to Beringia," a summary of a "flurry" of studies that suggest that ancient people lingered in the land bridge longer than previously thought: *Science* 343 (February 28, 2014): 961–63 and 979–80.
41. O'Neill, *The Last Giant*, p. 13.
42. Ibid.
43. Ibid., p. 17.
44. J. Louis Giddings, *Ancient Men of the Arctic*, London: Secker & Warburg, 1968.
45. O'Neill, *The Last Giant*, p. 114.
46. Ibid., pp. 145–47. Renée Hetherington et al., "Climate, African and Beringian subaerial continental shelves, and migration of early peoples," *Quarternary International* (2007), DOI:10.1016/j.quaint.2007.06.033.

Chapter 13: The Overlaps Between New Disciplines:
Ethology, Sociobiology, and Behavioral Economics

1. Robert A. Hinde, "Konrad Lorenz (1903–89) and Niko Tinbergen (1907–88)," in Ray Fuller (ed.), *Seven Pioneers of Psychology*, London: Routledge, 1994, pp. 76–77 and 81–82. Niko Tinbergen, *The Animal in Its World* (2 volumes), London: George Allen & Unwin, 1972, especially Volume 1, pp. 250ff; Hans Kruuk, *Niko's Nature: The Life of Niko Tinbergen and His Science of Animal Behavior*, Oxford: Oxford University Press, 2003, p. 74. See also: Richard W. Burkhardt, Jr., *Patterns of Behaviour: Konrad Lorenz, Niko Tinbergen and the Founding of Ethology*, Chicago: University of Chicago Press, 2005, which stresses the links between ethology and sociobiology, between psychology and biology, and between ethology and "behavioral ecology."

2. Jane Goodall, *Through a Window: Thirty Years with the Chimpanzees of Gombe*, London: Weidenfeld & Nicolson, 1990, p. 13.

3. Ibid., p. 12.

4. Ibid., p. 30.

5. Ibid., pp. 101ff.

6. Ibid., p. 242.

7. Robert Ardrey, *African Genesis*, London: Collins, 1961.

8. See for example: F. W. Marlowe, "Hunter-Gathers and Human Evolution," *Evolutionary Anthropology: Issues, News and Reviews* 14, No. 2 (2005): 54–67.

9. Clifford Geertz, *Local Knowledge*, New York: Basic Books, 1983, p. 8.

10. Ibid., p. 435.

11. Geertz's work continues as two lecture series published as books. See his: *Works and Lives*, London: Polity, 1988, and *After the Fact*, Cambridge, MA: Harvard University Press, 1995.

12. Noam Chomsky, *Language and the Mind*, New York: Harcourt Brace, 1972, pp. 13 and 100ff.

13. John Lyons, *Chomsky*, London: Fontana/Collins, 1970, p. 14; Noam Chomsky, *Syntactic Structures*, The Hague: Mouton, 1957.

14. R. L. Trivers, "The Evolution of Reciprocal Altruism," *Quarterly Review of Biology* 46, no. 1 (1971): 35–57; "Parent-Offspring Conflict," *American Zoologist* 14, no. 1 (1974): 249–64.

15. Daniel Kahneman and Amos Tversky. "Judgment Under Uncertainty: Heuristics and Biases," *Science* 185, no. 4157 (1974): 1124–31.

16. Jacques Monod, *Chance and Necessity: An Essay on the Natural Philosophy of Modern Biology*, New York: Alfred A. Knopf, 1970, p. 158.

17. Ibid., p. 168.

18. Ibid., p. 177.

19. Edward O. Wilson, *Sociobiology: The New Synthesis*, Cambridge, MA: Harvard University Press, 1975; abridged edition, 1980, p. 19.

20. Ibid., p. 17.
21. E. O. Wilson, *Consilience: The Unity of Knowledge*, New York: Vintage, 1998, pp. 125 and 218.
22. Wilson, *Sociobiology*, p. 253.
23. Ibid., p. 331.
24. Ibid., p. 340.
25. Wilson, *Consilience*, pp. 125 and 218.
26. John Alcock, *The Triumph of Sociobiology*, Oxford: Oxford University Press, 2001, pp. 177–78.
27. Ibid., pp. 200–201.
28. Ibid.
29. Noam Chomsky, *Language and Problems of Knowledge*, Cambridge, MA: MIT Press, 1988, p. viii.
30. Ibid., p. 152.

Chapter 14: Climatology + Oceanography + Ethnography Myth = Big History

1. John Savino and Marie D. Jones, *Supervolcano: The Catastrophic Event That Changed the Course of Human History*, Franklin Lakes, NJ: New Page, 2007, p. 125.
2. Ibid.
3. Ibid., pp. 132 and 144. See also: Michael D. Petraglia et al., "Middle Paleolithic Assemblages from the Indian Subcontinent Before and After the Toba Supereruption," *Science* 317, July 2007, pp. 114–16.
4. Michael Petraglia et al., "Middle Paleolithic Assemblages." See also: Kate Ravilious, "Exodus on the Exploding Earth," *New Scientist*, April 17, 2010, pp. 28–33.
5. Curiously, the map of the Toba explosion overlaps the course of the tsunami of Christmas 2004, which occurred due to an earthquake at Simeulue, off the west coast of northern Sumatra, which affected most of Sri Lanka and southern India.
6. Stephen Oppenheimer, *Eden in the East: The Drowned Continent of South East Asia*, London: Weidenfeld & Nicolson, 1998, p. 17.
7. Savino and Jones, *Supervolcano*, p. 47.
8. C. Leonard Woolley, *The Sumerians*, Oxford: Clarendon Press, 1929, p. 6.
9. Oppenheimer, *Eden in the East*, p. 17.
10. Ibid., p. 18.
11. Ibid.
12. Peter Bellwood, *First Farmers: The Origins of Agricultural Societies*, Oxford: Blackwell, 2005, pp. 130 and 133.
13. Oppenheimer, *Eden in the East*, p. 35.
14. Ibid., p. 24.
15. Ibid., p. 32.

16. Ibid., p. 39.
17. Ibid., p. 62.
18. Geoff Bailey, "World Prehistory from the Margins: The Role of Coastlines in Human Evolution," *Journal of Interdisciplinary Studies in History and Archaeology* 1, No. 1, summer 2004, p. 43.
19. David Frawley and Navaratna Rajaram, *Hidden Horizons: Unearthing 10,000 Years of Indian Culture*, Shahibaug, Amdavad-4: Swaminarayan Aksharpith, 2006, pp. 61 and 65.
20. Georg Feuerstein et al., *In Search of the Cradle of Civilisation*, Wheaton, IL, and Chennai, India: Quest Books, 2001, p. 91.
21. Oppenheimer, *Eden in the East*, p. 317.
22. Ibid., plate 1, facing p. 208.
23. Ibid., p. 378.
24. Ibid., p. 359.
25. Ibid., p. 373.
26. Johanna Nichols, *Linguistic Diversity in Space and Time*, Chicago and London: University of Chicago Press, 1992, pp. 9–10.
27. Mircea Eliade, *A History of Religious Ideas*, Volume 1, London: Collins, 1979, p. 20.
28. Ibid., p. 5.
29. Anne Baring and Jules Cashford, *The Myth of the Goddess: Evolution of an Image*, New York: Arkana/Penguin Books, 1991/1993, pp. 9–14.
30. Wayland Barber and Barber, *When They Severed Earth from Sky*, passim.
31. *Nature*, DOI:10.1038/nature07995.
32. Haim Ofek, *Second Nature: Economic Origins of Human Evolution*, Cambridge, UK: Cambridge University Press, 2011, especially the maps on pp. 185 and 188.
33. *Nature*, DOI:10.1038/nature08837.
34. Malcolm Potts and Roger Short, *Ever Since Adam and Eve: The Evolution of Human Sexuality*, Cambridge, UK: Cambridge University Press, 1995, p. 85.
35. Chris Scarre, "Climate Change and Faunal Extinction at the End of the Pleistocene," chapter 5 of *The Human Past*, edited by Chris Scarre, London: Thames & Hudson, 2006, p. 13. See also: Bellwood, *First Farmers*, p. 65.
36. David R. Harris (ed.), *The Origin and Spread of Agriculture and Pastoralism in Eurasia*, London: University College London Press, 1996, p. 135.
37. Peter Watson, *Ideas: A History from Fire to Freud*, London: Phoenix/Weidenfeld & Nicolson, 2006, p. 77; Harris (ed.), *The Origin and Spread*, p. 264.
38. Jacques Cauvin, *The Birth of the Gods and the Origins of Agriculture*, Cambridge, UK: Cambridge University Press, 2000 (French publication, 1994, translation: Trevor Watkins), p. 15.
39. Ibid., pp. 39–48.
40. Peter Watson, *The Great Divide: Nature and Human Nature in the Old World and the New*, New York: Harper Perennial, 2013, p. 133.

41. Brian Fagan, *The Long Summer: How Climate Changed Civilization*, New York: Basic Books, 2004, p. 103.

42. Michael Balter, *The Goddess and the Bull: Çatalhöyk: An Archaeological Journey to the Dawn of Civilization*, New York: Free Press, 2005, pp. 176ff; David Grimm, "Dawn of the Dog," *Science* 348, no. 6232 (April 17, 2015).

43. http://wholehealthsource.blogspot.com/2008/08/ life-expectancy-and-growth-of.html, posted August 5, 2008.

44. Elaine Pagels, *Adam and Eve and the Serpent*, London: Weidenfeld & Nicolson, 1988, p. 29.

45. Ibid., p. 27.

46. For the "refugium" of Beringia, see: "Welcome to Beringia," *Science* 343 (February 28, 2014): 961–63 and 979–89; and Heather Pringle, "Ancient Infant Was Ancestor of Today's Native Americans," *Science* 343 (February 14, 2014): 716–17; Timothy Taylor, *The Prehistory of Sex*, London: Fourth Estate, 1997, pp. 132 and 144.

47. Charles Seife, *Decoding the Universe: How the New Science of Information Is Explaining Everything in the Cosmos, from Our Brains to Black Holes*, New York and London: Viking/Penguin, 2007.

48. Ibid., p. 137.

49. See: *New Scientist*, April 25, 2015, pp. 8–9; *Meteoritics & Planetary Science*, doi. org/3vn.

Chapter 15: Civilization = The Orchestration of Geography, Meteorology, Anthropology, and Genetics

1. Jared Diamond, *Guns, Germs and Steel: The Fates of Human Societies*, New York and London: W. W. Norton, 2005, p. 47.

2. Peter D. Clift and R. Alan Plumb, *The Asian Monsoon: Causes, History and Effects*, Cambridge, UK: Cambridge University Press, 2008, p. 136.

3. Brian Fagan, *The Long Summer: How Climate Changed Civilisation*, London: Granta, 2004, p. 170.

4. Ibid.

5. Clift and Plumb, *The Asian Monsoon*, p. 214.

6. Andrew Sherratt, "Alcohol and Its Alternatives," in Jordan Goodman et al. (eds.), *Consuming Habits: Drugs in History and Anthropology*, London and New York: Routledge, 1995, pp. 16–17.

7. Ibid., pp. 17–18.

8. Peter T. Furst, *Hallucinogens and Culture*, Novato, CA: Chandler & Sharp, 1976/88, pp. 2–3.

9. Elisabeth Benson and Lisa Lucero (eds.), "Complex Polities in the Ancient Tropical World," *Archeological Papers of the American Anthropological Association* 9 (1999): 151–65.

10. Carl O. Sauer, *Agricultural Origins and Dispersals*, Cambridge, MA: MIT Press, 1952/1969, p. 73.

11. See the illustration on p. 431 of Peter Watson, *The Great Divide: Nature and*

Human Nature in the Old World and the New, New York: HarperCollins, 2011, in which book there is a much fuller account of the issues discussed in this chapter.

12. Stuart Piggott, *Wagon, Chariot and Carriage*, London and New York: Thames & Hudson, 1992, p. 16. Robert Drews, *The End of the Bronze Age: Changes in Warfare and the Catastrophe ca. 1200 BC*, Princeton, NJ: Princeton University Press, 1994, pp. 104, 106, 112, 119, and 125.

13. Sherratt, "Alcohol and Its Alternative," p. 392.

14. Gérard Chaliand, *Nomadic Empires: From Mongolia to the Danube*, translated by A. M. Berrett, Livingston, NJ: Transaction, 2005, pp. 8–10, and A. M. Khazanov, *Nomads and the Outside World*, translated by Julia Crookenden, Cambridge, UK: Cambridge University Press, 1984, p. 92.

15. Daniel Hillel, *The Natural History of the Bible: An Environmental Exploration of the Hebrew Scriptures*, New York: Columbia University Press, 2006, pp. 16–18 and 56–62. See also: Dafna Langgut, Israel Finkelstein, and Thomas Litt, "Climate and the Late Bronze Collapse: New Evidence from the Southern Levant," *Tel Aviv* 40 (2013): 149–75.

16. Guy G. Stroumsa, *The End of Sacrifice: Religious Transformation in Late Antiquity*, translated by Susan Emanuel, Chicago: University of Chicago Press, 2009, pp. 53–54.

17. Geoffrey W. Conrad and Arthur A. Demarest, *Religion and Empire: The Dynamics of Aztec and Inca Expansionism*, Cambridge, UK: Cambridge University Press, 1984, pp. 26 and 29; David Carrasco, *Quetzalcoatl and the Irony of Empire: Myths and Prophecies in the Aztec Tradition*, Revised Edition, Boulder, CO: University Press of Colorado, 2000, pp. 104ff.

18. Colin Barras, "Dawn of a Continent," *New Scientist*, July 4, 2015, pp. 28–33; See also: Watson, *The Great Divide*, pp. 512ff.

19. Hillel, *The Natural History*, pp. 56–62.

Chapter 16: The Hardening of Psychology and Its Integration with Economics

1. Gilbert Ryle, *The Concept of Mind*, London: Hutchinson, 1949, pp. 36ff and 319ff.

2. Miles Weatherall, *In Search of a Cure: A History of Pharmaceutical Discovery*, Oxford: Oxford University Press, 1990, p. 254.

3. Ibid., p. 257.

4. James Le Fanu, *The Rise and Fall of Modern Medicine*, London: Little, Brown, 1999, p. 68.

5. Ibid.

6. David Healy, *The Anti-Depressant Era*, Cambridge, MA: Harvard University Press, 1997, pp. 52–54, which discusses the influential *Nature* article of 1960 on this subject.

7. Michael Rutter, *Genes and Behaviour: Nature-Nurture Interplay Explained*, Oxford: Blackwell, 2006, pp. 71ff.

8. Ibid., pp. 71ff. See also: Nicholas Rose and Joelle M. Abi-Rached, *Neuro: The New Brain Sciences and the Management of the Mind*, Princeton, NJ: Princeton University Press, 2013, pp. 78–81.

9. Rose and Abi-Rached, *Neuro*, pp. 131–32.

10. Rutter, *Genes and Behaviour*, p. 147. Lucas A. Mongiat and Alejandro F. Schindre, "A Price to Pay for Adult Neurogenesis," *Science* 344 (May 9, 2014): 594–95; Emily Underwood, "Lifelong Memories May Reside in Nets Around Brain Cells," Science Weekly News, *Science*, October 30, 2015.

11. Clare Wilson, "Psychiatry: The Reboot Starts Here," *New Scientist*, May 10, 2014, pp. 10–12; J. Andrew Pruszynski and Jörn Diedrichsen, "Reading the Mind to Move the Body," *Science* 348, no. 6237 (May 22, 2015): 860–61, and Tyson Afialo et al., "Decoding Motor Imagery from the Posterior Parietal Cortex of a Tetraplegic Human," 906–10 of the same issue; "Paralysed Man Thinks to Walk," *New Scientist*, October 3, 2015, p. 17; *Journal of Neuroengineering and Rehabilitation*, doi.org/7wc.

12. Paul Gilbert and Kent G. Bailey (eds.), *Genes on the Couch: Explorations in Evolutionary Psychology*, London: Psychology Press, 2002, pp. 127–29 and 168–69. See also: Kalman Glantz and John D. Pearce, *Exiles from Eden: Psychotherapy from an Evolutionary Perspective*, New York and London: W. W. Norton, 1989.

13. Rutter, *Genes and Behavior*, pp. 178–210.

14. Ibid., p. 112.

15. Nessa Carey, *The Epigenetics Revolution: How Modern Biology Is Rewriting Our Understanding of Genetics, Disease and Inheritance*, London: Icon Books, 2011, pp. 67–72.

16. Ibid., pp. 247–49.

17. Daniel Kahneman, *Thinking Fast and Slow*, London: Allen Lane, 2011, Part 7, pp. 277ff.

18. Michael S. Gazzaniga, *Who's in Charge? Free Will and the Science of the Brain*, London: Robinson, 2012. Chapter 3, pp. 75–103 is devoted to The Interpreter. But see also pp. 108–14 and 168–72.

19. Richard Thaler, *Misbehaving: The Making of Behavioural Economics*, London: Allen Lane, 2015.

20. Ibid., p. 314.

21. Ibid., p. 339.

22. Ibid., p. 330.

23. David C. Parkes and Michael P. Wellman, "Economic Reasoning and Artificial Intelligence," *Science* 349, no. 6245 (July 17, 2015): 267–72; Samuel J. Gershman et al., "Computational Rationality: A Converging Paradigm for Intelligence in Brains, Minds and Machines," *Science* 349, no. 6245 (July 17, 2015): 273–78.

Chapter 17: Dreams of a Final Unification: Physics, Mathematics, Information, and the Universe

1. Charles Seife, *Decoding the Universe: How the New Science of Information Is Explaining Everything in the Cosmos, from our Brains to Black Holes*, New York: Viking, 2006, p. 59.
2. Ibid.
3. Ibid., p. 101.
4. George Dyson, *Turing's Cathedral: The Origins of the Digital Universe*, London: Penguin, 2012, p. 11.
5. Ibid., p. 9.
6. Katie Hafner and Matthew Lyon, *Where Wizards Stay Up Late: The Origins of the Internet*, New York: Simon & Schuster, 1996, pp. 253–54; John Naughton, *A Brief History of the Future: The Origins of the Internet*, London: Weidenfeld & Nicolson, 1999, pp. 92–119, passim.
7. Naughton, *A Brief History*, pp. 140ff; Hafner and Lyon, *Where Wizards*, p. 192.
8. Hafner and Lyon, *Where Wizards*, pp. 204 and 223–27.
9. Ibid., pp. 253 and 257–58.
10. Brian Winston, *Media, Technology and Society: A History from the Telegraph to the Internet*, London: Routledge, 1998.
11. Seife, *Decoding the Universe*, p. 58.
12. Ibid., p. 89.
13. Ibid., p. 95.
14. Ibid., pp. 116–17.
15. Michael White and John Gribbin, *Stephen Hawking: A Life in Science*, New York and London: Viking, 1992, pp. 137–38.
16. Ibid., pp. 154–55.
17. Ibid., pp. 208 and 274–75.
18. Ibid., pp. 292–301.
19. Martin Rees, *Just Six Numbers: The Deep Forces That Shape the Universe*, London: Weidenfeld & Nicolson, 1999; White and Gribbin, *Stephen Hawking*, pp. 216–17.
20. See, for example, *New Scientist*, March 11, 2014.
21. David Bohm and B. J. Hiley, *The Undivided Universe: An Ontological Interpretation of Quantum Theory*, London: Routledge, 1993, p. 358.
22. David Deutsch, *The Fabric of Reality*, London: Penguin Books, 1997, pp. 352ff.
23. Frank J. Tipler, *The Physics of Immortality: Modern Cosmology, God and the Resurrection of the Dead*, London: Macmillan, 1995.
24. Deutsch, *The Fabric of Reality*, p. 358.
25. Ibid.
26. Stephen Wolfram, *A New Kind of Science*, Champaign, IL: Wolfram Media, Inc., 2002, pp. 715–16.

27. See, for example; Steven Weinberg, "Is the Universe a Computer?," *New York Review of Books*, October 24, 2002; Philip Ball, "Life, the Universe and a Game of Chequers," *Guardian*, July 14, 2002; George Johnson, "A New Kind of Science: You Know That Space-Time Thing? Never Mind," *New York Times*, June 6, 2002; Judith Rosen, "Weighing Wolfram's 'New Kind of Science,'" *Publishers Weekly*, January 13, 2003.

28. Wolfram, *A New Kind of Science*, p. 474.

29. Ibid., p. 719.

30. Ibid., p. 772.

31. Lee Smolin and Roberto Mangabeira Unger, *The Singular Universe and the Reality of Time*, Cambridge, UK: Cambridge University Press, 2014, p. 331.

32. Ibid., p. 142.

33. Ibid., p. 275.

34. Ibid., p. xii. Compare: Lee Smolin, *The Trouble with Physics: The Rise of String Theory, the Fall of Physics and What Comes Next*, London: Penguin, 2013, which considers alternatives to string theory.

35. Smolin and Unger, *The Singular Universe*, pp. 102 and 112.

36. Ibid., p. 301.

37. Ibid., pp. 328 and 346.

38. Brian Greene, *The Elegant Universe: Superstrings, Hidden Dimensions and the Quest for the Ultimate Theory*, London: Jonathan Cape, 1998, pp. 174–76.

39. Ibid., pp. 10–13.

40. Ibid., pp. 329–31.

41. Seife, *Decoding the Universe*, p. 177.

42. Ibid., p. 178.

43. Ibid., p. 181.

44. Jacob Aron, "Quantum Technology to Hit the Streets," *New Scientist*, October 17, 2015, p. 10.

45. Freeman Dyson, *Infinite in All Directions*, New York: Harper and Row, Cornelia and Michael Bessie Books, 1988.

46. Steven Weinberg, *Dreams of a Final Theory: The Scientist's Search for the Ultimate Laws of Nature*, New York: Vintage, 1993, pp. 14 and 33. See also: John C. Taylor, *Hidden Unity in Nature's Laws*, Cambridge, UK: Cambridge University Press, 2001. There are good chapters on quantum theory and special relativity; order breaks symmetry; what hold quarks together; how do we tell the future from the past?; and string theory: beyond points. The strength of this book is less its originality than its organization and clarity.

47. Salam, *The Unification*, p. 128.

48. Weinberg, *Dreams of a Final Theory*, p. 136.

49. Ibid., p. 141.

50. Ibid., p. 50.

51. Salam, *The Unification*, pp. 29–30.

52. Sean Carroll, *The Particle at the End of the Universe: The Hunt for the Higgs and the Discovery of a New World*, New York and London: Oneworld, 2012, p. 4.
53. Ibid., p. 89.
54. Ibid., p. 99.
55. Ibid., p. 170. See *New Scientist*, June 20, 2015, pp. 5 and 33–35 for a discussion, and Adam Frank and Marcelo Gleiser, "A Crisis at the Edge of Physics," *International New York Times*, June 9, 2015.

Chapter 18: Spontaneous Order: The Architecture of Molecules, New Patterns in Evolution, and the Emergence of Quantum Biology

1. Simpson, *This View of Life*.
2. Ibid., p. viii.
3. Ibid., pp. 110–11.
4. Arthur Eddington, *The Nature of the Physical World*, London: Macmillan, 1928, pp. 172–73.
5. Ernest Nagel, *The Structure of Science: Problems in the Logic of Scientific Explanation*, London: Routledge & Kegan Paul, 1961. Quoted in Ilya Prigogine and Isabelle Stengers, *Order Out of Chaos: Man's New Dialogue with Nature*, London: Flamingo/Fontana, 1984, p. 8.
6. Ibid., pp. 362–63.
7. Ibid., pp. 367–80 and 433–35.
8. Ibid., p. 372.
9. Ibid., p. 380.
10. Philip Warren Anderson, "More Is Different: Broken Symmetry and the Nature of the Hierarchical Structure of Science," *Science* 177, no. 4047 (August 4, 1972): 393–96. See also: Ilya Prigogine, *The End of Certainty*, New York: Free Press, 1997, pp. 16 and 18.
11. John Taylor's examples are given in his *Hidden Unity in Nature's Laws*, Cambridge, UK: Cambridge University Press, 2001, pp. 290–95.
12. Prigogine, *End of Certainty*, p. 18.
13. Prigogine and Stengers, *Order Out of Chaos*, p. 162.
14. Ibid., p. 71.
15. Ibid., p. 87.
16. James Gleick, *Chaos: Making a New Science*, New York: Penguin, 1987.
17. Prigogine and Stengers, *Order Out of Chaos*, p. 13; Prigogine, *End of Certainty*, p. 173.
18. See, for example: Paul Davies (ed.), *The New Physics*, Cambridge, UK: Cambridge University Press, 1989, p. 321; and Prigogine and Stengers, *Order Out of Chaos*, p. 152.
19. Davies, *The New Physics*, pp. 322–23. For further details of the B-Z reaction, see: Philip Ball, *Designing the Molecular World*, Princeton, NJ: Princeton University Press, 1994, pp. 308–9.

20. Prigogine and Stengers, *Order Out of Chaos*, p. 152. For the amoebae reactions, see Ball, *Designing*, p. 310.
21. Stuart Kauffman, *The Origins of Order: Self-Organization and Selection in Evolution*, Oxford: Oxford University Press, 1993.
22. Ronald Fox, *Biophysics Journal* 65 (December 1993): 2698–99.
23. Kauffman, *The Origins of Order*, p. 154.
24. Ibid.
25. Ibid., p. 173.
26. Robert Wesson, *Beyond Natural Selection*, Cambridge, MA: MIT Press, 1991.
27. Ibid., p. 17.
28. Ibid., p. 50.
29. Ibid., p. 70.
30. Ibid., p. 89.
31. Ibid., p. 200.
32. Ibid., p. 297.
33. Al-Khalili and McFadden, *Life on the Edge*, p. x.
34. Ibid., p. 297.
35. Ibid., p. 128. See also: Ball, *Designing*, p. 99.
36. Ibid., p. 131.

Chapter 19: The Biological Origin of the Arts, Physics and Philosophy, the Physics of Society, Neurology and Nature

1. Wilson, *Consilience*, p. 162.
2. Ibid., pp. 125 and 218.
3. Ibid., p. 8.
4. Ibid., p. 12.
5. Ibid., p. 127.
6. Ibid., p. 218.
7. Ibid., p. 221.
8. Ibid., p. 225.
9. Ibid.
10. A. R. Sanders et al., "Genome-Wide Scan Demonstrates Significant Linkage for Male Sexual Orientation," *Psychological Medicine*, doi: 1017/SOO33291714002451. And: http://www.sciencedaily.com/releases/2014/09/140918141151.htm.
11. Sam Harris, *The Moral Landscape: How Science Can Determine Human Values*, New York: Free Press, 2010, p. 4.
12. Ibid., pp. 28–30.
13. Ibid., p. 73.
14. Ibid., p. 191.
15. Patricia Churchland, *Braintrust: What Neuroscience Tells Us About Morality*, Princeton and Oxford: Princeton University Press, 2011, p. 178.

16. Ibid., p. 9.

17. Ibid., p. 52.

18. Ibid., p. 77.

19. Philip Ball, *Critical Mass: How One Thing Leads to Another*, London: William Heinemann, 2004, p. 35.

20. Ibid., pp. 165–78.

21. Ibid., p. 195.

22. Ibid., p. 199.

23. Ibid., p. 398.

24. Ibid., p. 415.

25. Lynn Gamwell, *Mathematics + Art: A Cultural History*, Princeton, NJ: Princeton University Press, 2015.

26. Patricia Churchland, *Neurophilosophy: Toward a Unified Science of the Mind-Brain*, Cambridge, MA: MIT Press, 1986, especially chapters 3 (for the work of Andras Pellionisz and Rodolfo Llinás), 5, and 7.

27. Churchland, *Braintrust*, pp. 48–59, 77–78, and 201–2 for the oxytocin/vasopressin material.

28. Daniel Dennett, *Freedom Evolves*, London: Allen Lane, 2003; and see the interesting review of Dennett's book by Mary Midgley in the *Guardian*, March 1, 2003.

29. Dennett, *Freedom Evolves*, p. 89.

30. Hilary Putnam, *Philosophy in an Age of Science: Physics, Mathematics, and Skepticism*, edited by Mario de Caro and David Macarthur, Cambridge, MA: Harvard University Press, 2012, chapter 2, pp. 51 ff.

31. Ibid., p. 52.

32. Ibid., p. 59.

33. Ibid., p. 244.

34. Ibid., p. 131.

35. Ibid., p. 138.

36. Ibid., pp. 41–42. See also: Tim Maudlin, "The Tale of Quantum Logic," chapter 6 in Yemima Ben-Menahem (ed.), *Hilary Putnam*, Cambridge, UK: Cambridge University Press, and Nancy Cartwright, "Another Philosopher Looks at Quantum Mechanics," chapter 7 of the same book.

37. Putnam, *Philosophy in an Age*, p. 336.

38. Jonathan Glover, *What Sort of People Should There Be?*, London: Penguin, 1984; *Choosing Children: Genes, Disability and Design*, Oxford: Oxford University Press, 2006. See also: Julian Savulescu and Nick Bostrom, *Human Enhancement*, Oxford: Oxford University Press, 2013.

39. Jürgen Habermas, *The Future of Human Nature*, Cambridge: Polity Press, 2003, p. 38.

40. Ibid., pp. 39 and 48.

41. Ibid., pp. 56, 79–78, and 87.

Conclusion: Overlaps, Patterns, Hierarchies:
A Preexisting Order?

1. Robert Laughlin, *A Different Universe: Reinventing Physics from the Bottom Down*, New York: Basic Books, 2005, p. 211.
2. John Dupré, *The Disorder of Things: Metaphysical Foundations of the Disunity of Science*, Cambridge, MA: Harvard University Press, 1993, p. 1.
3. Ibid., p. 19.
4. Ibid., pp. 62–63.
5. Ibid., p. 89.
6. Ibid., p. 92.
7. Peter Galison and David J. Stump (eds.), *The Disunity of Science: Boundaries, Contexts, and Power*, Stanford, CA: Stanford University Press, 1996, p. 135.
8. Dupré, *The Disorder of Things*, p. 147.
9. Ibid., p. 167.
10. Galison and Stump, *The Disunity of Science*, p. 39.
11. Ibid., p. 68.
12. Ibid., p. 139.
13. Dupré, *The Disorder of Things*, p. 224.
14. Ibid., p. 115.
15. Ibid., p. 182.
16. Ibid., p. 199.
17. Galison and Stump, *The Disunity of Science*, p. 402.
18. Ibid., p. 67.
19. Dupré, *The Disorder of Things*, p. 224.
20. Richard Dawkins, *The Blind Watchmaker*, London: Longman, 1986/Penguin 1988 and 1991, p. 123.
21. Nick Lane, *The Vital Question: Why Is Life the Way It is?*, London: Profile Books, 2015, p. 39.
22. *Science* 348, no. 6237 (May 22, 2015): 860–61 and 906.
23. Stuart Kauffman, *Reinventing the Sacred: A New View of Science, Reason, and Religion*, New York: Basic Books, 2010, p. 36.
24. Ibid., p. 98.
25. Steven Strogatz, *Sync: The Emerging Science of Spontaneous Order*, London: Penguin, 2003/2004, passim but especially pp. 168–83.
26. Laughlin, *A Different Universe*, pp. xv, 4 and 12.
27. Ibid., p. xi.
28. Ibid., p. 5.
29. Ibid., p. 15.
30. Ibid., p. 50.
31. Ibid., p. 63. See also: Robert Laughlin, *The Crime of Reason and the Closing of the Scientific Mind*, New York: Basic Books, 2008, where he slays a few more scientific shibboleths (like black holes).
32. Laughlin, *A Different Universe*, p. 76.

33. Ibid., p. 109.
34. Ibid., p. 170.
35. Ibid., pp. 200–201.
36. Ibid., p. 219.
37. Lane, *The Vital Question*, p. 132.
38. Ibid., p. 133.
39. Ibid., p. 134; *Science* 348, no. 6235 (May 8, 2015): 615–66.
40. Lane, *The Vital Question*, p. 92.
41. Ibid., p. 107.
42. Ibid., p. 173. Lane introduced the importance of mitochondria and "proton" power in an earlier book, *Power, Sex, Suicide: Mitochondria and the Meaning of Life*, Oxford: Oxford University Press, 2005.
43. Lane, *The Vital Question*, pp. 103–4.
44. Ibid., p. 173.
45. Ibid., p. 134.
46. Adam Benforado, *Unfair: The New Science of Criminal Injustice*, New York: Crown, 2015; *Science* 345, no. 6213 (November 28, 2014): p. 1146.
47. Ibid., p. 1147.
48. Ibid., p. 1148.
49. Ibid., p. 1146.
50. Ball, *Critical Mass*, pp. 581–83.
51. Pedro Domingos, *The Master Algorithm: How the Quest for the Ultimate Learning Machine Will Remake Our World*, New York and London: Basic Books/ Penguin, 2015.

INDEX